The physiology and biochemistry of cestodes

T0269315

The physiology and biochemistry of cestodes

J. D. SMYTH

London School of Hygiene & Tropical Medicine
University of London

AND

D. P. McMANUS

Queensland Institute of Medical Research
and University of Queensland, Australia

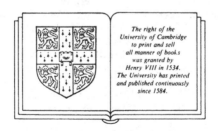

The right of the
University of Cambridge
to print and sell
all manner of books
was granted by
Henry VIII in 1534.
The University has printed
and published continuously
since 1584.

CAMBRIDGE UNIVERSITY PRESS

CAMBRIDGE

NEW YORK NEW ROCHELLE MELBOURNE SYDNEY

CAMBRIDGE UNIVERSITY PRESS
Cambridge, New York, Melbourne, Madrid, Cape Town, Singapore, São Paulo

Cambridge University Press
The Edinburgh Building, Cambridge CB2 8RU, UK

Published in the United States of America by Cambridge University Press, New York

www.cambridge.org
Information on this title: www.cambridge.org/9780521355575

First published 1989
This digitally printed version 2007

A catalogue record for this publication is available from the British Library

Library of Congress Cataloguing in Publication data
Smyth, J.D. (James Desmond), 1917–
The physiology and biochemistry of cestodes / J.D. Smyth and D.P.
McManus.
p.   cm.
Bibliography: p.
Includes index.
ISBN 0 521 35557 5
1. Cestoda – Physiology. 2. Platyhelminthes – Physiology.
I. McManus, D. P. (Donald Peter), 1949–. II. Title.
[DNLM: 1. Cestoda – physiology. QX 400 S667pb]
QL391.P7S57 1989
595.1′21041 – dcl9
DNLM/DLC
for Library of Congress   88–25596 CIP

ISBN 978-0-521-35557-5 hardback
ISBN 978-0-521-03895-9 paperback

Contents

Contents

Contents

Preface

The text of this book is based essentially on *The Physiology of Cestodes* (J. D. S. Smyth; W. H. Freeman/Oliver & Boyd, 1969) with an extended content and title to take into account the impact of biochemistry and molecular biology on the field. In addition, other new investigative techniques, such as transmission and scanning electron microscopy, cytochemistry, immunochemistry, population dynamics, immunobiology, and *in vitro* culture has greatly extended our understanding of cestode physiology and in this text we have attempted to review progress made up to 1986/87. Within the permissible space restrictions, it has not always been possible to quote work prior to 1970 and the reader is referred to the earlier volume for these data. Reference to early work of major fundamental importance, however, has been retained.

As well as being the causative organisms of a number of major human and animal diseases (e.g. cysticercosis, hydatidosis), cestodes serve as elegant experimental models for the study of fundamental biological phenomena. These include not only problems of specific parasitological interest, such as host-specificity, but also more basic problems such as enzyme dynamics, membrane transport and cell and tissue differentiation (especially asexual/sexual differentiation), common to many other biological fields.

An attempt has been made to give a representative worldwide coverage of the literature with major reviews being quoted where possible. Even with vigorous selection, the number of references has increased from 492 in the earlier volume to nearly 1000 in this version. Where appropriate, reference to papers in the less well-known foreign languages are supplemented with the relevant *Helminthological Abstract* number so that readers can readily consult an abstract in English. The number of figures has been increased by some 50 new diagrams or photographs and there are 71 tables, most of them new.

Preface to the second edition

One of us (D. P. Mc.) was responsible for Chapters 4, 5 and 6, and the other (J. D. S.) for the remaining chapters. Both of us were previously on the staff of the Department of Pure and Applied Biology, Imperial College, University of London, where much of our work reported here was carried out.

J. D. Smyth
D. P. McManus
April 1988

Acknowledgements

We are grateful to the following who have read and commented critically on various parts of the text: Dr R. Bell, Dr E. M. Bennett, Professor C. Bryant, Dr I. Fairweather, Professor D. W. Halton, Dr W. M. Hominick, Dr J. C. Shepherd, Professor L. T. Threadgold, Mrs M. M. Smyth.

We are especially indebted to Dr G. A. Conder and Dr Margaretha Gustafsson who contributed original micrographs. The typing of much of the extensive tabular material was carried out largely by Mrs P. Mill to whom our thanks are due. We are indebted to the London School of Hygiene and Tropical Medicine and Imperial College of Science and Technology for providing background facilities.

During the preparation of this book, Professor Smyth was the recipient of a Leverhume Emeritus Fellowship from the Leverhulme Trust and their generous support in this respect is gratefully acknowledged. Much of the authors' research work reported here has been supported by grants from the EEC, the Medical Research Council, the Wellcome Trust and the World Health Organization, to which due acknowledgement is made.

We are grateful to numerous authors and to the following editors or publishers for permission to use material: Academic Press Inc. (London) Ltd; *Acta Parasitologica Polonica*; Alan R. Liss Inc.; American Society of Parasitologists; *Angewandte Parasitologie*; *Annales Zoologici Fennici*; Australian Society for Parasitology; Cambridge University Press; Ciba Foundation; Edward Arnold; Elsevier Science Publishers B. V.; *Evolution*; Fisheries Society of the British Isles; *Gastroenterology*; George Allen & Unwin; *Japanese Journal of Parasitology*; *Japanese Journal of Veterinary Research*; John Wiley & Sons; *Journal of Biological Chemistry*; *Journal of Experimental Zoology*; Macmillan Publishers Ltd; Martin Nijhoff Publishers; *Molecular and Biochemical Parasitology*; Ohio University Press; Pitman Publishing Ltd; Raven Press, New York; Springer-Verlag; *Revista Iberica de Parasitologia*; Williams & Wilkins Co.

1

The cestodes: general considerations

General account

Cestodes represent a group of organisms which present many features of exceptional physiological interest. They are, for example, almost unique amongst parasites in that the adult worm occupies only one particular habitat, the alimentary canal, in one particular group of animals, the vertebrates. Moreover, the known exceptions to this generalisation occur in sites related to the alimentary canal – the bile duct, the gall bladder or the pancreatic ducts.

The only adult forms which occur in hosts other than vertebrates are members of the subclass Cestodaria, whose physiology is largely unknown (and is not considered here) and a few neotenic forms in oligochaetes (i.e. *Archigetes*).

The dominating morphological features of adult cestodes are: (*a*) an elongated tape-like body – a form elegantly adapted to the tubular habitat provided by the host gut; and (*b*) the absence of an alimentary canal, in either the adult or larvae. The latter feature is of major physiological importance, for it means that the external surface of the worm, which ultrastructural studies have revealed as a 'naked' cytoplasmic *tegument* unprotected by a resistant 'cuticle', has evolved as a metabolically active surface through which the transport of all nutrients into, and waste materials out of, the worm must take place. This characteristic makes the cestode tegument a superb model for membrane transport studies and, for these, several species, especially *Hymenolepis diminuta*, have been used extensively. In contrast to the adult worm, larval cestodes can occur in almost any location in the intermediate host, although many species show a predeliction for a particular organ. In tissue sites, an analogy has been drawn between cysticerci and the embyros of placental mammals (*796*). Both theoretically and physiologically, this analogy has much to recommend it, for it must be remembered that a mammalian embryo (which is of paternal as well as maternal origin) is genetically 'foreign' or 'non-self'

1

The cestodes: general considerations

tissue. Thus, it can also be considered to be 'parasitic' in the maternal uterus and essentially shares the same problems as a cestode.

Chief among these are (a) absorption of nutrients, (b) elimination of waste products, and (c) survival in an immunologically 'hostile' environment. In addition, a cestode faces a problem of access for its reproductive products (eggs or cysts) to the outside world and hence transmission to its next definitive or intermediate host.

It is self-evident that the immediate areas of contact of the cestode with its host – i.e. the *host–parasite interface* – is one of great physiological interest and its ultrastructure and biochemical activities have received much attention (*459, 462, 622, 623, 624*).

Problems of the life cycle

With rare exceptions (e.g. *Hymenolepis nana*), cyclophyllidean cestodes require at least one intermediate host, and pseudophyllidean cestodes require two, or more rarely, three such hosts. The physiological activities of cestodes have been studied largely in these two groups, with the result that the physiology of less well-known orders, such as the Trypanorhyncha, has been largely neglected.

In this text, the physiology of the egg, larval and adult stages of cestodes is reviewed and an attempt is made to examine those factors which may influence or determine the various development patterns encountered during the life cycles. The study of the physiology of an organism in depth involves investigations at molecular, cellular, tissue, organ, whole organism and ecological levels. Only when such wide-ranging studies are made is it possible to obtain an integrated picture of its physiology. Such studies frequently raise questions fundamental to whole areas of biology and, in this respect, cestodes can be regarded as valuable, if unusual, models for the investigation of basic biological phenomena and are so regarded in this text.

Problems of special interest are: activities at the host–parasite interface (especially the transport of substances across the tegument); the biochemistry of various species at different phases of development, in different hosts or in different habitats; the factors controlling growth and sexual differentiation of the adult strobila; the nature and formation of the egg and its protective membranes and the factors stimulating its hatching in different host habitats; the mechanism of penetration of larvae into different tissue sites; the factors which determine whether larvae differentiate in a sexual or asexual direction; and the immunological relationships of all stages of cestodes with their hosts and especially the mechanisms whereby they are able to survive in immunologically hostile environments (Chapter 11).

2

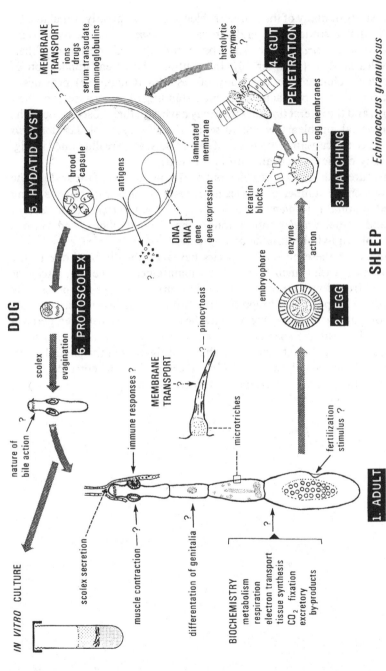

Fig. 1.1. Some physiological problems associated with the life cycle of the hydatid organism *Echinococcus granulosus*. (After Smyth, 1987a.)

3

The cestodes: general considerations

The study of many of the above problems has been greatly stimulated by the fact that cestodes of several genera (e.g. *Hymenolepis, Echinococcus, Mesocestoides, Spirometra*) can now be cultured through most or all of their life cycles *in vitro* (Chapter 10). In addition, the recent application of molecular techniques to the study of cestodes (Chapter 6) provides a fundamental new approach to our understanding of many aspects of their biology, and the interaction that they may have with their hosts. Some of the physiological problems which arise in the life cycle of *Echinococcus granulosus*, the causative organism of hydatid disease, are illustrated in Fig. 1.1; many of these are dealt with in this text.

A feature of the research carried out on the physiology of the group is the relatively small number of species investigated – most experiments being carried out on *Hymenolepis* spp., *Taenia* spp., *Moniezia* spp., *Echinococcus* spp. and *Mesocestoides* spp. amongst the Cyclophyllidea, and various species of Diphyllobothriidae amongst the Pseudophyllidea.

The use of these particular species has undoubtedly been due to the relative ease with which they can be maintained in the laboratory or obtained from local sources, such as abattoirs. This does not necessarily mean that the species used represent the *best* experimental material for the study of that particular problem – only the most *convenient*. Other, perhaps less readily available species, may be much better models. It cannot be stressed strongly enough, therefore, that investigation of the physiology of *any* cestode species, however rare or aberrant (and whether or not of economic or medical importance), is worthy of study.

2

The adult cestode: special structural features relevant to its physiology

General considerations

The general morphology of cestodes, including the ultrastructure, is now too well known to be reviewed at length here. It should be noted, however, that the traditional concept of the tapeworm strobila being divided into discrete 'segments' or 'proglottides' is largely illusionary, in that although the strobila is constricted at intervals between segments, the interior is filled with uninterrupted parenchymal cells and longitudinal muscles throughout the whole length of the body (519). Valuable reviews of the basic structure and/or ultrastructure are those of Arme & Pappas (27, 28) and Threadgold (878) on cestodes in general and that of Arai (19) on *Hymenolepis* spp. in particular. In this chapter therefore, with the exception of the tegument, which is discussed in some detail, only those morphological features of particular physiological interest are considered.

The tegument

General considerations

Because tapeworms lack a gut, all nutritive material must pass through the body surface and waste materials likewise be eliminated through it. The basic structure may, in fact, be compared with a gut turned inside out, with the body covering or *tegument* (Fig. 2.1) serving the absorptive functions normally associated with the intestinal mucosa (Fig. 2.2). The structure and physiology of the tegument is, therefore, of fundamental importance to the understanding of cestode physiology as a whole and has received considerable attention.

Although the tegument contains specific systems for molecular and ion transport – especially amino acids, hexose sugars, vitamins, purines, pyrimidines, nucleotides, and lipids – it probably also serves a number of other vital functions (647): (a) it is a major site of catalytic activity and

5

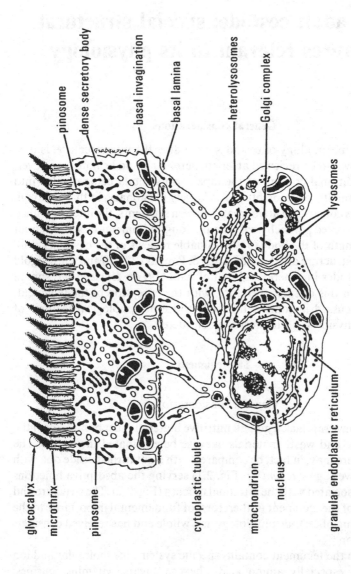

Fig. 2.1. Diagrammatic representation of the tegument of a typical adult cestode, based on several species. (After Threadgold, 1984.)

glycocalyx
microthrix
pinosome
cytoplastic tubule
mitochondrion
nucleus
granular endoplasmic reticulum
pinosome
dense secretory body
basal invagination
basal lamina
heterolysosomes
Golgi complex
lysosomes

The tegument

contains enzymes of parasite and possibly of host origin (p. 10); (*b*) it may be a site for volume regulation; (*c*) it serves a protective function both against the host's digestive enzymes (see p. 11) and the host's immune reactions; (*d*) it may also function as a site of metabolic transfer. Valuable reviews on various aspects of the ultrastructure and physiology of the tegument are those of: Arai (*19*), Arme & Pappas (*27, 28*), Halton (*289*), Kuperman (*419*), Lumsden (*456, 457*), Lumsden & Murphy (*459*), Lumsden & Specian (*462*), Pappas (*624*), Podesta (*647, 648, 649*) and Threadgold (*878*).

Recent studies by transmission (TEM) and/or scanning (SEM) electron microscopy on the adults of species (other than *Hymenolepis*) are those on: *Multiceps endothoracicus (342, 343), Proteocephalus ambloplitis (139, 369), Bothriocephalus acheilognathi (267), Taenia crassiceps (601, 410), T. taeniaeformis (519), Anomotaenia constricta (244), Paricterotaenia porosa (244), Proteocephalus tidswelli (869), Eubothrium salvelini (858), Hydatigera taeniaeformis (375), H. krepkogarski (742), Diphyllobothrium dendriticum (279), D. latum (974), Tetrabothrius* spp. (*14*), *Triaenophorus nodulosus (901), Eubothrium rugosum (901), Echinococcus granulosus (870), E. multilocularis (519), Grillotia dollfusi (954),* and *Spirometra erinacea (972).*

Properties of absorptive surfaces

Before discussing the functioning of the cestode tegument, however, it is essential to have some understanding of the properties of absorptive surfaces in general and the intestinal mucosa in particular. Such surfaces have a number of closely identifiable morphological, physiological and biochemical characteristics. A major morphological adaptation is the amplification of the surface area exposed to the intestinal milieu. In the mammalian intestine this amplification is achieved by a 'brush border' of microvilli (Fig. 2.2) which may amplify the exposed surface by as much as 26–30 times (*58, 459*), there being, perhaps, 3000 microvilli per mucosal cell (*877*). In cestodes, the amplification of the surface is achieved by the presence of delicate cytoplasmic extensions, or microtriches (Figs. 2.1 and 2.2), reminiscent of mucosal cell microvilli. The size and numerical density of the microtriches vary both between species and in different regions of the same worm, as does the degree of amplification of the surface.

The surface amplification factor (SAF) has been comprehensively investigated in adult *Hymenolepis diminuta, H. nana, Eubothrium crassum, Proteocephalus pollanicoli, Schistocephalus solidus* (plerocercoid) and *Taenia crassiceps* (cysticercus), using a computerised image analysis system (*883*). The SAF ranged from 2.2 (immature proglottides of *H. diminuta* to 16.3 (mature proglottids of *P. pollanicola*). If, however, the dense distal tip

7

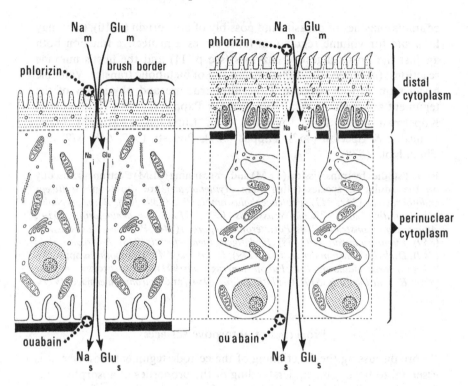

(a) INTESTINAL MUCOSA   *(b)* CESTODE TEGUMENT

Fig. 2.2. Structural analogy between the tegumental cells of cestodes (right) and the mammalian intestinal epithelium (left) both of which show surface amplification (see text).

Details have been omitted from the central regions of both systems to show schematically a model of a sodium-dependent glucose transport system. Phlorizin-sensitive glucose (Glu_m) is accumulative and coupled to the uptake of ambient sodium (Na_m) from the medium. In effect, sodium uptake is downhill due to a low internal concentration (Na_i) maintained by active extrusion (ouabain-sensitive) of sodium (Na_s) at the 'serosal surface'. Passage of glucose across the basolateral membrane is carried out by some mechanism, as yet unknown. (Right, based on Béguin, 1966. Left, Reprinted from Morphological and functional aspects of the cestode surface, by R. D. Lumsden and W. A. Murphy, in *Cellular interactions in symbiosis and parasitism*, ed. C. B. Cook, P. N. Pappas and E. D. Rudolph, 1980 by permission, © by the Ohio State University Press. All rights reserved.)

of each microthrix is excluded – as it may not be involved in absorption and transport – the *functional* amplification factor (FAF) shows a range 1.73–11.75; this compares with a FAF value for the mouse mucosa of 26 (*58*). As the analogy with the intestinal mucosa is so close, this amplified border is here also referred to as a 'brush border'.

The ultrastructure of microtriches has been extensively studied and is generally well known. For details of the modifications of the fine structure encountered, the various reviews (p. 7) should be consulted. As the microtriches of *Hymenolepis* have been described in detail (*462*) the account below will be confined to this species. In *H. diminuta*, microtriches have been quoted as having a maximum diameter of 0.14–0.19 μm and a maximum length of 0.9–1.08 μm, by different authors (*878*); these variations probably reflect differences in technique and in the region and age of the specimens examined. A microthrix has a bipartite structure, the distal point being an electron-dense tip separated from the basal shaft by what appears to be a flattened sac, which would allow for considerable flexibility of the tip. The core of each shaft contains numerous fine filaments which resemble the actin components of microvilli.

Within the interstices of the brush border is a hydrodynamic layer of 'unstirred water', which must act, to some extent, to modify the kinetics of the transport processes (Chapter 5). The significance of this layer as a diffusion barrier, is, however, a matter of dispute (*458*).

The glycocalyx

GENERAL PROPERTIES

The apical plasma membrane of the mucosal cell is further coated with a *glycocalyx* of mucopolysaccharide and glycoproteins. The membrane proper of the brush border contains pores through which molecules of amino acids, sugars etc. can pass. In cestodes, the glycocalyx, which, in common with other platyhelminths, appears to be produced by synthesis within the tegument itself, contains a preponderance of acidic groups which result in the membrane having a net electronegative fixed surface charge (*459*). Analysis reveals a predominance of glycoproteins, with molecular weights ranging from 12 000 to 237 000 (*387*) (Fig. 2.3; Table 2.1). The evidence suggests that free hexoses and amino acids absorbed by the worm are rapidly incorporated into macromolecules in the perinuclear cytoplasm and in turn are transferred to the brush border. It has been estimated that in *H. diminuta* the glycocalyx has a turnover rate of about 6 h (*457*). In *Echinococcus granulosus* the composition of the glycocalyx has

9

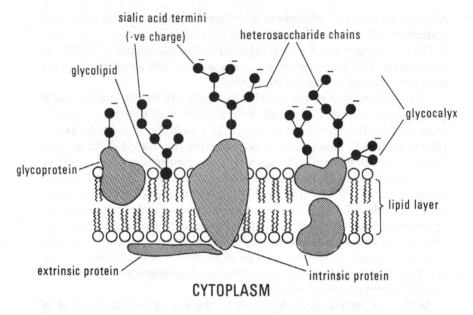

Fig. 2.3. Diagrammatic representation of the molecular organisation of the tegument plasma membrane (based on the fluid mosaic model of membrane structure of Singer & Nicolson (1972)). The carbohydrate moieties of the membrane glycoproteins and glycolipids are exposed on the external face as the glycocalyx. (After Smyth & Halton, 1983.)

been reported to be different in protoscoleces from cysts of sheep and horse origin and this character may prove to be of taxonomic value in separating different 'strains' of this species (*491*).

Amongst other activities, the glycocalyx has the capacity to concentrate both inorganic and organic ions. Some of these ions may serve to activate surface (i.e. membrane-bound) enzymes, a number of which have been reported in cestodes (Chapter 6).

The glycocalyx also has the potential of binding high molecular weight substances such as *host* enzymes (e.g. amylase), which may allow 'contact digestion' (= 'membrane' digestion) to take place (*674, 796*). The presence of such intrinsic enzymes could confer a 'kinetic' absorption advantage on the worm (in relation to the intestinal mucosa with which it is competing for nutrients) in that the spatial relationships at the worm surface would be such that the breakdown products of enzyme activity would be immediately available for absorption by the tegument (*622*). This aspect is considered further in Chapter 6.

The tegument

Table 2.1. *Molecular weights of proteins and glycoproteins associated with the tegumental brush border of* Hymenolepis diminuta. *(Data from Knowles & Oaks, 1979.) (See also Figs. 6.3 and 6.6)*

(Glyco) protein	$M_r \times 10^{-3a}$	(Glyco) protein	$M_r \times 10^{-3}$	(Glyco) protein	$M_r \times 10^{-3}$
1	237	10	128	19	56
2	218	11	119	20	49
(I)	214	12^b(III)	117	VI	45
3	198	13	103	21	44
(II)	195	14	94	22	36
4	184	15	79	23	35
5	169	16	75	24	30
6	162	(IV)	73	25^b(VII)	24
7	155	17	69	26^b(VIII)	22
8	148	18	64	27^b(IX)	17
9	130	(V)	58	28^b(X)	12

Note: Parentheses indicate glycoprotein represented by Roman numeral.
[a] Mean molecular weight determined by comparison to the relative mobility of the standards from five separate experiments.
[b] Indicates corresponding glycoprotein. Correlation coefficient of the linear regression analysis determined from mobility of the standards versus the log of their molecular weight is 0.992 ± 0.001 S.E.

RESISTANCE TO DIGESTION

The mechanism whereby the proteins and glycoproteins within the brush border plasma membrane are able to survive digestion by the intestinal proteolytic enzymes is unknown (*397*). The structure of the membrane proteins appears to have evolved in such a way that the protein-sensitive peptide sequences are protected from enzymic attack (*397*). Intact *Hymenolepis* have been shown to inhibit proteolytic enzymes such as trypsin and chymotrypsin, but inactivation is accomplished without the secretion of detectable inhibitor molecules into the medium. However, pepsin, subtilisin (EC 3.4.4.16, 'protease' VII) and papain are unaffected (*744*). In these experiments, although trypsin was inhibited by the worm, it retained its catalytic activity (as tested by its action on azoalbumin as a substrate) against some low molecular weight synthetic substrates (e.g. benzoyl-DL-arginine-*p*-nitroanilide and *p*-tosyl-L-arginine methyl ester). Examination of the trypsin after incubation with worms (pulse labelled) failed to demonstrate the presence of an inhibitor of worm origin associated with the inactivated enzyme (*744*). It has been suggested that inactivation

11

probably merely involves a small structural change, which alters the enzyme's activity towards the higher molecular weight substrates only (*744*). See also Chapter 6.

Although cestodes can clearly resist pepsin and intestinal enzymes, it has been found that when protoscoleces of *Echinococcus* are treated with pepsin (a prerequisite for *in vitro* culture) the activity of a number of enzymes is reduced (*491*), a result which, at present, cannot be explained.

Absorption

The passage of nutrient molecules, especially hexose sugars and amino acids, across surface membranes of mammalian tissues – and especially the intestinal mucosa and red blood cells – has been much investigated. In the former system, numerous hypotheses have been proposed to account for the transport of such materials against substantial concentration differences between luminal and cytoplasmic regions of the gut. The well-known Gradient Hypothesis developed by Crane (*155, 156*) indicates that the system involves a sodium-coupled accumulative transport system. This view was based on the observation that all cells appear to have in their membrane structure, specific sites through which sugars may enter and other sites out of which Na^+ is pumped. The exact mechanism of the sodium 'pump' is not clear, although it has received much attention in mammalian tissues.

That the sodium pump operates in cestodes has been clearly shown in *H. diminuta* and the same inhibitors (ouabain and phlorizin) which inhibit the system in the mucosal cells are also effective in cestodes (Figs 2.2 and 2.4). One major difficulty, however, in applying the intestinal cell model to the cestode tegument is that the enzyme Na^+/K^+-ATPase, which is involved in the former, has only been identified with certainty in *E. granulosus* (*491*) and not in any other species (*878*).

The various mechanisms involved in the uptake of low molecular substances by the cestode surface and the enzyme activities at the host–parasite interface are discussed further in Chapters 5 and 6. The uptake of macromolecules by endocytosis is discussed on pp. 14 and 44.

Cellular organisation

GENERAL STRUCTURE

A major structural difference between the intestinal mucosa and the cestode tegument is that the latter is a syncytial epithelium organised into two zones

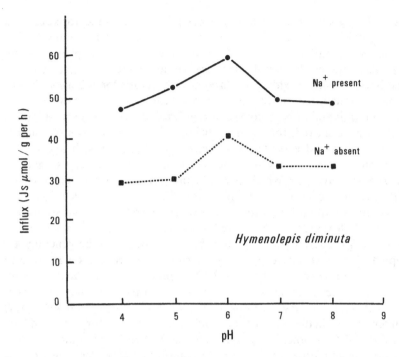

Fig. 2.4. Uptake of methionine by *Hymenolepis diminuta* in Na^+ or Na^+-free fluids of varying pH. (After Lussier *et al.*, 1979.)

– a superficial, anucleate zone and a nucleated cytosomal zone. These two zones correspond to the 'cuticle' and 'sub-cuticular zones' of classical helminthology.

The terminology in the literature is confusing. Terms used for the outer (nucleate) zone include: *syncytial layer, surface layer, distal cytoplasm, tegument*; and for the inner (nucleated) zone: *perikarya, proximal layer, tegumental cells, cytons, perinuclear cytoplasm.* Bearing in mind that descriptive terms are more likely to be remembered and used correctly than non-descriptive words, the terms *distal cytoplasm* and *proximal cytoplasm*, which are well established in the literature, are used in this text, for these zones. The term *tegumental cytons* is used for the basic cells which make up the syncytial epithelium (Fig. 2.1).

The distal cytoplasm contains an abundance of granules, membrane-bound vesicles and mitochondria (Fig. 2.1). The majority of vesicles and granules appear to have their origin in the Golgi apparatus of the tegumental cytons, from where they are passed to the distal cytoplasm. These granules probably play a major role in the formation and replacement of the glycocalyx and the microtriches. Whether endocytosis occurs in the tegument of adult cestodes has long been a matter of considerable dispute and, until recently, all efforts to demonstrate it had proved unsuccessful. Using the electron-dense tracers ruthenium red, lanthanum nitrate and horseradish peroxidase as markers, endocytosis 'on a huge scale' has been demonstrated in the pseudophyllidean cestodes *Schistocephalus solidus* (adult and plerocercoid) and *Ligula intestinalis* (plerocercoid) (*335, 882*), ruthenium red, for example, being sequestered, transported across the tegument and exocytosed within 6 min (*882*). Smooth and coated micropinocytotic vesicles have also been found in the cysticercus tegument of *Taenia crassiceps*, which provides indirect evidence of endocytosis, although the process has not been confirmed experimentally (*188, 880, 881*). It is curious that all early experiments to demonstrate endocytosis in *adult* cyclophyllideans (*H. diminuta*) (*463*) proved negative, in spite of its demonstration in pseudophyllideans, and this view appears to be generally accepted (*458, 878*). However, using the ruthenium red marker referred to above, this process appears to have been demonstrated in *H. nana (fraterna)* (*652*). In these experiments, 30–35-day-old worms from mice were incubated in Hédon–Fleig medium for 1 h and transferred to 0.44% (w/v) ruthenium red in the same medium and incubated again for 5 min and then fixed for TEM. Ruthenium grains were clearly identified in pinosomes but *only* in the immature proglottides. The pinosomes appeared beneath each other, giving the impression of the inward movement forming an apparent 'pinocytotic track'. The authors speculated that there may be fundamental differences between the tegument in different strobilar regions and that nutrition in the posterior regions may be carried out by membrane digestion (p. 123), which may not operate in immature regions. It is possible that the failure of other workers to demonstrate endocytosis in cyclophyllidean cestodes may be due to examining only mature regions of the strobila. These recent observations are of great fundamental interest and significance and clearly require further investigation.

FREEZE-FRACTURE STUDIES

The freeze-etch technique has been applied to the tegument of adult *Hymenolepis microstoma*, larval *Echinococcus granulosus*, *Taenia taeniaeformis* and *T. crassiceps* (*66, 147*) and the results are in general agreement with the structural characterisation envisaged from TEM studies (Fig. 2.5). One new observation to emerge from such studies is the presence of vertically orientated channels, membrane-bound channels passing from the proximal tegumental membrane to the distal surface where they appear to open by pores (*147*). These may be the 'pore canals'

The scolex

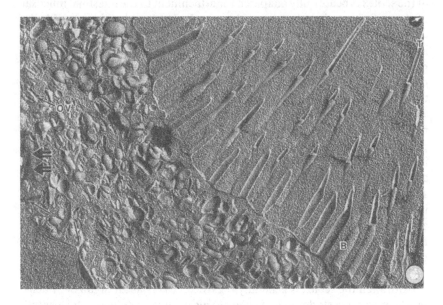

Fig. 2.5. *Taenia taeniaeformis* (strobilocercus): freeze-etch preparation, × 23 800. The tegument contains numerous oblong vesicles (OV) and round, irregularly shaped inclusions. Microtriches have short, broad bases (B) and long, thin tips (T). Note also the double-bound intrusions in the base of the tegument (arrow) and channel (double arrows) within the tegument. The arrow at top right shows the direction of the external surface. (After Conder *et al.*, 1983.)

postulated by Threadgold (*876*) but now thought by him and later workers probably to represent poorly fixed dendritic processes of sensilla.

The scolex

General account

THE SCOLEX–HOST INTERFACE

Cestodes live in what might be termed a 'hazardous' environment, in which the peristaltic movement of the gut and the passage of partly digested food make the possession of an efficient form of attachment an essential prerequisite for survival.

The various forms of cestode scolex or holdfast are well known and are described in standard texts (*32, 345, 800, 933*). In general, the morphology of the scolex is beautifully adapated for attachment to the intestinal mucosa of a specific host. Some species with a poorly developed scolex (e.g. *Ligula intestinalis*) – i.e. one not adapted to the topography of any particular host intestine – may have a wide host spectrum. Since the scolex forms the major point of contact between the adult worm and the host tissue – the *host– parasite interface* – its structure and function are clearly of major importance to an understanding of the physiology of this group. Yet, surprisingly, although various aspects of it have been studied in some detail, e.g. the morphology, the fine structure, the presence or absence of glands – the basic physiology of the scolex as an entity (i.e. independent of that of the strobila) is largely unknown. It is not known, for example, whether the metabolism and/or nutrient requirements of the scolex is different from that of the strobila. However, it is known that the scolex plays an important role in the induction of host immunity, at least in *Hymenolepis* (*334*). Little work appears to have been carried out in this area. In *Echinococcus*, at least, the metabolic pathways and nutritional requirements of the evaginated protoscolex (which is essentially a scolex) appear to be the similar to those of the adult worms (i.e. scolex plus strobila) (*488*).

Although the scolex is generally regarded largely as an organ of attachment, in some cestodes, e.g. *Echinococcus*, it may also have a 'placental' function and absorb nutrients directly from the mucosal wall – a condition which occurs in some trematodes (*810*). The microtriches covering the scolex region often show a structure different from that of the strobila, a situation presumably related to the topography of the host mucosa.

Scolex form and function

There are three main types of scoleces – the acetabulate, the bothriate and the bothridiate. The *acetabulate* type, which is characteristic of the Cyclophyllidea, can be divided broadly into the 'non-penetrative' and the 'penetrative' types. In the former, an attached sucker encloses a group of villi but the scolex does not penetrate deeply into the mucosa. In the 'penetrative' type, attachment is more intimate and the crypts of Lieberkühn are invaded. In the Taeniidae, this type of scolex often bears a *rostellum* – a dome-like or finger-like extension of the scolex – present in such well-known forms as *E. granulosus, Dipylidium caninum* and *H. nana*. In some cases, e.g. *E. granulosus* and *H. nana*, the rostellum carries a *rostellar pad* (Fig. 9.5, p. 240), which, when expanded, forces the hooks outward and makes withdrawal difficult. Rostellar glands are present in some species (see below).

The *bothriate* type of scolex consists typically of a pair of shallow, elongate, sucking *bothria*. Each bothrium may take the form of a groove, slit or saucer or, by fusion of the margins, a tube. In some pseudophyllids, e.g. *Schistocephalus* (Fig. 10.1, p. 261), the bothria occur as shallow grooves which can have little, if any, attachment function. This species has a progenetic plerocercoid and maturation (in the bird gut) is achieved very rapidly, within 36–48 h – an adaptation undoubtedly related to the weak mode of attachment and the associated difficulty of maintaining a position in the gut for more than a short period. In this case, maturation is triggered by the rapid rise in temperature (to 40 °C) on entering the bird gut or *in vitro* (Fig. 10.2, p. 263).

The *bothridiate* type of scolex is characteristic of the Tetraphyllidea, and typically bears four, leaf-like outgrowths or *bothridia* (phyllidea), whose morphology is closely adapted to the topography of the host mucosa (*958*). Many scoleces are also armed with hooks; in the tetrarhynchid cestodes, four retractable spiney *proboscoides* may be protruded. In some instances (e.g. *Spathebothrium*) there is no scolex or any other form of attachment, and in some (e.g. *Parabothrium gadi-pollachi*), the scolex degenerates to form a 'scolex deformatus' (*796*). Some tetraphyllids possess a large protrusible muscular mass or *myzorhynchus*. Although this appears to act as an organ of attachment, it may also have a nutritive (i.e. placental) and/or sensory function.

Scolex glands

Most glands in adult cestodes are found in the scolex. The interproglottidal glands which occur in species of *Moniezia* are an exception; these give positive reactions for the periodic acid–Schiff (PAS) test, RNA, esterase and phosphatases, but the nature and function of their secretion is unknown (*340*). Scolex (or head or frontal) glands have been investigated by Kuperman & Davydov (*422*) in a large number of species, especially Pseudophyllidea (Table 2.2). Three kinds of secretions have been recognised (Fig. 2.6): (*a*) *apocrine*, in which the secretion accumulates in the tegument as projections or 'tumules' – the discharge of the secretion is brought about by the partial or complete destruction of these; (*b*) *eccrine*, in which the secretory ducts penetrate the distal cytoplasm, open without any projection, and secretion is released through membrane fusion (exocytosis); (*c*) *microapocrine*, in which the secretion enters the distal tegument and fills it – the secretory discharge occurs by means of small evaginations of the outer plasma membrane and their subsequent detachment.

Glands in the rostellar region of the scolex of a number of Cyclophyllidea, e.g. *Hymenolepis* spp., *Taenia solium*, *T. crassiceps*, *Davainea*

The adult: special structural features

Table 2.2. *Species of adult cestodes with scolex glands, investigated by Kuperman & Davydov* (1982*b*).

Cestode species	Host species
Eubothrium rugosum	*Lota lota*
E. salvelini	*Salvelinus alpinus*
E. crassum	*Oncorhynchus tschawytscha*
E. acipenserinum	*Acipenser stellatus*
Bothriocephalus gowkongensis[a]	*Cyprinus carpio*
B. scorpii	*Myxocephalus quadricornis*
B. claviceps	*Anguilla anguilla*
Diphyllobothrium latum	*Mesocrisetus auratus*
Schistocephalus solidus	fam. Laridae
Ligula intestinalis	fam. Laridae
Triaenophorus nodulosus	*Esox lucius*
T. crassus	*E. lucius*
T. meridionalis	*E. lucius*
Cyatocephalus truncatus	*O. kisutch*
Proteocephalus percae	*Perca fluviatilis*
P. exiguus	*Coregonus albula*
Khawia sinensis	*C. carpio*
Caryophyllaeus laticeps	*Abramis brama*
Archigetes sieboldi	*Tubifex tubifex*
Choanotaenia porosa	fam. Laridae

Note: For larval cestodes examined, see Table 8 of Kuperman & Davydov (1982*b*).
[a] Synonym of *B. acheilognathi* (Pool & Chubb, 1985).

proglottina, Aploparaxis furcigera and *E. granulosus*, have also been described (*410, 796*). In the last species, the secretion (unusually!) originates in the nucleus of the gland cells and droplets of secretion can be seen at the rostellar surface (*794, 796, 813, 866*). Intranuclear bodies have also been observed in the rostellar tegumental cell bodies of *H. microstoma* (*938*). Ultrastructure studies suggest that the secretion takes place by a *holosecretory* mechanism (*866*), i.e. release of secretory products involves the destruction of the gland cells which produce it. This secretion has proved to be unusually labile and chemically unreactive. Histochemical evidence points to it being a lipoprotein (*813*). The common occurrence of glands in the scolex region suggests that their secretions must play an important role in the activities at the host–parasite interface, but, to date, no function for them has been identified.

18

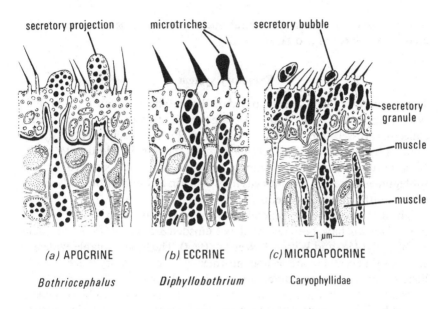

secretory projection   microtriches   secretory bubble

secretory granule

muscle

muscle

—1 μm—

(a) APOCRINE   *(b)* ECCRINE   *(c)* MICROAPOCRINE

Bothriocephalus   *Diphyllobothrium*   Caryophyllidae

Fig. 2.6. Three types of secretory mechanisms found in the scolex glands of adult cestodes. (After Kuperman & Davydov, 1982*b*.)

Parenchyma

As in other parasitic platyhelminths, parenchymal cells and fibrous interstitial material occupy most of the space between the various organs. The nature of the cestode parenchyma was long a mystery to helminthologists and was often described as being 'vacuolar'. This appearance was undoubtedly due to the harsh fixatives used in early studies which often dissolved out lipids and failed to preserve glycogen, which – although easily fixed in the small pieces used for electron microscopy – is notoriously difficult to fix if present in quantity in large tissue pieces. The major question was whether the 'parenchymal space' represented an *intra*cellular or *inter*cellular space. These spaces are frequently loaded with glycogen and in the case of progenetic plerocercoids, such as those of *Schistocephalus* or *Ligula*, may amount to more than 50% of the dry weight of the worm. Prominent lipid droplets too, are present (*462*). In *H. diminuta*, the close association of these cells with myofibrils has led some workers to the conclusion that the 'parenchyma cells' are, in fact, myocytons (see Fig. 2.7) of which two or more may be cytoplasmically interconnected (*462*). If this interpretation proves to be correct, then the musculo-parenchymal tissue could be considered, to some extent, to be syncytial.

Excretory system

The cestode excretory system is based on the platyhelminth protonephridial system, with flame cells and collecting vessels; two to four longitudinal collecting vessels are common but up to 20 may occur (*345*). In the pseudophyllids *Ligula* and *Schistocephalus*, there is, in addition to some 16 longitudinal canals, a complex network of vessels situated in the subtegumental region. Flame cells may occur in any part of the body, even in the central nervous system.

The ultrastructure and general features, which have been studied largely in *H. diminuta*, have been reviewed by Lumsden & Specian (*462*), Lumsden & Hildreth (*458*) and Wilson & Webster (*964*). The little available evidence suggests that cestodes are osmoconformers whose 'excretory' system plays little or no role in osmoregulation and appears to be, in fact, largely excretory (*462*). In cestodes, the excretory tubules appear to be syncytial, with subadjacent nucleated cytons and lumenal cytoplasm bearing microvilli. This situation is in contrast to that in trematodes, where excretory tubules tend to be cellular structures with lumenal surfaces amplified by lamellae (*810*). The composition of the excretory canal fluid in *H. diminuta* has been studied in detail (*947–951*). The extent to which the excretory system functions in the absorption or reabsorption of electrolytes or organic compounds or is involved in their circulation is a matter that clearly requires further investigation.

Muscular system

Ultrastructure

The morphology and ultrastructure of the cestode musculature has been reviewed by Lumsden & Hildreth (*458*) and Lumsden & Specian (*462*). A cestode muscle cell consists of two portions (*a*) the non-contractile cytoplasmic portion or *myocyton* containing the nucleus and usually packed with glycogen and (*b*) the *myofibril* portion containing myofilaments (presumably actin and myosin) (Fig. 2.7). The myocyton is generally positioned some distance from the myofibrils to which it is connected by tendrillar processes (*462*). The latter appear to be connected to

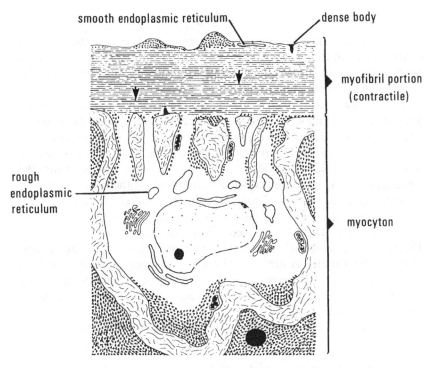

Fig. 2.7. Cortical muscle element consisting of a contractile myofibril portion and a mycyton portion. Arrows indicate two of the thick myofilaments distributed among the more numerous thin myofilaments. (After Lumsden & Hildreth, 1983.)

more than one myocyton and, in places, the myocytons are cross-connected by nexus junctions (= gap junctions). Cestode myofibrils closely resemble those of turbellarians and most trematodes in being generally non-striated and containing thick and thin myofilaments; dense bodies (Fig. 2.7), which represent attachment sites for the thin myofilaments, are associated with the sarcolemma. Although a sarcoplasmic reticulum is generally poorly developed, its cisternae are arranged consistently beneath the sarcolemma. However, a T-tubule system (tubular invaginations of the sarcolemma) which serves, in vertebrate striated muscle, to transport stimuli to the myofibrils, appears to be lacking, as is the case in trematodes (*810*). In cestodes, striated myofibrils are, however, found in some situations, e.g. the tentacular bulbs of trypanorhynchids. For detailed descriptions of muscle ultrastructure, the reviews cited above should be consulted.

The adult: special structural features

Physiology

Comparatively little is known regarding the physiology of cestode muscle contraction, but it is assumed that a two filament (actin–myosin) sliding mechanism operates as in all invertebrate and vertebrate systems so far studied (*462*). The interaction of the thin filaments with the thick would exert tension against the cell surface at the points of dense body insertion. This force, when transmitted to the surrounding connective tissue, would result in body movements. The smooth muscle fibres are characteristic of slow-contracting fibres in other invertebrates, and, being able to sustain contractions for long periods, would be appropriate to maintain the strobilar tonus against the peristaltic movements of the host intestine.

Although some species of cestodes (e.g. *Hymenolepis diminuta*; p. 236) undergo diurnal migrations within the intestine – and therefore their suckers may not always be in a state of sustained contraction – others (e.g. *Echinococcus granulosus*) apparently remain in the same place and appear to be able to maintain contraction. How this is achieved physiologically is not known, but it has been speculated that this may operate through a 'catch' muscle mechanism (as in lamellibranchs) or by the involvement of special stretch receptors (*796*).

The neurophysiology of cestode muscle is discussed on pp. 32–4.

Nervous system

General account

The nervous systems of helminths, including cestodes, have been reviewed by Falkmer *et al.* (*207*), Gustafsson (*278*), Halton (*289*), Rohde (*707*) Shisov (*758*) Sukhdeo & Mettrick (*828*) and Lumsden & Hildreth (*458*). It has been stressed that 'the borderline between the nervous system and the endocrine system has become indistinct and the two parts are now dealt with under the common name "neuroendocrine system"' (*278*).

Until recently, the structure, ultrastructure and physiology of the cestode nervous system was a very neglected area of research. The reasons for this are not hard to find and relate chiefly to the fact that the sheath and glial elements, which, by selective staining, are used to visualise the nervous system in other groups, are completely lacking in cestodes. The problem of visualisation, however, has been revolutionised by two techniques, one based on the cytochemical demonstration of cholinesterase and the other, more recently introduced, based on highly specific immunocytological techniques for the demonstration of neuropeptides and serotonin

22

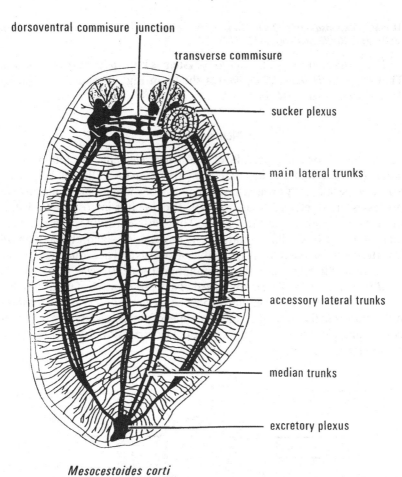

Mesocestoides corti

Fig. 2.8. Nervous system of the tetrathyridium larva of *Mesocestoides* sp. based on histochemical (for cholinesterase) and histological techniques. (After Hart, 1967.)

(5-hydroxytryptamine (5-HT)). The latter, in particular, has provided much valuable data on cestode neurobiology (*278, 283*). The fluorescent histochemical technique for catecholamines has also been used with some success (*280, 755*).

Representative species investigated by these and other techniques such as TEM include the following: *Dipylidium caninum* (*755, 756*), *Diphyllobothrium dendriticum* (*277, 280–283*), *Echinococcus granulosus* (*570*), *Hymenolepis diminuta* (*462, 966*), *H. microstoma* (*936, 937, 941, 943*), *H. nana* (*202, 206, 966*), *Mesocestoides* sp. (*304*),

The adult: special structural features

Moniezia expansa (*408*), *Oochoristica sigmoides* (*411*), *Pelichnobothrium speciosum* (*260*), and *Raillietina* spp. (*411, 717*).

The gross anatomy of the nervous system is known for very few species. That of adult *H. nana* (*206, 966*) is shown in Fig. 2.9 and that of larval *Mesocestoides* sp. (*304*) in Fig. 2.8.

Neurocytology

In cestodes, most, perhaps all, of the neural elements are found in the ganglia of the scolex and the main nerve tracts. In these, the cell bodies are arranged peripherally around a core made up of a tangled mass of neuronal processes often referred to as 'neurites' because it is impossible to distinguish between axons and dendrites. This mass constitutes the so-called 'neuropile'. Like all differentiated cells in cestodes – and probably in all invertebrates – nerve cells are incapable of mitosis and hence new nerve cells must arise directly from the germinative (=stem) cells (*276*). In *D. dendriticum*, the neuropile is composed of a dense network of unmyelinated nerve fibres with no extracellular stroma between the fibres (*277*). The fibres differ mainly in the content of different types of vesicles, which can be dense-core, small clear or large clear vesicles. The structure of synapses is discussed further on pp. 25–7.

Hymenolepis nana

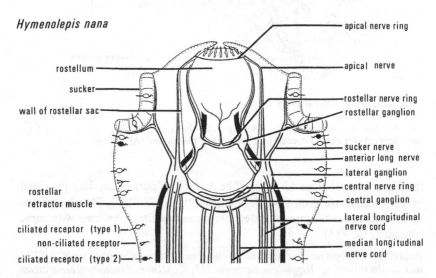

Fig. 2.9. Nervous system of adult *Hymenolepis nana*. (After Fairweather & Threadgold, 1983.)

24

Nervous system

Neurotransmitters

The message substances operating in neuroendocrine systems are amines, amino acids, acetylcholine and peptides. It should also be pointed out that our present understanding of neurotransmission in general may still be very limited, for according to Bloom (*86, 87*), in vertebrates at least, 'modern methods of chemical analysis are providing an ever-increasing list of new transmitter molecules'. This situation is likely to be reflected in invertebrates also.

SYNAPSES

For a basic account of synapses in general, see the text by Threadgold (*877*). Synapses have been examined in a number of genera, e.g. *Diphyllobothrium*, *Echinococcus* and *Hymenolepis* (*277, 726, 936, 941, 943*). In the plerocercoid of *Diphyllobothrium dendriticum*, synapses are formed between neurites in which the presynaptic neurite contains (*a*) *both* dense-core vesicles *and* clear vesicles or (*b*) small clear vesicles only (Fig. 2.10). The former correspond closely to the dense-core vesicles of aminergic neurones, and have been tentatively classified as aminergic synapses (*277*), whilst the latter are classified as cholinergic synapses. These neurones are discussed further below.

In the scolex ganglion of the tetraphyllid *Pelichnobothrium speciosum*, where unipolar and multipolar neurones are found, the nerve cell processes have been reported as forming 'tight junctions which cannot be interpreted as synaptic contacts' (*260*); this anomalous observation clearly requires confirmation.

AMINERGIC NEURONES

5-HT is the only aminergic neurotransmitter so far detected in cestodes (*278*). The widespread occurrence of 5-HT and acetylcholine (ACh) in the cestode nervous system suggests that, as in trematodes, 5-HT plays an important role as an excitatory neurotransmitter and that ACh functions as an inhibitory transmitter. The presence of 5-HT in the plerocercoids of *Diphyllobothrium dendriticum* has been demonstrated elegantly by immunocytology by Gustafsson *et al.* (*283*) (Fig. 2.11). Early evidence suggested that cestodes may have little or no ability to synthesise 5-HT, this substance being taken up largely from the host (*297, 528*). However, it has since been demonstrated that *H. diminuta* has the capacity to synthesise 5-HT, both from tryptophan and 5-hydroxytryptophan (Fig. 6.12, p. 137) (*681, 683*).

It is especially interesting to note that *host* 5-HT may play a part in the

Diphyllobothrium dendriticum

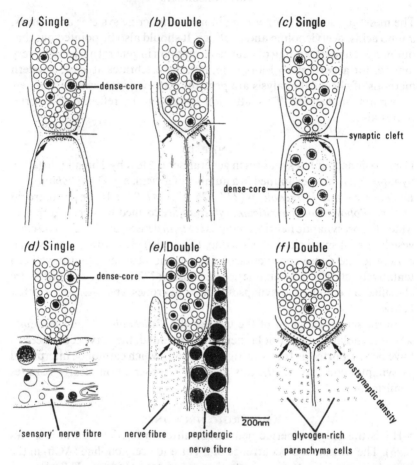

Fig. 2.10. *Diphyllobothrium dendriticum*: types of single and shared synaptic contacts. The nerve terminals are filled with dense-core and small clear vesicles. (*a*) Single synapse. (*b*) Shared synapse. (*c*) Single synapse on nerve fibre with dense-core and small clear vesicles. (*d*) Single shared synapse on large lucent nerve with mixed vesicle content and of presumed sensory nature. (*e*) Shared synapse on peptidergic nerve fibre and fibre lacking special characteristics. (*f*) Shared synapse on glycogen-rich parenchyma cells. The postsynaptic densities (bold arrows) and the synaptic clefts (thin arrows) are well developed in all synapses. (After Gustafsson, 1984.)

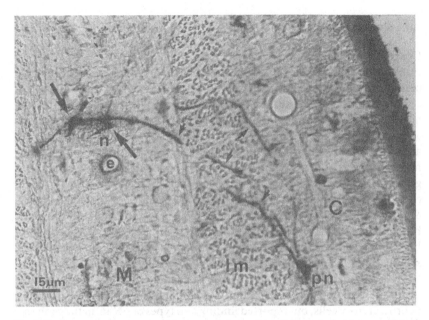

Fig. 2.11. Serotonin-immunoreactive neurones (large arrows) in the main nerve cord (n) of *Diphyllobothrium dendriticum* plerocercoid. M, Medullary parenchyma; C, cortical parenchyma; lm, longitudinal muscle layer; e, main excretory duct. The nerve fibres extend (small arrows) through the longitudinal muscle layer out to the peripheral nerve ring (pn). Sections stained with Sternberger's immunoperoxidase-antiperoxidase (PAP) technique. (Courtesy Dr Margaretha K. S. Gustaffson.)

curious diurnal migrations which some species (e.g. *H. diminuta*) make up and down the intestine (p. 236). It has been shown that this circadian rhythm can apparently be correlated with similar changes in the 5-HT levels in the small intestine which, in turn, are related to the pattern of host feeding (*132, 529, 827*).

AMINERGIC SYNAPSES

In *D. dendriticum*, a number of types of presumptive aminergic synapses have been described (*277*) (Fig. 2.10). Evidence that the dense-core granules in this and other species may contain storage granules of a monoamine such as noradrenaline (= norepinephrine), dopamine or 5-HT is based on microspectrofluorometric analysis, histochemical reactions and biochemical analysis (*206, 277, 295–297*).

27

The adult: special structural features

General account

Presumptive neurosecretory cells (pNSC) have been described in both cestodes and trematodes, identification being based on the staining of contained granules by paraldehyde-fuchsin (PAF). It is now widely recognised that this stain is non-specific so that conclusions based on its use alone are equivocal. If, however, a cycle or pattern of activity in the cell can be associated with a corresponding pattern of mitosis, differentiation or morphogenesis, the case for identification of such cells as neurosecretory is greatly strengthened. Potential peptidergic cells were first demonstrated in cestodes in the (rudimentary) rostellum of *Hymenolepis diminuta* by Davey & Breckenridge (*169*) although the identity of these particular cells has since been challenged (see below). Presumptive neurosecretory cells have also been reported in *H. nana* (*201–203*), *H. microstoma* (*937*), *Bothriocephalus scorpii* (*372*), *Echinococcus granulosus* (*570*), *Diphyllobothrium dendriticum* (*280, 281*). Later, Specian *et al.* (*819*) concluded that the PAF-positive cells in the scolex of *H. diminuta* were uniglandular endocrine cells and *not* neurosecretory cells, but reported finding two types of pNSC in the cephalic ganglia and lateral nerve cords. It was speculated that the neurosecretory material from these cells regulated the activities of the gland cells.

Experimental activation

Perhaps the most striking evidence for concluding that such PAF-positive cells were, in fact, neurosecretory cells, came first from the studies of Gustafsson & Wikgren (*280, 281*) on *D. dendriticum*. In the plerocercoid of this species, peptidergic neurones were tentatively identified using, in addition to PAF, chromalum-haematoxylin/phloxine, paraldehyde-thionin, rescorcin-fuchsin and Alcian blue/Alcian yellow; this conclusion was later confirmed by immunocytological tests (see below). Aminogenic neurones were identified by the Falck–Hillarp fluorescent technique (which emits yellow fluorescence with 5-HT). These workers further showed that when the plerocercoids of *D. dendriticum* (from fish) were cultured *in vitro* at 38 °C (i.e. as in the natural or experimental warm-blooded host, a bird or a hamster) increased cellular activity began after only 5 min of cultivation and was marked after 1 h. After this period the number of pNSC increased at least 10-fold over the number in untreated plerocercoids.

Such a sudden increase in activity would be compatible with a physiological 'switch' occurring during the larval/adult transformation process triggered by the change from a cold-blooded to a warm-blooded environment; in this case, the stimulating factor could be the release of

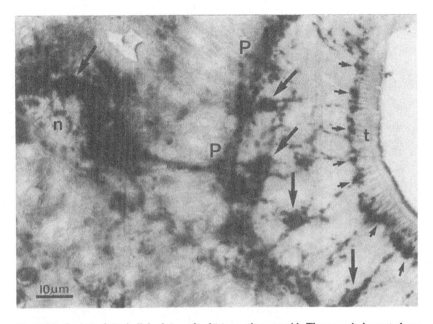

Fig. 2.12. Scolex of *Diphyllobothrium dendriticum* plerocercoid. The neural elements have been labelled with growth hormone releasing factor (GRF), a vertebrate neuropeptide. In the main nerve cord (n) one GRF-immunoreactive cell body can be seen (large arrow); in the peripheral nervous system (P) several GRF-immunoreactive cells bodies occur (large arrows). The small arrows point to nerve terminals beneath the basal lamina of the tegument along the inner border of the bothridia. Sections stained with Sternberger's immunoperoxidase-antiperoxidase (PAP) technique. (Courtesy Dr Margaretha K. S. Gustaffson.)

neurosecretory material. Presumably, this material could reach the target cell either (*a*) via synaptic junctions which allow direct delivery to the target cell or (*b*) by an indirect route, probably by extracellular diffusion, a process which appears to occur in coelenterates and free-living platyhelminths.

Immunological identification
Using immunocytological and immunofluorescent methods, nine verte-brate neuropeptides (Table 2.3) have since been detected unequivocally in the neuroendocrine system of *D. dendriticum* (*278, 283*): bovine pancreatic polypeptide, growth hormone release factor (GRF), peptide histidine isoleucine, gastrin, gastrin releasing peptide, Leu-enkephalin, neurotensin, vasotocin and oxytocin and one neuropeptide, FMRF-amide (Phe-Met-Arg-Phe-NH$_2$), a neuropeptide first isolated from the bivalve clam *Macrocallista nimbosa*. This unequivocal immunological identification

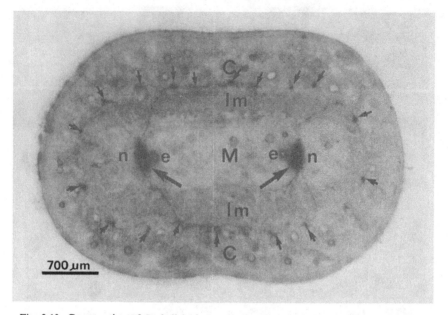

Fig. 2.13. Cross-section of *Diphyllobothrium dendriticum* plerocercoid with FMRF-amide immunoreactive nerve fibres in the main nerve cords (n) (thick arrows) and in the peripheral nerve cords (small thin arrows) in the cortical parenchyma (C). M, Medullary parenchyma; e, main excretory duct; lm, longitudinal muscle layer. Note the thin fibres extending through the muscle layer. Sections stained with Sternberger's immunoperoxidase-antiperoxidase (PAP) technique. (Courtesy Dr Margaretha K. S. Gustaffson.)

means that the criteria for identification of peptidergic neurones as postulated by Bern (*69*) are now satisfied.

The distribution of neuropeptides in *D. dendriticum*, which is very extensive, has been described in great detail by Gustafsson (*278, 283*); Figs. 2.12 and 2.13 illustrate the elegant cytological pictures obtained typically with the immunocytological technique.

CHOLINERGIC NEURONES

The presence of cholinesterase has been detected by chemical or histochemical means in a number of species: e.g. *Diphyllobothrium latum, D. dendriticum* (*796*); *Taenia saginata, T. taeniaeformis, Hymenolepis* spp., *E. granulosus, Dipylidium caninum* (*202, 435, 755, 756, 796*). Putative cholinergic synapses are characterised by their content of small clear vesicles tightly packed on the presynaptic side (*278*). Acetylcholine is thought to serve as an inhibitory neurotransmitter and has been shown to have a

Nervous system

Table 2.3. *Use of immunocytological techniques to demonstrate the nature of neurotransmitters in cestodes and turbellarians. The presence of neurones in the nervous system which are immunoreactive with sera raised against some neurohormonal peptides and against serotonin (5-HT).* (*Data from Falkmer* et al., *1985*)

Substance	Free-living		Parasitic	
	Microstomum lineare	*Polycelis nigra*	*Diphyllobothrium dendriticum*	*Schistocephalus solidus*
Somatostatin	−	−	−	−
FMRF-amide	+	+	+	+
VIP	−	−	−	−
Neurotensin	−	−	+	nt
Leu-enkephalin	−	nt	+	+
Met-enkephalin	−	+	+	+
Vasotocin	+	nt	+	nt
Serotonin	−	+	+	+

Note: −, no immunoreactive cells observed; +, immunoreactive neurones and/or nerve fibres found; FMRF, Phe-Met-Arg-Phe-NH$_2$; VIP, vaso-active intestinal polypeptide; nt, not tested.

potent inhibitory effect on the musculature of *H. diminuta* and *H. microstoma* (see later).

OTHER NEUROTRANSMITTERS

Evidence is lacking for the presence of the amino acids, γ-aminobutyric acid (GABA), glutamic acid, aspartic acid and glycine, which are believed to play a role as transmitter substances in vertebrates and other invertebrates.

Sense organs

The tegument of cestodes contains a wide variety of (presumed) sense organs which terminate in nerve processes. These can be divided broadly into two types: (*a*) ciliated receptors and (*b*) non-ciliated receptors. The former have been described in the adult and developmental stages of cestodes. They consist essentially of a bulb-like expansion of the dendrite from which a cilium-like process extends through the tegument. Their function has yet to be established, but various authors have suggested that they might be chemoreceptors (*85*), tactile mechanoreceptors, rheoreceptors (*12, 570, 940*) or osmoreceptors (*324*).

The adult: special structural features

Non-ciliated receptors show much variation. Some resemble ciliated receptors in that they have a terminal dendritic bulb with a basal body but no distal cilium (*85, 942*). Others, as in *H. nana* (Fig. 2.9), show little structural specialisation and appear to be little more than free nerve endings which may have a single, double or triple structure. Endings which do not penetrate the tegument to the external environment may have a mechanoreceptive or proprioceptive (i.e. receptive to internal stimuli) – rather than chemoreceptive – function (*12, 206, 942*).

Neuromuscular physiology

The difficulties of studying the nervous system are reflected in the fact that very little is known regarding the neuromuscular physiology of cestodes. Apart from an early study by Rietschel (*693*) on *Catenotaenia pusilla*, the

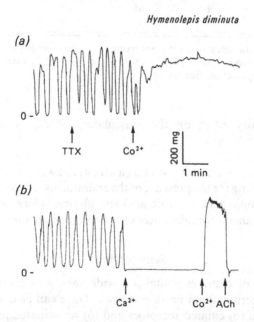

Fig. 2.14. *Hymenolepis diminuta*: effects of calcium and cobalt on contractions of the longitudinal musculature in strips of tissue cut from the adult worm. In normal Ringer's solution, periodic contractions of the longitudinal muscles were generated spontaneously (*a*) and (*b*). (*a*) Tetrodotoxin (TTX, 5×10^{-6} M) did not affect spontaneously generated muscle contractions. The application of $CoCl_2$ (5×10^{-3} M) resulted in sustained muscle contractions. (*b*) The application of $CaCl_2$ (5×10^{-3} M) caused muscle relaxation (final calcium concentration 6.2×10^{-3} M). $CoCl_2$ (5×10^{-3} M) elicited sustained contraction. Acetylcholine (ACh, 5×10^{-4} M) caused relaxation. (After Thompson & Mettrick, 1984.)

neuromuscular physiology of only the longitudinal muscles of the cyclophyllideans *Hymenolepis diminuta* and *H. microstoma* (*862*) and of the cestodarian *Gyrocotyle* spp. (*384, 404*) have been studied in any detail.

Experimental studies

Rietschel (*693*) found that when the longitudinal nerve cords of *C. pusilla* were severed there was no effect on the rhythmic contractions of the longitudinal strobilar muscle – a result suggesting that the latter are independent of direct neural control.

The studies on *Hymenolepis* spp. utilised *in vitro* systems using adult worms and strips of worm body wall (*862*). Although intact worms were found to be insensitive to ionic changes in the supporting medium, strips of body wall were extremely sensitive to ionic manipulation – a result presumably due to penetration differences. In whole strobila (0.5–1.0 cm), the tension generated (as measured by a force transducer) was made up of two components – a small brief tension (in *H. diminuta* of 500 mg amplitude and 4 s duration) superimposed on larger, longer peaks (up to 1200 mg in amplitude and 30 s duration). Excising the neck had generally no effect, but in some strobila there was an apparent reduction in tension amplitude.

In split-worm preparations, the level of Ca^{2+} in the bathing solution affected both spontaneous and evoked contractions in both species; the addition of $CaCl_2$ greatly reduced the amplitude and frequency of the contractions. $CoCl_2$ (which acts as a calcium-channel blocker in cell membranes) was found to be the most effective salt in stimulating muscle contractions; barium, cadmium, manganese, magnesium and nickel ions were less effective. Acetylcholine was found to have a strong inhibitory effect (Fig. 2.14). This result is in keeping with the view that 5-HT is the major stimulating neurotransmitter affecting muscle contractions, and preliminary experiments have confirmed the stimulating action of this substance (*862*). The concentration of extracellular Ca^{2+} may play a role in acetylcholine release.

The electrophysiological properties of cestode nerves appear to have been studied only in the cestodarian *Gyrocotyle* spp. (*384*). Some worms were found to display spontaneous electrical activity – either small potentials (6–16 μV in amplitude, lasting 10–20 ms) or larger spikes (16–20 μV of 8–10 ms duration) being recorded. An interesting result arising from this work was that the response to a single electrical stimulus usually evoked a single, large compound potential (LCP) which was graded and showed a decremental conduction, especially in a posterior direction. Thus a loss of 50% amplitude of the LCP was recorded over a few millimetres of nerve cord. This has been interpreted as indicating that 'much of the information processed in the longitudinal nerve cords may be concerned with local regions of the animal and only a small portion is invoked with conduction over appreciable distances' *(384)*.

It is evident from the above that the electrophysiology of cestodes – and indeed the neuroendocrine systems of the group in general – is an area which calls particularly for further detailed investigation.

Reproductive system

This is considered in Chapter 7.

3

The adult cestode in its environment

General considerations

Cestodes differ from trematodes and nematodes in that the adults, with a few exceptions, occupy one type of habitat – the alimentary canal. Even the exceptions occur in sites related to the gut. Examples of some aberrant genera are: *Stilesia*, *Thysanosoma* (bile ducts of sheep), *Porogynia* (bile ducts of guinea fowl), *Atriotaenia* (pancreatic ducts of *Nasua*), *Progamotaenia* and *Hepatotaenia* (bile ducts, gall bladder and liver of marsupials). *Hymenolepis microstoma*, which is found in the bile ducts of rodents, is widely used as a laboratory model (*446*).

An understanding of the physiology and morphology of the alimentary canal is thus of particular importance to the study of cestode physiology. Knowledge in this field has expanded greatly within recent years and a number of valuable reviews are available (*161, 213, 370, 507, 602, 730, 970*). The alimentary canal, considered specifically as an environment for parasitic helminths, has also been reviewed (*54, 56, 157, 527, 530*). The immune responses of the gut are reviewed in Chapter 11 and it must be emphasised that, as a result of these responses, the structure and function of the intestine can undergo profound changes when infected with cestodes (Table 3.1).

The alimentary canal as a biotope

Intestinal parameters

MORPHOLOGY

It is not intended to give here a detailed description of the vertebrate alimentary canal, accounts of which are given in the reviews referred to above. Some parameters of the intestinal environment which are likely to be important for the establishment and growth of a cestode are shown in Fig. 3.1. It is important to appreciate, however, that the physiology of a

35

Table 3.1. *Potential changes in intestinal structure and function induced by parasitic infections. (Data from Wakelin, 1986)*

Cellular and structural changes
Altered epithelial cell kinetics
Villous atrophy, crypt hyperplasia
Infiltration by eosinophils, macrophages, neutrophils
Mastocytosis
Increased goblet cells
Increased plasma cells
Increased intra-epithelial lymphocytes
Membrane changes in epithelial cells
Dedifferentiation
Physiological changes
Increased secretion of mucus
Increased vascular and epithelial permeability
Increased net fluid secretion across mucosa
Fluid accumulation in lumen
Increased motility
Decreased transit time
Mediator changes
Increased levels of Ig in lumen
Increased levels of myeloperoxidase and phospholipase
Increased levels of histamine, 5-HT, leukotrienes, prostaglandins

Note: Ig, immunoglobulin; 5-HT, 5-hydroxytryptamine.

particular cestode species is likely to be related not only to the physico-chemical conditions within the gut but also to the actual topography of the gut surface and the nature of its related glands. This question, which can be considered as part of the micro-ecology of cestodes, with a few exceptions (796), has not received much attention from parasitologists. It does not appear to be generally appreciated that, even in closely related hosts, the microstructure of the gut may show variation in such features as the size of villi and the width and depth of crypts – characters which may be of fundamental importance for the attachment, establishment and survival of a cestode. The morphology of the scolex of a particular species must clearly be very closely adapted to that of the gut of its host and this could play a role in determining host specificity. This is perhaps best seen in fish hosts such as *Raja montagui*, *R. clavata* and *R. naevus*, whose intestines show much variation. Species of the tetraphyllid genus *Echeneibothrium*, which parasitise these hosts, possess scoleces adapted to the depths of the villi, crypts or reticulations of the mucosa in each host species (957, 958). Again,

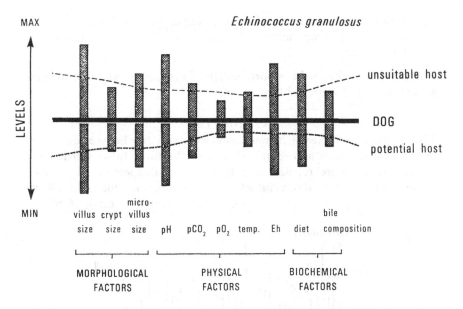

Fig. 3.1. Some parameters of the intestinal environment which play a part in the determination of the host specificity of a cestode; based on *Echinococcus granulosus*. The levels of the various factors are only notional, as the limits of tolerance have not been defined precisely. A host with an intestine with parameters ranges outside those shown cannot act as a suitable host. (After Smyth 1969.)

the crypt size of the dog, fox and cat differ substantially, which could possibly account for the fact that *Echinococcus granulosus* can grow in the dog but not in the cat and rarely in the fox, whereas *E. multilocularis* can be grown in all these hosts. As was pointed out earlier, other factors, such as the physico-chemical characteristics of the gut and its immunological responses, must also, of course, be taken into account. (Figs. 3.1 and 11.2, p. 287).

Mucosa

The basic cytology and ultrastructure, physiology and biochemistry of the mucosa, has been reviewed in detail by Henry (*318*) and Nugent & O'Connor (*602*). This topic can be considered broadly under four headings.

(*a*) The villous epithelial cells concerned with absorption: enterocytes.

(*b*) The generative epithelium of the crypts.

(c) Various specialised cells of the villi and crypt epithelium.
(d) The cellular components of the connective tissue of the intestinal lamina propria.

Only (a)–(c) are likely to be important in relation to cestode attachment and growth and are considered further below.

ENTEROCYTES

The morphology of the villi in man shows some variation between different races but also varies according to anatomic location, those villi in the jejunum being more regular than those in the duodenum. The most numerous cells of the villous epithelium are the *columnar absorptive cell* or *enterocyte* (Fig. 3.2) and the *goblet cell* (considered later). The amplification

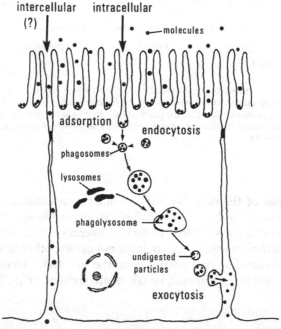

Fig. 3.2. General mechanisms for the uptake and transport of macromolecules by an enterocyte. *Intracellular uptake*: after absorption and endocytosis by the microvillous membrane, macromolecules are transported in small vesicles and larger phagosomes. Intracellular digestion occurs when lysosomes combine to form phagolysosomes. Intact molecules that remain after digestion are deposited in the intercellular space by exocytosis. *Intercellular uptake*: alternatively, macromolecules may cross the tight junction barrier between cells and diffuse into the intercellular space. (After Walker, W. A. & Isselbacher, K. J. Uptake and transport of macromolecules by the intestine: possible role in clinical disorders. *Gastroenterology*, **67**: 531–50, © by Williams & Wilkins (1974).)

of the surface of the enterocyte has already been described (p. 7). Its ultrastructure reveals a high mitochondrial content, a prominent Golgi system and a heterogeneous collection of lysosomes; the smooth and rough endoplasmic reticulum are well developed. What has been termed the 'biochemical anatomy' of the enterocyte, i.e. the subcellular fractionation and biochemical study of the various organelles, has been studied in some detail and has thrown much light on the pathology of the duodenum in man. This approach does not appear to have been used in examining the effect of adult cestodes on the biochemistry of the enterocytes in parasitised animals and such an approach could prove to be an interesting one for further research.

CRYPT GENERATIVE EPITHELIUM

Since the scolex of many species (e.g. *E. granulosus*) penetrates into the crypts of Lieberkühn, it is clearly important to know something of the physiology of this region. The crypts are a region of intense mitotic activity and the undifferentiated cells there give rise to the mature enterocytes, which are constantly being lost and replaced. In all mammals, the 'turnover time' (i.e. the time taken for replacement of the number of cells equal to that in the total population) was found to be less than 3 days. The turnover time in the duodenum of the rat is 1.6 days, the cat 2.3 days and man 2–6 days (*785*). The new cells formed in a crypt move up along the surface of a villus to the tip as cells are shed. Thus, the contents of the duodenum are constantly being supplemented by shed mucosal cells, and the autolysis of these undoubtedly releases a significant amount of nutritional material, which would be available to a cestode. This turnover of mucosal cells could also account for the fact that the immunological responses against adult cestodes are, with a few exceptions, rather weak (Chapter 11).

SPECIALISED CELLS AND SECRETIONS

The goblet cell
This is a simple mucus-secreting cell which does not undergo division. Mucus secreted by the goblet cells makes the intestinal surface viscid, a condition which must greatly assist the close adhesion of tapeworm strobila. Adhesion to the mucosa is probably essential to the normal nutritional efficiency of cestodes, but the degree of dependence may vary widely with species. At least in *E. granulosus*, as demonstrated by experiments *in vitro* (Fig. 10.6, p. 271), contact with a solid proteinaceous substrate appears to be essential for strobilisation. Rather surprisingly, this requirement is not essential for *E. multilocularis*, which strobilates *in vitro* in

a monophasic medium (i.e. in the absence of a substrate). *In vitro* experiments with the pseudophyllidean cestodes *Schistocephalus solidus* and *Ligula intestinalis* have also demonstrated that compression of the strobila against a soft surface such as the mucosa is essential for insemination and fertilisation (*788, 802*). *In vitro*, this is achieved by culturing these species within cellulose tubing (Fig. 10.2, p. 263).

Mucus
The structure, function, biochemistry, physico-chemical properties and secretion of mucus has been reviewed in detail by Nugent & O'Connor (*602*). Gastrointestinal mucus is a water insoluble, weak, viscoelastic gel adherent to the mucosal surface and is part of the mucosal defence against mechanical damage, acid and pepsin digestion (*10*). The macromolecules of mucus glycoproteins have proved to be much larger than previously recognised, an M_r of $10 \times 10^6 - 45 \times 10^6$ being quoted (*116*). These glycoproteins have a polymeric structure of subunits consisting of a linear array of oligosaccharide clusters interspersed with 'naked' stretches of protein; the subunits are joined by disulphide bridges between non-glycosylated regions of the proteins core (Fig. 3.3) (*10, 116*).

Mucus plays a major role in maintaining the mucosal pH, by possessing a standing gradient of HCO_3^- secreted by the epithelium at its lumenal surface. Use of glass or antimony electrodes have shown that the extracellular pH at the epithelial surface is maintained at pH 7.0, in spite of the lumenal pH being as low as 2.0 (experimentally produced) for periods of more than 1 h (*222*). This situation may have profound physiological implications for a cestode for it could imply that the scolex could be at pH 7.0 whereas the strobila might be at a strongly acid pH.

Other mucosal cells
In addition to enterocytes and goblet cells, the mucosa contains *enteroendocrine* cells, which are specialised neurosecretory cells, *M cells*, which may be involved in macromolecular uptake, and *Paneth cells*, which have recently been recognised as mononuclear phagocytic cells containing lysozyme (*318*) whose release may be cholinergically regulated. Nothing appears to be known regarding the possible interactions of these cells in cestode infections. Paneth cells, for example, could play a major role in phagocytosing oncospheres (e.g. of *Taenia saginata*) hatching in the gut of their intermediate host (cattle) and attempting to cross the gut barrier. The immune responses of the intestine are discussed further in Chapter 11.

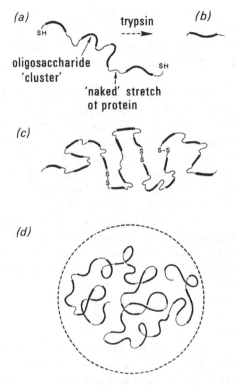

(a) trypsin *(b)*
----→

oligosaccharide
'cluster'

'naked' stretch
of protein

(c)

(d)

Fig. 3.3. Schematic illustration of the proposed architecture of mucus glycoproteins. (*a*) Subunits constitute a linear array of oligosaccharide clusters interspersed with naked stretches of protein. (*b*) Trypsin digestion affords glycopeptides corresponding to the oligosaccharide clusters. (*c*) The 'whole mucins' are formed by an end-to-end association of subunits via disulphide bonds. (*d*) The hydrodynamic model of mucus glycoproteins conforms to a random coil within a spheroidal solvent domain. (After Allen *et al.*, 1984.)

Intestinal physiology

General concepts

DIGESTION AND ABSORPTION

Concepts of vertebrate intestinal function have changed continuously over the past few years and the processes of digestion and absorption are now recognised as being very complex. Broadly, the present state of knowledge indicates the following:

41

(*a*) Carrier molecules are involved in the movement of sugars, peptides and amino acids across the mucosal membrane. For example, as explained on p. 12, the transport of glucose appears to be coupled to an inward Na^+ gradient through a common carrier, forming a ternary complex (Na^+–carrier–solute.)

(*b*) Concepts of protein digestion have undergone a radical change. It was formally believed that intestinal absorption of proteins required their complete breakdown to free amino acids. There is now overwhelming evidence that absorption of dipeptides and tripeptides can take place 'on a large scale' and an uptake system for small peptides occurs in the brush border membranes of enterocytes. Unexpectedly, the uptake concentration of peptides was found to be much larger for amino acid residues in the form of peptides than in the free form. Thus, peptides are quantitatively more important substrates for absorption than are free amino acids. The final breakdown of the peptides to amino acids (in which form they reach the bloodstream) takes place in the brush border membrane of the cytoplasm of the enterocytes. The peptide carrier system apparently shows some preference for peptides with bulky side-chains and L-stereoisomer amino acid residues at both the amino and carboxyl terminals (*2, 577*).

(*c*) There is some evidence that the molar ratios of the lumenal amino acid pool remain fairly constant, independent of the quantity and composition of protein digested in the host diet (*796*). This view, however, has been challenged (*158, 523*). In ducks, for example, it was shown (*158*) that the amino acid mixture in the intestine was markedly affected by diet (Table 3.2) and that the dilution by endogenous protein (from shed enterocytes, secretions etc.) was insufficient to mask the amino acid pattern of ingested protein.

(*d*) As already pointed out (p. 9), the mucosal surface is covered by a highly organised, filamentous glycocalyx rich in ionised sugar moieties and has many functional correlates. The role of the 'unstirred water' layer (p. 9) within the interstices of the glycocalyx in modifying mucosal transport kinetics is uncertain (but see pp. 81–2, Chapter 5).

Absorption mechanisms

MEMBRANE TRANSPORT

Mechanisms

Some aspects of membrane transport have been discussed in Chapter 2. The mechanisms whereby cells – especially those of the mucosa – take up

Table 3.2. *Effect of two different diets on the amino acid composition in the intestine of ducks. (Data from Crompton & Nesheim, 1969)*

Amino acid	Soya-bean diet		Maize gluten diet	
	Intestinal lumen	Diet	Intestinal lumen	Diet
Arginine	0.85	0.76	0.34	0.22
Glycine	0.53	0.64	0.15	0.27
Histidine	0.35	0.27	0.11	0.14
Isoleucine	0.47	0.58	0.23	0.30
Lysine	0.72	0.59	0.19	0.13
Methionine	0.24	0.18	0.15	0.15
Phenylalanine	0.64	0.57	0.47	0.38
Threonine	0.86	0.45	0.42	0.22
Valine	0.57	0.59	0.27	0.19

substances from their environment have been much investigated and the following mechanisms are recognised:

(*a*) passive diffusion,
(*b*) active transport,
(*c*) facilitated diffusion,
(*d*) endocytosis.

Although it is beyond the scope of this text to deal with these mechanisms in detail, because they operate at both surfaces of the host–parasite interface, i.e. at the mucosal and cestode surfaces, they have a special significance in the study of cestode physiology. For this reason they are summarised briefly below.

Passive diffusion
This is the movement of substances across membranes by mechanisms which apparently follow the simple laws of diffusion, i.e. the rate of movement of a given substance is proportional to the concentration difference across the membrane. In the intestine, this appears to be the uptake mechanism of relatively few materials, among which are water-soluble vitamins, some nucleic acid derivatives and many lipid-soluble substances.

Active transport
As uptake by simple diffusion is slow, it is not surprising to find that the intestinal mucosa has developed mechanisms which result in a more rapid

transport of the food materials required for synthetic purposes. Salts, glucose, amino acids and lipids are all absorbed by such mechanisms. The use of the term active transport is often restricted to 'those processes in which a substance moves across a membrane *against an electrochemical gradient* and consequently requires energy supplied by cellular mechanisms'. The process is further characterised by two features: (*a*) it shows stereospecificity, which involves competitive inhibition by chemically similar compounds; (*b*) it is inhibited by poisons of energy metabolism. The role played by Na^+ in active transport (the so-called sodium pump) has been discussed (p. 12). Amino acids, dipeptides, sugars etc. each have their own specific carrier system and uptake site. It has already been pointed out that during protein digestion only some 25% of ingested proteins are broken down to amino acids, the rest are broken down to peptides. Rather unexpectedly, it has been found (2) that the peptide carrier system is more efficient than the amino acid carrier system; for example, glycine absorption from diglycine (*Gly-Gly*) (Fig. 3.4) and triglycine (*Gly-Gly-Gly*) solutions is greater than that of the free amino acid (*Gly*) itself. This appears to be because the glycine absorption from a free *Gly* solution is solely dependent on the amino acid carrier system, whereas glycine absorption from a tetraglycine solution can use *both* amino acid and peptide carrier systems. For example, absorption from 50 mM tetraglycine solution is greater than that from a 200 mM free *Gly* solution. At a *Gly* concentration of 200 mM, the amino acid carrier system is well saturated; on the other hand, 50 mM tetraglycine, which contains essentially the same amount of *Gly* ($4 \times Gly \times 50 = 200$ *Gly*), the concentrations of *Gly* and *Gly-Gly-Gly* released by the brush border hydrolysis of tetraglycine fall far below the saturation concentrations of these systems (2). The affinity of dipeptides for uptake sites are known to be influenced by stereoisomerism, the length of side-chain, substitutions of amino and carboxy terminals and the number of amino acid residues.

Facilitated diffusion
This is a term coined to denote a transport mechanism in which the rate of attainment of diffusion is accelerated without any direct expenditure of energy. It differs from active transport in being unable to operate against an electrochemical gradient. The absorption of D-xylose by the intestine comes under this heading.

Endocytosis
It is well known that the intestine of many neonatal mammals, including man, has the capacity to ingest macromolecules (such as immunoglobulin in colostrum). In man, this capacity markedly decreases as the endocyte

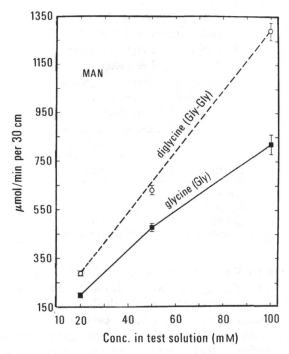

Fig. 3.4. Rates of Gly uptake from glycine and diglycine test solutions which are equivalent in Gly content (e.g. 20 mM Gly v. 10 mM Gly-Gly). The perfusion studies were done in the jejunums of five healthy human volunteers. (After Adibi, S. A. & Kim, Y. S. Peptide absorption and hydrolysis. In *Physiology* of *the gastrointestinal tract*, ed. L. R. Johnson, © 1981, with permission from Raven Press, New York.)

membrane in the mucosa matures. Within recent years, however, it has been demonstrated that uptake of macromolecules is still possible in the adult intestine (Fig. 3.2). The exact cells involved in this uptake are not clear but M cells may be involved. These cells appear also to operate as a system which facilitates the uptake of antigens into intestinal lymphoid tissue. Hence this uptake system may be important in the IgA-producing cell cycle (Fig. 11.4, p. 288) and may facilitate the local immune response (*924*). The transport mechanisms of cestodes are discussed further in Chapters 5 and 6.

Physico-chemical characteristics

pH

The pH of the alimentary canal has been reviewed by a number of authors (*370, 796*) most of the data referring to man and the more commonly used

Table 3.3. *pH and HCO_3^-*
concentration in intestinal
secretions in the dog. Secretions
from Thiry-Vella loops of
unanaesthetised dogs, unexposed
to air during collection. (Data
from Wilson, 1962)

Location	pH	'Total CO_2' (HCO_3^- + CO_2) (mM)
Jejunum	6.8	19
Ileum	7.6	83
Colon	8.0	90

laboratory animals. Results from different workers show some divergence, probably due to different protocols of investigation. It was formerly thought that the pH of the duodenum was alkaline, but more critical work in dogs showed that it is generally just on the acid side – about 6.7–6.8 (Table 3.3). It has been pointed out earlier that the mucus coat of the mucosa plays a major role in buffering the pH at the enterocyte surface at about neutrality, although the lumenal pH may be much lower. The pH in the crypts of Lieberkühn, in which the scolex of some species may be embedded, is likely to be that of the adjacent tissues, about 7.4.

In the rat, the presence of *Hymenolepis diminuta* has a marked lowering effect on the pH; 1 h after feeding, the lowest pH reached in the intestine of uninfected rats was found to be 6.4, but that in infected rats fell to < 6 (*524, 796*), probably due to secretion of organic acids by the worm (Chapter 5).

Oxidation–reduction potential

This characteristic, which may have considerable significance for cestode metabolism (especially in relation to electron transport) has been relatively little studied. In the rat, perhaps the most studied laboratory host, the stomach contents are reported to have an Eh value of + 150 mV and that of the upper and lower small intestine about − 100 mV (*796*). That this negative potential is largely due to the presence of microflora is seen from the fact that, in germ-free rats, the Eh of the large intestine was found to be + 30 to + 200 mV, but this became negative after contamination with various coliforms (*970*). The Eh may prove to be an important factor in *in vitro* culture systems (p. 258).

Physico-chemical characteristics

Table 3.4. *The percentage of the gases in the small intestine of various vertebrates. (Data from Smyth, 1969)*

Animal	CO_2	O_2	CH_4	H_2	N_2
Rabbit	13–75	0–0.2	2–2.8	7.7–18	6–75
Dog	15.9	0.3	—	26.5	57.3
Goose	2–87	0–3.6	0–13.5	0.7–20	68–85
Pig	2.2–80	0.1–8	0–28	0–40	2–92
Horse	15–43	0.6–0.8	0	20–24	37–60
Cattle, sheep and goats	62–92	0	0–6.6	0–37	1.0

Oxygen tension

The oxygen tension of the vertebrate intestine has long been a matter of interest to parasitologists on account of its significance to the aerobic or anaerobic metabolism of intestinal parasites (Chapter 5). A high pO_2 in the environment does not necessarily mean that a parasite which lives in such an environment will have an aerobic metabolism. The trematode *Schistosoma mansoni*, for example, although it lives in the portal blood vessels (an aerobic site), appears to have an essentially, although not exclusively anaerobic metabolism (490). The composition of intestinal gas has proved to be technically extremely difficult to measure and the data are 'widely' divergent (800). Some available data are given in Table 3.4. The main controversy has centred on whether differences exist between the pO_2 in the lumen and that on the mucosal surface. Early work on the rat (800) found that the pO_2 on the mucosal surface was three times higher than in the lumen, where it was believed to approach zero. In the duck intestine the same order of result was obtained (159, 800). In contrast, using a tonometric method, it was concluded that, in the dog, the level of lumenal pO_2 (Fig. 3.5) was a true reflection of the mucosal pO_2 (291). The pO_2 may also be affected by the nutritional state of the host. For example, in a dog, the O_2 level fell from 4–6% during fasting to zero after a meat meal (800).

Carbon dioxide

CO_2 is frequently the predominant duodenal gas and plays an important role in the physiology of cestodes in two ways: (*a*) as a source of carbon atoms for metabolic functions via CO_2 fixation (Chapter 5); (*b*) as a 'trigger' in the hatching of cyclophyllidean eggs (p. 192) or the excystment of larval cysts. Production of CO_2 results from the interaction of HCO_3^- and H^+, the latter being derived from secretions from the gut, biliary and pancreatic

47

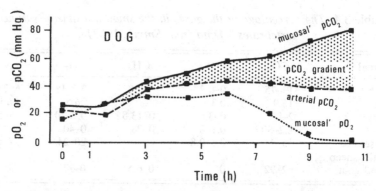

Fig. 3.5. The change of mucosal pO_2 and of mucosal to blood pCO_2 gradient throughout a period of general anaesthesia in a dog. (1mmHg $\approx$ 133.3 Pa.) (After Hamilton, J. D., Dawson, A. M. & Webb, J. P. W. Observations on the small gut 'mucosal' pO_2 and pCO_2 in anesthetized dogs. *Gastroenterology*, **55**; 52–60, © Williams & Wilkins (1968).)

secretions. In some cases the lumenal pCO_2 has been reported to reach levels as high as 600 mmHg (1 mmHg $\approx$ 133.3 Pa) (*56*). Detailed consideration of the complex reactions between HCO_3^- and H^+ which are beyond the scope of this book have been reviewed by Levitt *et al.* (*444*) and Befus & Podesta (*56*).

Other gases

In addition to O_2 and CO_2, N_2, H_2 and CH_4 are found in various concentrations in different vertebrates and these gases make up 99% of the intestinal gas in man. The source of H_2 and CH_4 appears to be bacterial metabolism and, in cattle rumen, CH_4 is known to be produced by *Methanobacterium rumentium*. The composition and origin of intestinal gas in man has been reviewed in detail by Levitt *et al.* (*444*).

Bile

General account

IMPORTANCE TO INTESTINAL PARASITES

Bile is of particular importance to intestinal parasites, for in many cases it is involved in the 'trigger' mechanisms concerned in the hatching of helminth eggs, excystation of protozoan cysts, evagination of cestode scoleces and excystment of trematode metacercaria. In relation to cestodes, its role in

48

Table 3.5. *Qualitative data on the composition of bile acids in man and common laboratory and domestic animals. (Data from Smyth, 1969)*

Species	Rate of secretion ml/kg body wt per 24h	Bile acids	Nature of conjugation
Man	2.6–15.0	Cholic, chenodeoxycholic, deoxycholic, lithocholic, ursodeoxycholic	Taurine, glycine
Ox	15.4	Cholic, deoxycholic, chenodeoxycholic, lithocholic, 3α-hydroxy-12-oxocholanic, 3α, 12α-dihydroxy-7-oxocholanic, 7α, 12α-dihydroxy-3-oxocholanic, 3α-hydroxy-7, 12α-dioxocholanic, stereocholic, sapocholic	Taurine, glycine
Sheep	12.1	Cholic, deoxycholic, chenodeoxycholic	Taurine, glycine
Guinea-pig	228	Chenodeoxycholic; 3α-hydroxy-7-oxocholanic, cholic, ursodeoxycholic	Taurine, glycine
Rabbit	118	Deoxycholic, lagodeoxycholic, cholic, lithocholic	Glycine
Fox	?	Cholic, deoxycholic	Taurine
Dog	5–52	Cholic, deoxycholic, chenodeoxycholic	Taurine
Cat	14.0	Cholic	Taurine
Rat	28.6–47.1	Cholic, chenodeoxycholic, ursodeoxycholic, α- and β-muricholic	Taurine, glycine
Mouse	?	Cholic, α- and β-muricholic, chenodeoxycholic	?
Pig	25.2	Cholic, hyocholic, chenodeoxycholic, hyodeoxycholic, 3α, 6β-dihydroxycholanic, 3β, 6α-dihydroxycholanic, 3α-hydroxy-6-oxocholanic (or allocholanic), lithocholic	Taurine, glycine

scolex evagination and egg hatching is dealt with later (pp. 192, 233). It is not always appreciated that the composition of bile varies greatly between species (Table 3.5) and that this difference may play a part in determining host specificity (796).

Bile is secreted continuously by the cells of the liver and stored in the gall bladder. In a few mammals, e.g. the rat, a gall bladder is lacking. Although small amounts of bile may enter the alimentary canal regularly, the ingestion of food normally stimulates the emptying of the gall bladder and bile enters the gut in quantity at that time. It is important to note that, during the period of storage in the gall bladder, certain substances, such as water and

Table 3.6. *Composition of hepatic and gallbladder bile; values are averages for man, dog and cat. (Data from Rose, 1981)*

	Hepatic	Gall-bladder
Na$^+$ mequiv./l	160	270
K$^+$	5	10
Cl$^-$	90	15
HCO$_3^-$	45	10
Ca^{2+}	4	25
Mg^{2+}	2	—
Bilirubin (mM)	1.5	15
Protein (mg/100 ml)	150	—
Bile acids (mM)	50	150
Phospholipids (mM)	8	40
Cholesterol (mM)	4	18
Total solids (mg/l)	—	125
pH	8.0	6.5

organic salts, may be absorbed so that the composition of 'gallbladder bile' and 'hepatic fistula bile' may differ substantially (Table 3.6).

Chemical composition

Although bile contains pigments, cholesterol, fatty acids, some sugars and bile salts, it is these last which are of particular importance to the physiology of cestodes.

Bile salts are substances derived from sterols, which make up a substantial part of the solid matter in bile and which play a central role in lipid absorption, by virtue of their surface-active properties. The structure and properties of these salts have been reviewed by Haslewood (*305*) and Heaton (*316*). Bile salts essentially have molecules of 'detergent' type hydrocarbon, with a fat-dissolving part and a polar, water-attracting part. The fat-dissolving part consists of the bulk of the steroid nucleus. The hydroxyl groups are so distributed that hydration can readily take place; the remainder of the molecule will dissolve the fatty phase. Emulsification of fat/water complexes can thus occur easily. The terms bile acid and bile salt are used somewhat interchangeably in the literature.

Types of bile salts

There are three types of bile salts (*a*) C_{27} (or C_{28}) alcohols, (*b*) C_{27} (or C_{28}) acids, and (*c*) C_{24} acids. The alcohols have been isolated only from amphibia and fishes (especially elasmobranchs). The C_{27} (or C_{28}) bile acids are little known, and occur mainly in reptiles. The C_{24} acids are much better known and are mostly derived from cholanic acid, $C_{24}H_{40}O_2$. The best known salts are those of cholic acid, hyocholic acid, chenodeoxycholic acid, deoxycholic acid and lithocholic acid (Table 3.5).

Conjugation of bile salts

Bile alcohols and acids normally occur in conjugated form. Alcohols are conjugated with sulphate, giving substances of the type $R:CH \cdot O–SO_3$. Bile salts conjugate with the amino acids taurine ($NH_2CH_2CH_2SO_3H$) and glycine (NH_2CH_2COOH). This is only a broad generalisation, however, as information on this point is incomplete. Cat and dog bile are conjugated almost exclusively with taurine, whereas rabbit bile acids are conjugated almost exclusively with glycine. It has been shown (*796*) that the tegument of some species (e.g. *Echinococcus granulosus*) is especially sensitive to some bile salts, such as deoxycholic acid, which lyses it rapidly. This property has proved to be of value in lysing oncospheres to release excretory/secretory (E/S) antigens without their having to be cultured *in vitro* (Chapter 11).

Osmotic relationships

General account

MOST CESTODE SPECIES

Many cestodes appear to behave like osmometers, and when placed in an environment with a different osmotic pressure can only adjust their internal osmotic pressure by varying their body volume. *Moniezia*, *Hymenolepis*, *Schistocephalus* and *Callibothrium* change their weight almost arithmetically in response to a change in external osmotic pressure (*796*).

The question of the osmotic relationship between a cestode and its environment is complicated by the fact that some substances can pass through membranes by means other than diffusion (p. 42). Hence the actual osmotic pressure of a solution as measured by physico-chemical means may not be as significant to a worm as the actual content to which the tegument of the worm is 'permeable' – using the word in its widest sense as indicated above. The reason for this is that substances in the medium which (theoretically) contribute to the total osmotic pressure of the medium do not actually exert osmotic pressure across the tegument of the worm, which separates the worm from its environment.

The adult cestode in its environment

Thus, differences in 'permeability' of solute can create curiously anomalous situations. In the case of the avian cestode, *Tetrabothrius erostris*, both sucrose at 0.192 M ($\Delta = 0.36$ °C) and NaCl at 0.140 M ($\Delta = 0.56$ deg. C) appear to be isosmotic; this result can be interpreted as showing that this species is less permeable to sucrose than to Na$^+$ and Cl$^-$.

CESTODES OF ELASMOBRANCHS

It is interesting to note that, in the case of cestodes of elasmobranchs, urea – which is well known to contribute substantially to the osmotic pressure of blood in many species of elasmobranchs – occurs in quantity in some cestodes of these fishes. The tetraphyllidean *Calliobothrium verticillatum* contains urea concentrations of $1.03(\pm 0.46)\%$ of the wet weight, the equivalent of 3.7% dry weight (796). When worms were incubated in solutions lacking urea, the urea in the worm tissues rapidly disappeared, but this did not occur if urea was present in the external medium. Similar results were obtained with the tetraphyllideans *Phyllobothrium foliatum* and *Inermiphyllidium pulvinatum*. Later studies with ^{14}C-labelled urea on the kinetics of urea entry into *C. verticillatum* showed that equilibrium between urea in worm tissues and in the external medium was reached in 60–90 min.

It is surprising to find that *Lacistorhynchus tenuis*, from the same host as *C. verticillatum*, does not behave osmotically in a similar manner. The amount of free urea in this species is low, and later work showed that urea was metabolised by this cestode (Chapter 6).

4

The adult: general metabolism and chemical composition; lipid metabolism

General background

Basic problems

Metabolic studies on cestodes present a number of challenging problems. A basic difficulty is to relate results from studies carried out *in vitro* to the processes which actually occur *in vivo*. Cestodes, with their complex tegumental transport mechanisms (Chapters 5 and 6), are in a state of dynamic equilibrium with their hosts, and removal from the host environment tends to destroy this balance. The most favourable *in vitro* culture conditions (Chapter 10) can partially mimic some of the complicated interactions which occur between host and parasite, but, nevertheless, it is clear that most metabolic experiments with cestodes *in vitro* take place under suboptimal, unphysiological conditions. Many (but not all) parasite biochemists are aware of these limitations and appreciate that these artificial systems produce data which must ultimately be tested *in vivo*, although the technical problems involved in such studies may prove difficult, if not insurmountable.

Another well-recognised complication in the study of cestode metabolism is the fact that a number of species (e.g. *Echinococcus granulosus, Hymenolepis diminuta, Taenia crassiceps*) have now been shown to exist as complexes of different strains, which may, often quite considerably, differ in their biochemistry. This important aspect is considered, in depth, in Chapters 5, 6 and 10. Furthermore, there is evidence that parasites from different host species or different strains of host show differences in metabolism, and the sex and circadian rhythm of the host can also influence the biochemistry of the parasite under study (*39*).

In the majority of species, enzyme assays and biochemical analyses have to be performed on homogenates. The spatial distribution of enzymes or substrates in the whole organism may be such that certain molecules, which may occur in different cellular sites (e.g. in the nucleus or in mitochondria)

53

in different cells or even in different tissues in the cestode, may never come into contact with one another in the living cell. Under the artificial conditions created within tissue homogenates, these materials may make intimate contact and undergo metabolic reactions which bear no relation to those which occur *in vivo*.

Respiration

The perplexing problem of oxygen utilisation by cestodes raises a number of questions of significant physiological interest, many of which are unanswered. A number of studies on oxygen consumption have been made, and all species of cestodes examined have been found to utilise oxygen *in vitro* when available. It is difficult to assess the significance of such data, however, largely because the oxygen tension at the establishment site is generally not known precisely (see Chapter 3), although all cestodes, both larvae and adults, are likely to have at least some oxygen available to them. There is, however, the additional problem of oxygen supply to all tissues, as cestodes have neither a circulatory system nor a respiratory pigment. Consequently, a concentration gradient from the peripheral to the central tissues will result and, in large species (e.g. *Moniezia*, *Hymenolepis*, *Taenia*), the oxygen tension in the central region may be zero.

As indicated above, the fact that oxygen is consumed *in vitro* does not imply that oxygen is utilised *in vivo*, unless evidence is presented that a similar level of oxygen is available *in vivo*. Some species (*Spirometra mansonoides, Mesocestoides corti, Hymenolepis nana, H. diminuta*) can be successfully cultured under strict anaerobic conditions, whereas others (*H. microstoma, T. crassiceps, Echinococcus* sp.) thrive best under air (*796*; Chapter 10). The significance of oxidative processes in the energy balance of cestodes is discussed in Chapter 5.

Chemical composition

General comments

A considerable quantity of data is now available on the chemical composition, including trace elements, of cestodes. Much of this information has been summarised by Barrett (*39*), Smyth (*796, 800*) and von Brand (*911*).

Some recent studies include those on: *Taenia taeniaeformis* and *Dipylidium caninum* (*578*); *Proteocephalus exiguus* and *Diphyllobothrium latum* (*18*); *Caryophyllaeus laticeps, Eubothrium rugosum, Proteocephalus* spp., *Triaenophorus crassus, D. latum* and *Ligula intestinalis* (*660*); *Bothriocephalus gowkongensis, Triaenophorus*

Chemical composition

Table 4.1. *The biochemical composition (in μg/mg dry wt ± S.E.) of protoscoleces and adults of* Echinococcus granulosus *from Kenya and the UK. (Data from McManus & Smyth, 1978; McManus, 1981)*

Host origin	Protein	Polysaccharides	Lipids	RNA	DNA
Horse[a]	550 ± 10	177 ± 2	109 ± 4	60 ± 4	5 ± 0.1
Sheep[a]	625 ± 12	169 ± 1	88 ± 3	89 ± 4	5 ± 0.1
Sheep	592 ± 16	166 ± 4	76 ± 4	74 ± 6	7 ± 1
Cattle	547 ± 14	146 ± 3	118 ± 6	52 ± 4	6 ± 1
Goat	569 ± 16	166 ± 4	95 ± 3	54 ± 4	6 ± 1
Camel	550 ± 17	215 ± 6	113 ± 4	53 ± 4	5 ± 1
Human	668 ± 16	164 ± 4	122 ± 4	57 ± 5	4 ± 1
Dog[b]	712 ± 18	139 ± 5	143 ± 4	96 ± 6	6 ± 1
Dog (natural)	584 ± 16	213 ± 8	122 ± 4	203 ± 7	4 ± 1

Notes:
[a] From the UK.
[b] Experimental infection with protoscoleces of human origin.

nodulosus and *L. intestinalis* (*826*); *L. intestinalis* (*823*); *Diphyllobothrium macroovatum* and *Diplogonoporus balaenopterae* (*975*); *Thysanezia giardi* (*773*); *Stilesia globipunctata* (*658*); *S. globipunctata, Avitellina centripunctata* and *Moniezia expansa* (*133, 853*); *Dioecocestus asper, Taenia pisiformis, Diploposthe laevis, Paricterotaenia porosa, Dubinolepis furcifera* and *L. intestinalis* (*355–358*); *Taenia cerebralis* (*752, 753*); *Echinococcus granulosus* (*67, 68, 136, 232, 488, 498, 583, 676, 830, 899*); *Raillietina tetragona* (*48*); *R. echinobothridia* (*584*); *M. expansa, T. hydatigena* and *T. solium* (*897*); *Hymenolepis diminuta* (*468, 469*) and *Moniezia benedini* (*900*).

Data on the chemical analysis of cestodes are of limited value unless the nutritional status of the host is known, as significant fluctuations in individual parasite components can occur. Furthermore, the chemical composition may vary with the 'strain' of both parasite and host, the host species, the age of the cestode and its degree of maturation. This is partly illustrated in Table 4.1, which shows the variation in biochemical composition of *Echinococcus granulosus* obtained from different hosts. Some of the available data must also be accepted with caution on technical grounds, because a number of the older analytical methods have been shown, by more modern workers, to be unreliable.

Major chemical constituents

In general, the proportions of the main tissue constituents, proteins, lipids, carbohydrates and nucleic acids (Table 4.2) show a somewhat different

The adult: general metabolism

Table 4.2. *Chemical composition of cestodes.[a]* (*Data from Smyth, 1976*)

Species	Dry wt as % of fresh wt	Glycogen	Lipid	Protein	Inorganic substances	Stage
Anoplocephala magna (*Taenia plicata*)	27.5	6	33.1	—	1.22	Adult
Cittotaenia perplexa	27.1	—	—	20.60	—	Adult
Diphyllobothrium latum	—	17.9	—	—	—	Plerocercoid
D. latum	9.0	20	16.6	60	4.8	Adult
Diphyllobothrium sp.	29.9	31.5	—	41	—	Plerocercoid
Diphyllobothrium sp.	30.8	36.2	—	48	—	Adult
D. dendriticum	27.0	36.2	—	—	—	Plerocercoid
Dipylidium caninum	20.4	—	—	—	—	Adult
Echinococcus granulosus[b]	14.8	19.8	13.6	6.25	13.5	Protoscoleces
Eubothrium rugosum	—	22.8	—	—	—	Adult
Hydatigera (Taenia) taeniaeformis	22.3	19.7	4.2	27.1	29.0	Strobilocercus (mice)
H. (T.) taeniaeformis	28	43.3	5.3	26.3	18.1	Strobilocercus (mice)
H. (T.) taeniaeformis	29.5	24.9	3.1	28.9	28.4	Strobilocercus (rats)
H. (T.) taeniaeformis	20	—	6.9	40.6	—	Adult
H. (T.) taeniaeformis	26.7	23.2	6.3	45.0	22.0	Adult
Hymenolepis diminuta	22.4	45.7	20.1	31.0	—	Adult
H. citelli	—	—	16.1	—	—	Adult
Ligula intestinalis	29.0	38–52	—	35–45	—	Plerocercoid
Moniezia expansa	9.2–11.0	24–32	30.1	21.8	10.5	Adult
Multiceps multiceps (*Coenurus cerebralis*)	25.3	—	—	—	27.4	Scolex
M. multiceps	12.4	—	—	—	4.1	Membranes
Raillietina cesticillus	20.5	31.8	15.5	36.4	11.5	Adult
Schistocephalus solidus	31.8	50.9	—	35.8	5.8	Plerocercoid
S. solidus	38	28.0	—	—	—	Adult
Taenia crassiceps	20.0	27.5	—	—	—	Larva
T. hydatigena (*marginata*)	16–26	28	4.9	—	—	Adult
T. saginata	12.2	48.8	11.2	32.0	5.3	Adult
T. solium	8.7	25.4	16.2	46	6.4	Adult
Thysanosoma actinioides	16.3	—	—	29.0	—	Adult
Triaenophorus nodulosus	—	13.8	—	—	—	Adult

Note:
[a] % Dry wt.
[b] See also Table 4.1.

pattern in cestodes from that in most other invertebrates: the protein content tends to be relatively low and the carbohydrate content (largely in the form of glycogen) high. Larval cestodes are especially rich in glycogen, which may reach extraordinary levels (greater than 50% of the dry weight) in plerocercoids of pseudophyllideans, such as *Ligula* or *Diphyllobothrium* (*800, 823*). The nature of the lipid (see below), protein and nucleic acid constituents (Chapter 6) are discussed later, but the high carbohydrate content is worthy of some special comment here.

POLYSACCHARIDES

Glycogen
In both larval and adult cestodes, the carbohydrate occurs mainly as the polysaccharide glycogen. Stored primarily in the parenchyma, glycogen serves typically as the most important energy reserve in cestodes and its metabolic role will be dealt with in Chapter 5. Glycogen synthesis (glycogenesis) has been investigated in detail only with *H. diminuta*, and the same steps, involving the enzymes phosphoglucomutase, glucose-1-phosphate uridylyl transferase and glycogen synthase, occur in cestodes as in mammals (*39, 534, 535, 698, 731*). Similarly, the properties of cestode glycogens resemble those of vertebrate glycogen and, for example, that of *Moniezia expansa* is a highly branched structure consisting of α-1,4- and α-1,6-linked glucopyranose units; it has an average chain length of 12.9 glucose units, with an average inner-chain length of 2.9 units and an average outer-chain length of 9.0 units (*608*). Glycogens isolated from various parasites by mild methods show a high degree of polydispersity, i.e. they exhibit a broad spectrum of molecular weights. About 30% of *H. diminuta* glycogen is very high in molecular weight (average 900 million), 60% averages 25–60 million, while intermediate molecular weight components are present in much smaller amounts (*698*). Whether these different molecular forms of glycogen can be correlated with the different types of morphologically distinguishable glycogen particles (α, β and γ) observed under the electron microscope (*698*) is open to question.

Other polysaccharides
Mucopolysaccharides are some of the most common structural carbohydrates in cestodes, although little is known of their biochemistry or function. They are heteropolymers and contain amino sugars (e.g. glucosamine, galactosamine) and uronic (glucuronic, galacturonic) acids. Often, mucopolysaccharides are complexed with proteins to form mucoproteins or glycoproteins, which, as discussed in Chapter 2, are major components of

the cestode glycocalyx. A series of histochemical studies has shown the presence of proteoglycans (acid mucopolysaccharides) and/or glycoproteins in the microtriches, tegument and parenchyma of cestodes and a valuable review is available of some of the early work (*457*).

The laminated layer of the hydatid cyst of *E. granulosus* has been shown to be a periodic acid–Schiff (PAS)-positive polysaccharide–protein complex, with the carbohydrate component being composed of glucose, galactose, galactosamine and glucosamine (*395, 684*). Partial structural similarity occurs between this polysaccharide and that present in the cyst membrane of the related species *Taenia hydatigena* (*860*) and it may be that these carbohydrates afford some protection to the parasites against the effector arms of the host immune response (Chapter 11). A number of other complex carbohydrate-containing components (glycoconjugates) have been identified in the tegument and exposed at the surface of *Hymenolepis microstoma* and *H. nana* by gold-labelled lectins and electron microscopy (*738*), and these molecules may well prove to be similarly protective. Furthermore, a surface carbohydrate of *T. taeniaeformis*, which is apparently secreted by the worm, has been shown to inhibit complement activation; this anti-complementary factor has been characterised as a glycosaminoglycan (a sulphated polysaccharide) (*292–294*). There is evidence that other molecules from this species, especially lipids (see below), also interact with complement. Such anti-complementary activity may be important in interfering with the effectiveness of complement-fixing anticestode antibodies.

The adults and eggs of *H. diminuta* possess four and two different kinds of polysaccharides, respectively; gas–liquid chromatographic analysis of the hydrolysed polysaccharides (Table 4.3) has revealed several sugars, with glucose and galactose predominating (*702*). Both these stages are refractory to proteolysis and this has lead to the interesting suggestion (*701*) that these complex carbohydrates, in association with specially resistant proteins, serve to protect *H. diminuta* against the numerous proteases present in the host gut.

This survival strategy may be more complex, however, as a number of studies (*376, 377, 515, 516, 591–593, 744, 832, 891*) have indicated that pseudophyllidean and cyclophyllidean cestodes produce their own protease inhibitors which can inactivate trypsin and chymotrypsin *in vitro*. Such protease inhibitors are also produced by nematodes and they appear to be proteins or protein–carbohydrate complexes (*592*). Precise chemical characterisation of these protease inhibitors has not yet been achieved in cestodes, although the active molecule from the strobilocercus of *T. taeniaeformis* has been shown to be a polypeptide (*taeniaestatin*) with an M_r of 19 500 under reducing conditions (*832*). The effectiveness of this molecule at inhibiting chymotrypsin *in vitro* is illustrated in Fig. 4.1. The mechanism whereby this and

Chemical composition

Table 4.3. *Gas-liquid chromatographic determination of neutral and amino sugars of* Hymenolepis diminuta *eggs and ten-day-old adults. (Data from Robertson* et al., 1984)

Sugar	Eggs	10-day-old worms
Rhamnose	0.47	0.05
Fucose	0.40	0.04
Ribose	0	0.04
Xylose	0.24	0
Mannose	0.03	0.01
Galactose	3.93	0.12
Glucose	9.34	4.12
Glucosamine	6.35	0.15
Galactosamine	0.31	0.11

Note: Results are given in μmol monosaccharide/mg lipid.

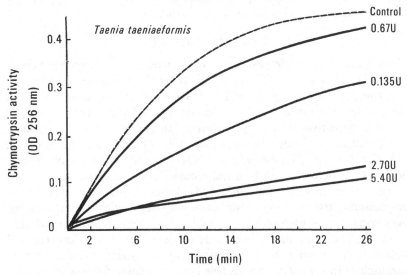

Fig. 4.1. Kinetic and dose–response inhibition of chymotrypsin by *Taenia taeniaeformis* proteinase inhibitor. (Reprinted with permission from *International Journal for Parasitology*, **14**: Suquet, C., Green-Edwards, C. & Leid, R. W. Isolation and partial characterization of a *Taenia taeniaeformis* metacestode proteinase inhibitor, © 1984 Pergamon Journals Ltd.)

Table 4.4. *Phosphate content (% fresh body wt) of calcareous corpuscles isolated from different cestodes. (Data from von Brand et al., 1969)*

Metacestodes	Phosphate (%)	Adults	Phosphate (%)
Cysticercus pisiformis	1.9	Taenia pisiformis	11.9
C. bovis	1.7	T. saginata	18.5–22.8
C. fasciolaris	4.0–4.2	T. taeniaeformis	9.6–11.3
C. cellulosae	1.1–1.4	Raillietina cesticillus	9.9
Taenia crassiceps	4.1	Priapocephalus spp.	12.1
Echinococcus granulosus	3.0–4.1	Diphyllobothrium latum	39.7
E. multilocularis	2.6–3.3		
Mesocestoides corti	1.5–1.8		
Spirometra mansonoides	5.3–6.2		
Ligula intestinalis	5.8		

other cestode protease inhibitors inactivate proteolytic enzymes is poorly understood (see Chapter 2), however, and their effectiveness in helping to prevent cestodes from being digested *in vivo* has yet to be ascertained.

Calcareous corpuscles

Many species of cestodes, especially the larvae, contain large numbers of curious concretions, termed *calcareous corpuscles*, made up of an organic base together with inorganic material. They are composed of concentric lamellae and vary in size; in some species they are very large – 16–32 μm (*Spirometra mansonoides*, *E. granulosus*) – but, in most species, they measure about 12 μm. They have been the subject of a number of studies (*36, 135, 351, 385, 386, 594, 796, 912, 913, 914*), but their composition, formation and, particularly, function still remain poorly understood.

The major inorganic components of calcareous corpuscles are calcium, magnesium, carbonate and phosphorus, with traces of other elements, but these constituents, especially phosphate, can vary considerably in relation to metabolic conditions and to cestode species (Table 4.4). Ultrastructural studies have shown that the corpuscles are produced intracellularly in corpuscle-forming cells, with a single corpuscle being formed in one cell, which is apparently destroyed in the process (*594*). The organic material contains RNA, DNA, protein, glycogen, mucopolysaccharides, various lipids and alkaline phosphatase (*796*). The mechanism of mineralisation of calcareous corpuscles is unknown although incorporation of calcium (Fig. 4.2), phosphate and trace elements into calcareous corpuscles has been shown in various cestode species (*39*). In addition, tetrathyridia of

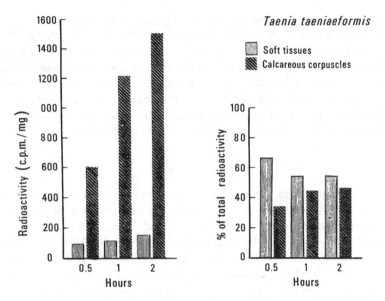

Fig. 4.2. *In vitro* accumulation of $^{15}Ca^{2+}$ by soft tissues and calcareous corpuscles of larval *T. taeniaeformis*. c.p.m., counts/min. (After Von Brand & Weinbach, 1975.)

Mesocestoides corti concentrate a variety of cations – chromium, copper, gallium, indium, thallium, zinc and zirconium – into calcareous corpuscles *in vitro* (*36*).

The role of calcareous corpuscles is not clear, although a number of speculations have been made. It has been shown that corpuscular material of *T. taeniaeformis* is removed more quickly under anaerobic than under aerobic conditions, and that it is also lost in acid but not alkali media. Consequently it has been suggested (*796*) that calcareous corpuscles buffer anaerobically produced acids (Chapter 5). Another possible protective function suggested for corpuscles is that of helping to buffer gastric hydro-chloric acid, which is encountered by cestode larvae during their passage through the host stomach (*135*). It is probable, however, that the role of the corpuscles is more complex than this, and the high level of inorganic ions has led to the suggestion that corpuscles may act as major reserves of these ions and CO_2. These materials may be called on suddenly – phosphates for phosphorylation reactions, the ions to act as enzyme cofactors, and the CO_2 for CO_2-fixation (Chapter 5) – such as when a cestode larva enters a host intestine and immediately requires rapid synthesis of ATP for muscular attachment, rapid initial growth and active transport of nutrients. That they play an important role in the metabolism of early developing intestinal

61

The adult: general metabolism

worms is reflected in the fact that the first sign of strobilar differentiation of *E. granulosus* and *E. multilocularis in vitro* is the disappearance of the calcareous corpuscles (p. 251; Fig. 9.10, p. 252). Other suggestions for the function of calcareous corpuscles include a role in lipid metabolism, tissue repair and osmotic balance, and they may act as rudimentary skeletal structures; they may also have an immunological role, as it has been suggested that anti-complementary factors are associated with the calcareous corpuscles of *E. granulosus* and *E. multilocularis* (*381*).

Nucleotides

Another important group of biochemical constituents in cestodes are nucleotides. Nucleotides (ribonucleotides and deoxyribonucleotides) are the basic building blocks of nucleic acids (see Chapter 6), but ribonucleotides, especially the adenine nucleotides (ATP, ADP and AMP) play a central role in other metabolic processes (Chapter 5), being intermediates in energy exchange, and they also act as coenzymes and enzyme modulators. A knowledge of the physiological concentrations of these nucleotides is critical to an understanding of their various functions in cestodes and the content of the adenine nucleotides, being relatively high, has been measured in several species (Table 4.5).

The energy status of a cell can be calculated from the levels of adenylate nucleotides and is expressed as the energy or adenylate charge:

$$\text{adenylate charge} = \tfrac{1}{2}\frac{[\text{ADP}] + 2[\text{ATP}]}{[\text{AMP}] + [\text{ADP}] + [\text{ATP}]}.$$

The adenylate charge integrates the metabolic pathways which produce and utilise high-energy phosphate. The mid-point for control by the adenylate charge is 0.85, this being the adenylate ratio at which ATP production in the cell equals energy utilisation. In mammalian tissues, the adenylate charge normally lies between 0.70 and 0.95, and, generally, cestodes conform (Table 4.5) and maintain a fairly high energy charge in their tissues. This is, perhaps, somewhat surprising in view of their essentially anaerobic energy metabolism (Chapter 5), which results in a low ATP production per mole of glucose catabolised. Use has been made of the adenylate charge to assess the metabolic integrity of cestodes *in vitro* and to monitor the effects of anthelmintics (*39*).

Nicotinamide nucleotide (NAD, NADP, NADH, NADPH) levels (Table 4.6) have been measured only in one cestode, *H. diminuta* (*42*), which probably reflects the difficulty of carrying out such an analysis. In this

Chemical composition

Table 4.5. *Adenine nucleotide content (μmol/g fresh wt) and adenylate charge in cestodes*

Species	AMP	ADP	ATP	Adenylate charge	Stage	Ref.
Moniezia expansa	0.35	0.72	0.65	0.59	Adult	*168*
Echinococcus granulosus	0.03	0.16	0.86	0.90	Protoscoleces (horse strain)	*500*
E. granulosus	0.04	0.36	1.11	0.86	Protoscoleces (sheep strain)	*500*
E. multilocularis	0.04	0.24	1.59	0.92	Protoscoleces	*500*
Hymenolepis diminuta	0.17	0.52	1.09	0.74	Adult (UT strain)[a]	*531*
H. diminuta	0.09	0.45	0.55	0.69	Adult (ANU strain)[b]	*612*
H. microstoma	0.20	0.63	1.02	0.72	Adult	*531*
Ligula intestinalis	0.15	0.26	1.67	0.86	Plerocercoid	*502*
L. intestinalis	0.30	0.49	1.12	0.72	Adult	*502*
Schistocephalus solidus	0.12	0.55	1.07	0.77	Plerocercoid	*63*

Notes:
[a] Toronto strain.
[b] Canberra strain.

species, NADP is mainly in the reduced form, whereas NAD is mainly in the oxidised form. The NAD content is higher than the NADP, and this is consistent with the role of NAD in energy metabolism, whilst NADP is involved in reductive synthetic pathways.

Phosphagens

Muscle and other tissues from both vertebrates and invertebrates usually contain a reserve of high-energy phosphate in the form of phosphagens. The phosphagens are all guanidines and they react reversibly with ATP:

$$\text{guanidine} + \text{ATP} \underset{\text{phosphotransferase}}{\overset{\text{phosphagen}}{\rightleftharpoons}} \text{phosphagen} + \text{ADP}.$$

The major phosphagens found in nature are arginine and creatine but parasitic helminths are unusual in that they possess no detectable phosphagens and none occurs in *Hymenolepis diminuta*, *Moniezia expansa*, *Ligula intestinalis* or *Schistocephalus solidus* (*44*). This absence of phosphagens has implications for control of metabolism in cestodes. If

Table 4.6. *The levels of nicotinamide nucleotides in* Hymenolepis diminuta *compared with* Fasciola hepatica *and* Ascaris lumbricoides. (*Data from Barrett & Beis, 1973*)

	Steady-state content (nmol/g fresh wt)					
	[NAD]	[NADH]	[NADP]	[NADPH]	[NAD]/[NADH]	[NADP]/[NADPH]
Hymenolepis diminuta	236	77	17	32	3	0.4
Ascaris lumbricoides	265	35	14	19	8	0.7
Fasciola hepatica	266	77	11	28	3	0.6

64

there is no reserve of high-energy substrate, a sudden energy demand should lead to a transient drop in ATP/ADP ratio. This has led to speculation (*44*) that, because cestodes live in a sheltered environment, either they are not subject to sudden energy demands or they may be able to accelerate their metabolism sufficiently quickly that a reserve of high energy phosphate is unnecessary. Curiously, *S. solidus*, unlike the other species, possesses an active phosphagen (taurocyamine) phosphotransferase (*44*) but its function remains enigmatic.

Lipid metabolism

General comments

Lipids are a highly diverse and heterogeneous group of compounds, with a variety of cellular functions. They are generally important energy reserves, although not in cestodes (see below), and are major components of cell membranes. Lipids also play important roles in enzyme regulation, cell surface recognition, cell interaction, glycoprotein synthesis, in the expression of surface antigenic determinants and in membrane transport. In addition, certain lipids (the quinones, see Chapter 5) form part of the electron transport chain, and lipids often occur in association with carbohydrates and proteins as glycolipids and lipoproteins. Lipoproteins may be highly antigenic in cestodes, examples being antigen 5 and antigen B, the two major antigens of *Echinococcus granulosus* hydatid cyst fluid (*690*) (see Chapter 6). Furthermore, there is evidence that glycolipids associated with the cestode tegument can interact directly with complement (*538*) and these molecules, in combination with surface glycoproteins (Chapter 6) are thought to play an important role in parasite protection *in vivo*.

Early research on cestode lipids has to be treated with some caution as the introduction of newer analytical methods, which include thin-layer and gas chromatography and infrared, ultraviolet and mass spectroscopy, has completely revolutionised this field. In spite of the advent of these new techniques, however, recent investigations on cestodes are few, being restricted mainly to analyses of lipid composition and studies of lipid synthesis. The valuable reviews of Smirnov (*780*), Smirnov & Bogdan (*781*), Barrett (*40*) and Frayha & Smyth (*233*) have comprehensively surveyed the field.

Some recent studies include those on: *Cotugnia digonopora* and *Raillietina fuhrmanni* (*595, 596, 597*); *R. cesticillus* (*635*); *R. tetragona* and *R. echinobothrida* (*915, 917*); *Hymenolepis diminuta* (*37, 371*); *Echinococcus granulosus* (*231, 341, 522, 685*); *Taenia taeniaeformis* (*537–539*); *Taenia crassiceps* (*540*); *Moniezia expansa*

The adult: general metabolism

Table 4.7. *Total lipid content and neutral lipid fractions identified in mature and larval cestodes. (Data from Frayha & Smyth, 1983; Barrett, 1983)*

	Fresh tissue (%)	Dry tissue (%)	Free fatty acids	Acylglycerols	Sterols, sterol esters	Waxes	Hydrocarbons
Mature cestodes							
Taenia saginata	3.3–3.8	31.1	+		+		
T. solium	14	–					
T. marginata	1.1	4.9					
T. plicata	9.1	33.1					
T. taeniaeformis	3.8	10.6			+		
Echinococcus granulosus	–	7.5–12.2					
Hymenolepis diminuta	5.1–5.8	21.2–34.6	+	+	+		+
H. citelli	3.9	16.1	+				
Thysaniezia giardi	–	15.8–41.3					
Dipylidium caninum	2.4	–					
Moniezia denticulata	1.3	16.2					
M. expansa	3.4	30.1			+		
Raillietina fuhrmanni	–	10.8			+		
Cotugnia digonopora	–	7.6			+		
Spirometra mansonoides	–	20–24	+	+	+		
Diphyllobothrium latum	1.6	–					
Calliobothrium verticillatum	7	25	+	+	+	+	
Proteocephalus exiguus	–	35.6	+	+	+		
Triaenophorus nodulosus	–	22.5	+	+	+		
Orgymatobothrium musteli	2.5–10	–	+	+	+		
Lacistorhynchus tenuis	1.2–10	–	+	+	+		
Eubothrium crassum		37.3	+		+		
Larval cestodes							
Taenia hydatigena							
Cysticerci membrane	–	18.2	+	+	+		+
Cysticerci fluid	–	4.7	+	+	+		+
Taenia taeniaeformis							
Strobilocerci	2.3	6.9					
Echinococcus granulosus							
Cyst membrane	–	1.3					
Cyst fluid	0.02	–	+	+	+		+
Protoscoleces	1.05–2.0	8.8–13.6	+	+	+		+
Echinococcus multilocularis							
Protoscoleces	–	16.1			+		+
Spirometra mansonoides							
Coracidium	–	25	+	+	+		

Lipid metabolism

Table 4.7. (*cont.*)

	Fresh tissue (%)	Dry tissue (%)	Free fatty acids	Acylglycerols	Sterols, sterol esters	Waxes	Hydrocarbons
Procercoid	–	10	+	+	+		
Plerocercoid	–	16–17	+	+	+		
Schistocephalus solidus							
Plerocercoid	2.6	9.8	+	+	+		
Ligula intestinalis							
Plerocercoid	–	14.2	+	+			

Note: +, present; –, not determined.

(*520*); *Diphyllobothrium dendriticum*, *Eubothrium crassum* and *Triaenophorus crassus* (*766, 782*); *D. dendriticum*, *D. latum* and *D. vogeli* (*272–274*); *Aploparaxis polystictae* (*49, 409*).

Lipid content

The relative total lipid and phospholipid content of mature and larval cestodes is given in Tables 4.7 and 4.8, from which it is clear that there is considerable variation between species. Not surprisingly, all the major neutral lipid (Table 4.7) and phospholipid (Table 4.8) fractions have been demonstrated in a variety of cestodes and, in general, the lipid composition of cestodes appears to be similar to that of other organisms. Histochemical studies have shown that the parenchyma of cestodes is the most important tissue for the storage of lipids (*233*). Lipids have also been demonstrated in calcareous corpuscles and tubular organs such as the uterus and excretory canals, and they have been found universally in cestode eggs, usually located between the embryo and the shell. Lipids also occur in the cestode tegument, and discrete droplets have been identified in the tegument of developing larvae of *T. taeniaeformis* (*537*). These lipids have been isolated and their composition is shown in Table 4.9. It appears that sterols, sterol esters and fatty acids are absorbed by the young parasite and organised into lipid droplets by the subtegumental cells. These are then transported to the tegument via cytoplasmic bridges, further processed and used for membrane synthesis. Relatively little is known regarding the lipid composition of the external plasma membrane in cestodes although lipids comprise 39% of the dry weight of the brush border plasma membrane of *H. diminuta*, with

The adult: general metabolism

Table 4.8. *Phospholipids of mature and larval cestodes. (Data from Frayha & Smyth, 1983; Barrett, 1983)*

	% Phospholipids of total lipids	Phosphatidic acid	Lysophosphatidic acid	Phosphatidylethanolamine (cephalin)	Phosphatidylserine	Lysophosphatidylethanolamine	Phosphatidylcholine (lecithin)	Lysophosphatidylcholine	Phosphatidylinositol	Sphingomyelin	Cardiolipin	Plasmalogens	Glycolipids (cerebrosides)
Mature cestodes													
Taenia saginata	4–10						+		+				
Taenia taeniaeformis	50			+	+		+				+		+
Hymenolepis diminuta	22–37	+	+	+	+		+		+		+	+	+
H. citelli	25			+	+		+			+	+	+	
Dipylidium caninum	21			+	+	+	+	+	+				+
Moniezia expansa	15–23												+
Spirometra mansonoides	53–58	+		+	+	+	+	+	+	+			+
Diphyllobothrium latum	31						+						+
Calliobothrium verticillatum	39			+	+		+				+		
Proteocephalus exiguus	49												
Triaenophorus nodulosus	52												
Orygmatobothrium musteli	25			+	+		+						
Lacistorhynchus tenuis	50			+	+		+						
Eubothrium crassum	65												
Larval cestodes													
Taenia hydatigena													
Cysticerci membrane	37			+	+		+	+	+	+			
Cysticerci fluid	39			+	+		+	+	+				
Taenia taeniaeformis													
Strobilocerci	47–50			+	+		+			+			+
Echinococcus granulosus													
Cyst fluid	0.1												
Protoscoleces	57			+	+		+	+	+	+			
Spirometra mansonoides													
Coracidium	64	+		+	+		+				+		+
Procercoid	58	+		+	+		+				+		+
Plerocercoid	64	+		+	+		+				+		+
Schistocephalus solidus													
Plerocercoid	52												
Ligula intestinalis													
Plerocercoid	27												

Lipid metabolism

Table 4.9. *Percentage distribution of neutral and polar lipids (glycolipids and phospholipids) in lipid droplets from 21-day old larvae of* Taenia taeniaeformis. *(Data from Mills et al., 1983)*

Neutral lipids	(%)	Polar lipids	(%)
Hydrocarbons	8	Phosphatidylethanolamine	9
Sterol esters	16	Phosphatidylcholine	25
Triglycerides	46	Phosphatidylserine	6
Free sterols	10	Glycosphingolipid-1	18
Diglycerides	4	Glycosphingolipid-3	15
Monoglycerides	5	Glycosphingolipid-4	14
Free fatty acids	11	Glycosphingolipid-5	13

cholesterol, phosphatidyl ethanolamines, cerebrosides and cardiolipins being the dominant components (*112*). That so few data are available on the lipid components of the surface membrane in cestodes is surprising, as these membranes constitute a primary component of the host–parasite interface (Chapter 2).

Fatty acids

Very little information is available regarding fatty acid uptake in cestodes, but it appears that both short- and long-chain fatty acids are absorbed by a mixture of diffusion and mediated transport (*39*). The major fatty acids of cestodes are usually C_{16} and C_{18} acids. Oleic ($C_{18.1}$) is the major C_{18} acid, although linoleic acid ($C_{18.2}$) or stearic acid ($C_{18.0}$) predominate in some species. Of fatty acids 50–60% are generally unsaturated in cestodes, which is similar to the situation in mammals. In certain species, however, including *Hymenolepis diminuta*, *Raillietina cesticillus*, *Diphyllobothrium dendriticum*, *Eubothrium crassum* and several shark tapeworms (e.g. *Calliobothrium verticillatum*, *Thysanocephalum cephalum*, *Lacistorhynchus tenuis*), the lipids are richer in unsaturated fatty acids, with 70–80% of the fatty acids being unsaturated (*40*). The types of fatty acids found in cestodes are often very similar to those of their hosts. This is shown most strikingly in tapeworms from sharks, which, like their elasmobranch hosts, contain large amounts of C_{20} and C_{22} polyunsaturated acids (*51, 110*). A gas–liquid chromatograph of the total fatty acid methyl esters of *C. verticillatum* (Fig. 4.3) shows the predominance of the C_{20} and C_{22} acids. The marked qualitative similarities between the fatty acid composition of the parasite and of the host fluids and tissues is shown in Table 4.10.

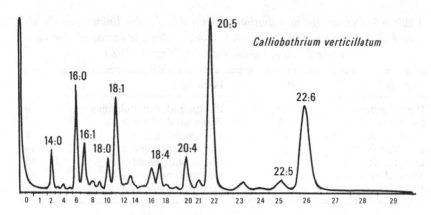

Fig. 4.3. Gas–liquid chromatograph of the total fatty acid methyl esters of *Calliobothrium verticillatum*, temperature-programmed from 150 to 200 °C at 2 deg. C/min. (After Beach *et al.*, 1973.)

Metabolism of fatty acids

In marked contrast to their hosts, cestodes are unable to synthesise long-chain fatty acids *de novo* from acetyl-CoA. Neither can they desaturate preformed long-chain fatty acids. Instead, fatty acid synthesis in cestodes is restricted to chain lengthening of host-derived fatty acids by the sequential addition of acetyl-CoA, although the mechanism for chain elongation is not known. Thus, *H. diminuta* can convert palmitate and stearate into saturated fatty acids with up to 26 carbons (*354*), while *Spirometra mansonoides* can elongate C_{16}, C_{18}, $C_{18.1}$, $C_{18.2}$ and $C_{18.3}$ acids to C_{20} and C_{22} acids (*532*). Prostaglandins, which are formed by the cyclisation of C_{20} unsaturated fatty acids, have profound physiological effects in vertebrates. They have been little studied in invertebrates but PGE_2 and PGI_2, both well-known immunomodulatory lipids, have been identified in the larvae of *T. taeniaeformis* (*436*). Furthermore, PGE_2 is released by this species (*436*) and it is possible that this prostanoid, by helping to suppress host cellular reactivity, may account for the lack of host cellular response around the living parasite in the liver and be a factor in its long-term survival.

In mammals, complex lipids are catabolised by lipases to yield fatty acids, glycerol and other components. The fatty acids are then broken down by beta-oxidation to give NADH, reduced flavoprotein and acetyl-CoA, which then enters the tricarboxylic acid cycle. Lipase activity has been detected in several cestodes (*513, 772*) but there is no evidence for an active beta-oxidation sequence in any species investigated to date. Somewhat surprisingly, however, some of the beta-oxidation enzymes have been

70

Lipid metabolism

Table 4.10. *Fatty acids of* Calliobothrium verticillatum *and of its host, the dogfish* Mustelus canis. *Samples were saponified, their fatty acids were extracted and methylated, and the fatty acid methyl esters were characterised and quantified by gas–liquid chromatography.* (*Data from* Beach *et al.,* 1973)

Fatty acid methyl esters	% Total fatty acid methyl esters						
		Mustelus canis					
	C. verticillatum; total	Intestinal contents; unesterified fatty acids	Intestinal contents; entire	Bile	Blood plasma	Intestinal wall	Liver
12:0	0.4	0.9	1.0	0.4	0.5	tr	tr
13:0	tr	—	tr	0.4	tr	—	—
i-14:0	0.6	0.5	1.0	2.0	0.5	tr	—
14:0	2.4	3.6	3.3	3.1	2.8	1.0	2.1
ai-15:0	0.2	1.2	1.0	2.1	0.7	tr	tr
15:0	0.7	1.2	1.4	2.0	0.8	0.3	0.5
i-16:0	0.8	1.7	2.2	1.7	3.1	0.3	0.3
16:0	6.7	14.6	13.1	14.4	16.0	15.1	20.1
16:1 (i-17:0)	4.3	7.1	5.8	6.3	6.4	3.9	10.9
17:0 (ai-17:0; 16:u)	1.2	2.3	2.4	2.2	1.3	1.0	1.4
17:1 (16:u, i-18:0)	1.1	1.9	2.1	1.0	1.8	0.8	0.8
18:0 (16:u, 17:u)	3.2	5.4	5.5	5.7	3.6	8.8	3.9
18:1 (16:u)	8.8	14.1	11.5	13.9	13.1	12.0	21.7
19:0 (16:u)	0.8	tr	tr	0.5	1.0	tr	tr
18:2	1.8	1.7	1.4	1.0	0.9	0.8	0.8
19:1 (i-20:0)	0.7	1.5	1.0	1.2	0.6	1.0	0.4
18:3 ω6 (20:0)	1.0	0	tr	1.1	tr	3.6	0.4
18:2 ω3 (20:1)	3.4	4.2	2.8	4.9	3.2	1.7	4.0
18:4 ω3 (19:u, 21:0)	3.0	1.9	2.4	0.6	2.0	0.5	0.7
20:2 ω6	1.1	1.6	1.9	0.8	0.9	0.6	0.6
20:2 ω6 (22:0, 19:u)	1.0	tr	1.0	1.6	tr	0.5	tr
20:4 ω6 (22:1, 19:u)	4.6	4.5	3.8	3.8	3.2	6.9	2.5
20:4 ω3	2.0	2.2	2.6	1.0	1.7	0.9	1.1
20:5 ω3 (22:2, 21:u)	21.1	13.4	10.2	7.0	11.1	11.0	7.3
22:4 ω6 (21:u, 24:1)	3.2	2.4	3.4	2.2	1.9	2.6	2.2
22:4 ω3 (24:u)	2.1	1.0	2.8	1.0	1.2	1.2	0.8
22:5 ω3	3.7	2.8	4.3	3.0	3.7	6.3	4.2
22:6 ω3	20.1	8.3	12.1	15.1	18.0	19.2	13.3

Note: tr, trace.

Table 4.11. *The activities of the beta-oxidation enzymes in cestodes (Data from Barrett, 1983)*

Enzyme	Activity (nmol/min per mg protein at 30°C)		
	Hymenolepis diminuta (adult)	*Schistocephalus solidus* (plerocercoid)	*Ligula intestinalis* (plerocercoid)
Acyl-CoA synthetase (short chain)	—	5.3	3.0
Acyl-CoA synthetase (long chain)	1.9	0.5	0.4
Acyl-CoA dehydrogenase	1.5	0.7	1.2
Enoyl-CoA hydratase	0	11.6	0
3-Hydroxyacyl-CoA dehydrogenase	0	2.2	0
Acetyl-CoA acyltransferase	12	9.6	4.7

demonstrated in several species and a complete sequence occurs in the plerocercoids of *Schistocephalus solidus* (Table 4.11). The function of these beta-oxidation enzymes in cestodes is far from clear.

Steroids

The dominant steroid present in cestodes is cholesterol. It enters hydatid cysts of *E. granulosus* by simple diffusion (*33*), and adult *H. diminuta*, in part, via a mediated system which is specific for sterols (*624*). Cestodes appear to be unable to synthesise this or any other steroid *de novo*. This inability to synthesise steroids is not peculiar to cestodes, as steroid synthesis is also absent in other platyhelminths, nematodes and many other invertebrate groups (*40*). Nevertheless, cestodes readily incorporate exogenous steroids and fatty acids into their steroid esters, although the pathways involved have not been studied. In addition, cestodes are able to synthesise polyisoprenoids such as farnesol, an intermediate in steroid synthesis. *H. diminuta* can synthesise the 2-*trans*, 6-*trans* isomer of farnesol (*218*), while the larvae of *E. granulosus* and *T. hydatigena* (*229, 230*) produce a similar compound with acetate as precursor. Moreover, biosynthetic labelling of worms followed by two-dimensional thin-layer chromatography has shown that *H. diminuta* can synthesise dolichols and a prenoid-linked quinone, probably rhodoquinone, in addition to farnesol (*371*). The block to sterol synthesis in cestodes appears to be the conversion of farnesol to the long-chain hydrocarbon squalene, but the reasons for this are unclear.

Farnesol mimics juvenile hormone activity in insects but there is no evidence for this compound having any hormonal role in cestodes. Another group of steroids which could function as hormones in cestodes is the ecdysteroids. These control moulting and metamorphosis in insects and in some crustacea. Free and conjugated ecdysteroids have been detected in *H. diminuta* (*521*) and *E. granulosus* (*522*), while ecdysone, 20-hydroxyecdysone and 20,26-dihydroxyecdysone have been identified in *M. expansa* by high-pressure liquid chromatography (HPLC), radioimmunoassay, and by gas chromatography/mass spectrometry (*520*). HPLC fractionation of the free ecdysteroid fraction from *M. expansa* is shown in Fig. 4.4. However, the origin of these steroids and their exact physiological function in cestodes remain obscure.

Other lipids

There is nothing particularly unusual about the composition or metabolism of the other lipid classes found in cestodes. Rather high levels of

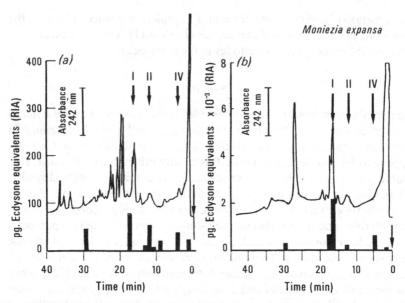

Fig. 4.4. HPLC fractionation of the free ecdysteroid fraction from *Moniezia expansa* (*a*) sample 1, 3.3% portion; (*b*) sample 2, 10% portion on a chromatographic system with collection of fractions every minute for the radioimmunoassay (RIA) (ICT-1 antiserum). The positions of authentic ecdysteroids are shown: IV, 20, 26-dihydroxyecdysone; II, 20-hydroxyecdysone; I, ecdysone. (After Mendis *et al.*, 1984.)

phospholipids, about 35% of the total, are commonly found, however (Table 4.8), which is rather more than occurs in most invertebrates (*37*). Given the precursors, cestodes can synthesise many of their own complex lipids and incorporation of label from ^{32}P- and ^{14}C-labelled fatty acids, glycerol and glucose into phospholipids and other lipid components has been demonstrated in several species. For example, plerocercoids of *S. solidus* (*43*) and *L. intestinalis* (*407*) absorb and incorporate [^{14}C]palmitate into neutral and phospholipid fractions (Table 4.12). More specific experiments with [^{32}P]orthophosphate have shown that *H. diminuta* possesses mechanisms for *de novo* synthesis of phospholipids (*944*). In addition, *H. diminuta* can incorporate [^{14}C]glucose into the glycerol moiety of acylglycerols and phosphoglycerides, the inositol moiety of phosphatidylinositol, the galactose moiety of glycolipids, and phospholipids (*945*).

Despite the fact that cestodes appear to be capable of considerable lipid synthesis, the actual pathways and enzyme systems involved have been little studied. Sphingomyelin synthesis has been investigated in *H. diminuta* (*37*), where it probably involves the same five-step pathway thought to occur in vertebrates involving the enzymes serine palmitoyl-transferase,

Lipid metabolism

Table 4.12. *The incorporation of exogenous[(U-^{14}C)]palmitate into lipids and carbon dioxide by plerocercoids of* Ligula intestinalis. *(Data from Körting & Barrett, 1978)*

	c.p.m./mg dry wt per h	
Fraction	Whole worms	Minced worms
Carbon dioxide	112	44
Neutral lipids	272 979	126 216
Free fatty acids	91 523	106 830
Phospholipids	325 237	452 196

Note: c.p.m., counts per minute.

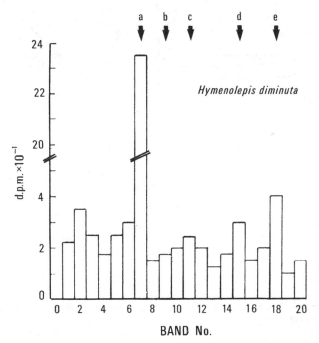

Fig. 4.5. Sphingomyelin synthesis in *Hymenolepis diminuta*: distribution of label in various intermediates, separated by thin-layer chromatography, after incubation with cytidine-5'-diphospho [methyl-^{13}C]choline. The position of the lipid standards is indicated by the arrows. (*a*) sphingomyelin; (*b*) dihydrosphingosine; (*c*) sphingosine; (*d*) ketosphingosine; (*e*) ceramide. The origin is at band O. d.p.m., disintegrations/min. (After Bankov & Barrett, 1985.)

3-oxosphinganine reductase, flavoprotein dihydrosphingosine reductase, sphingosine acyltransferase and ceramide choline-phosphotransferase. The presence in *H. diminuta* of the last enzyme in this pathway is demonstrated in Fig. 4.5, where parasite extracts rapidly incorporate label from cytidine diphosphocholine, the immediate precursor, into sphingomyelin. It is likely that synthesis of other lipids will also involve the same general pathways as exist in mammals, although there may well be differences in detail. A proposed scheme (*233*) for the predominant metabolic pathways of the major classes of lipids in cestodes is presented in Fig. 4.6.

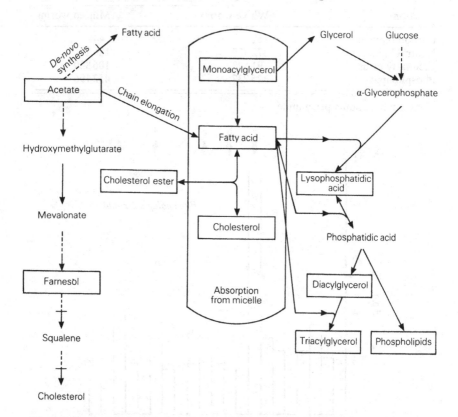

Fig. 4.6. Predominant metabolic pathways of the major classes of lipids in cestodes. (After Frayha & Smyth, 1983.)

5

The adult: carbohydrate metabolism

General considerations

Basic problems

The carbohydrate metabolism of cestodes has been more extensively studied than any other aspect of metabolism; it has been the subject of a number of reviews (*39, 41, 101, 102, 104, 129, 221, 398, 490, 492, 673, 698, 759, 816, 911, 931*). Much work in this area has been aimed at the identification of biochemical steps which might differ substantially from the host metabolism and which may, therefore, be of value for rational drug design. Unfortunately, in practice, the approach has not proved to be very rewarding and virtually all useful drugs against cestodes have been developed empirically rather than rationally. The chemotherapy of cestodes will not be covered here as this aspect has been comprehensively reviewed (*113, 114, 190, 735, 895, 946*).

Interest has focussed on four main areas: (*a*) the nature of exogenous carbohydrates utilised by cestodes, (*b*) the mechanisms by which these carbohydrates are absorbed, (*c*) the pathways of carbohydrate catabolism and (*d*) the regulatory control of these pathways.

These investigations have shown that cestodes, in common with other parasitic helminths, utilise carbohydrate as the major, and possibly only, energy substrate. In addition, carbohydrate catabolism in cestodes is characterised by the excretion of reduced end-products (e.g. lactate, propionate, succinate) even under aerobic conditions.

Much of the information accumulated has been obtained from studies *in vitro*. The pitfalls inherent in trying to relate data from *in vitro* experiments to the situation *in vivo* are well recognised (*490*). The time of incubation *in vitro*, the physiological condition of the cestode under study and differences in experimental conditions, especially the pO_2, pCO_2, pH, Eh, the presence of glucose, serum or bile and the concentration of ions such as Na^+, K^+ or Mg^{2+} in the incubation medium, may substantially affect the

Table 5.1. *End-products*, nmol/g wet wt *(means ± S.D.), of carbohydrate metabolism produced by Hymenolepis microstoma under different in vitro conditions. (Data from Rahman & Mettrick, 1982)*

End-product	Air-CO$_2$ (6)	95%O$_2$ + 5%CO$_2$ (6)	95%N$_2$ + 5%CO$_2$ (7)
Lactate	26 364 ± 4400[a]	19 427 ± 3037[b]	13 430 ± 1227
Succinate	37 982 ± 7768[a]	64 607 ± 4383[a]	101 805 ± 14 178
Acetate	13 055 ± 2218[a]	25 304 ± 4539	30 006 ± 3466
Propionate	3786 ± 653[b]	8505 ± 1432	6586 ± 1021
Total end-products	84 230 ± 12 937[a]	117 433 ± 8460[b]	154 051 ± 14 447
Total end-products excreted[c]	56 840 ± 11 464[a]	85 896 ± 6251[d]	110 931 ± 15 974

Notes:
[a] $P < 0.0005$ indicate significant difference from N$_2$ + CO$_2$ results.
[b] $P < 0.0025$
[c] Total of lactate, succinate, acetate and propionate in the incubation media.
[d] $P < 0.01$

Figures in parentheses denote numbers of determinations.

results obtained. For example, the presence or absence of CO_2 during incubation has a significant effect on the quantity of metabolites produced as a result of carbohydrate breakdown in *Hymenolepis microstoma* (Table 5.1). This effect is related chiefly to CO_2-fixation, which, as we shall see later, is a vital step in the metabolic pathways of this and other cestodes.

An important point which should be re-emphasised in relation to *in vitro* studies is the now substantial evidence that certain cestodes exist as a complex of intraspecific variants or strains (pp. 97–98). These strains may exhibit considerable quantitative and qualitative differences in carbohydrate metabolism, thereby complicating the interpretation of results. This particular aspect is elaborated on later. Furthermore, there is the additional problem of differing protocols used by independent research workers which can often make *in vitro* data comparisons difficult.

Carbohydrate utilisation *in vivo*

It is now well recognised that carbohydrate must be present in the host diet for cestode establishment and, in most species, for normal development and reproduction. This has been shown for *H. diminuta*, where lack of, or restriction in, host dietary carbohydrate results in reduced establishment and stunting in worm size (*697, 796*). Similar effects have been reported for *H. citelli*, *H. nana* and *Oochoristica* in rodents, *Raillietina* in birds and *Lacistorhynchus* in dogfish (*527, 698, 796*).

The level of infection in a host has a direct effect on the carbohydrate content of cestodes (*796*) and has been shown by Henderson (*317*) to affect glucose absorption by *H. diminuta in vitro* (Fig. 5.1). These are manifestations of the 'crowding' effect, reported in *Raillietina*, *H. microstoma*, *H. nana* and, most notably, *H. diminuta* (*697*), which results in an inverse correlation between worm size and increasing population density. Competition for host dietary carbohydrate presumably plays a role in the crowding effect, but the phenomenon is complex and a number of other factors are probably involved (*348, 697, 979, 980*). This phenomenon is addressed further in Chapter 9.

The main carbohydrate reserve found in cestodes is glycogen, as the level of free glucose is low, being usually less than 1 % of the dry weight. Another potential energy substrate, the disaccharide trehalose, occurs in some species (*39*) and a trehalase, which splits trehalose into two glucose moieties for subsequent phosphorylation to glucose-6-phosphate via hexokinase, has been reported in at least one cestode, *Stilesia globipunctata* (*581*).

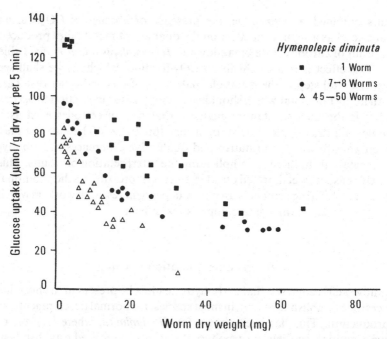

Fig. 5.1. The effect of the number of worms in an infection on the absorption of glucose from a 6 mM solution by *Hymenolepis diminuta*. If worms of similar size (i.e. dry weight) are compared, those from low density infections show greater rates of absorption than those from high. (After Henderson, 1977.)

Carbohydrate utilisation *in vitro*

A substantial amount of experimental work has been performed on the utilisation of exogenous carbohydrate, particularly glucose, during short periods of maintenance *in vitro*. All cestodes appear to be capable of taking up glucose and the mechanism of uptake, by active mediated transport, seems to be fairly similar in *H. diminuta, H. microstoma, Calliobothrium verticillatum, Taenia crassiceps* and *T. taeniaeformis* (*624*), being characterised by Michaelis–Menten kinetics and occurring against a concentration gradient. In addition, the process requires ATP, is specific for D-glucose, is inhibited by a variety of similar-structure monosaccharides (Table 5.2), can be competitively inhibited by glucose analogues such as 1-deoxyglucose or 3-*O*-methylglucose and is non-competitively inhibited by protein precipitants. The theoretical and practical considerations concerning mediated transport and other uptake mechanisms have been described in Chapter 3 and by Barrett (*39*). It should be pointed out that few studies on

Table 5.2. *The actions of various monosaccharides and other compounds as inhibitors of glucose uptake in various species of cestodes. (Data from Roberts, 1983)*

Inhibitor	Hymenolepis diminuta	Hymenolepis microstoma	Taenia crassiceps larvae	Calliobothrium verticillatum
D-Glucose	+	+	+	+
α-Methylglucoside	+	+	+	+
β-Methylglucoside	+	+	+	ND
1-Deoxyglucose	+	ND	ND	ND
2-Deoxyglucose	ND	−	ND	ND
3-O-Methylglucose	+	−	−	−
6-Deoxyglucose	+	ND	ND	ND
Glucosamine	ND	ND	−	−
N-Acetylglucosamine	ND	−	−	−
Galactose	+	+	+	+
2-Deoxygalactose	ND	−	ND	ND
6-Deoxygalactose	−	−	+	ND
Galacitol	ND	−	−	ND
Allose	+	ND	ND	ND
Mannose	ND	−	−	−
1,5-Anhydro-D-mannitol	−	ND	−	ND
Fructose	−	−	ND	−
Phlorizin	+	+	+	+
Phloretin	−	+	ND	−
Ouabain	−	−	−	−

Notes: +, the inhibitor inhibits glucose uptake; −, indicates no inhibition. ND, not determined.

nutrient uptake in cestodes have taken into account the syncytial nature of the tegument, the compartmentalisation of organic solutes and ions, the possible effect of bound host immunoglobulin and the 'unstirred water layer' at the surface of the tegument in the precise evaluation of transport mechanisms and in the determination of transport kinetic data. These factors will not be considered in detail here, because they have been discussed elsewhere (645, 647) but, for example, an unstirred water layer at the solution–membrane interface can lead to an underestimation of passive permeability coefficients and to a distortion of Michaelis constants for active transport processes (Table 5.3).

Glucose and galactose transport in cestodes appears to occur via a common carrier system, as these sugars are mutually competitive inhibitors. This uptake mechanism is potassium and phlorizin sensitive and sodium

Table 5.3. *The effect of the unstirred water layer on the kinetics of glucose uptake in* Hymenolepis diminuta. *(Data from Podesta, 1977)*

Rate of stirring (rpm)	Apparent Michaelis constant (mM)	Apparent maximal transport rate (μmol/g per h)	Apparent permeability coefficient (μmol/g per h per mM)	Michaelis constant in absence of unstirred layer (mM)	Maximal transport rate in absence of unstirred layer (μmol/g per h)
0	2.49	133	2.16	—	—
525	1.41	123	2.50	0.21	106
710	1.13	117	2.25	—	—
950	1.01	115	2.10	—	—
1550	0.71	111	2.50	—	—
2040	0.54	110	2.50	—	—

and chloride dependent (reviewed: *460, 624*). It has been suggested that the energy required for hexose uptake in cestodes may be derived partially from the maintenance of the sodium gradient across the external plasma membrane (Fig. 2.2, p. 8) but it is more likely to be supplied as metabolic energy, i.e. ATP (*646, 892*). In several respects, hexose uptake in cestodes is similar to that occurring in the mammalian intestine, although distinct differences are also apparent in the two systems (*39*). In both, glucose and galactose share a common sodium-dependent carrier which is inhibited by phlorizin (which probably competes with the sugars for the carrier site) and potassium. In contrast, glucose and galactose are the only sugars shown to be accumulated against a concentration gradient in cestodes, but, in mammals, in addition to glucose and galactose, some 12 different glucose analogues are actively transported. Furthermore, fructose is not metabolised in cestodes and enters by passive diffusion, but, in the mammalian intestine, this sugar is transported by a specific carrier. *H. diminuta*, it should be noted, is virtually impermeable to fructose (*39*), although why this should be the case is not known.

Intermediary carbohydrate catabolism

General account

It is assumed that the reader is familiar with the general pattern of the pathways involved in carbohydrate metabolism in mammalian tissues. Details of these pathways are given in most elementary texts in biochemistry. The metabolic pathways best known in mammals occur also in cestodes, although generally in a modified form; these are (i) the Embden–Meyerhof pathway of glycolysis, (ii) the Krebs or tricarboxylic acid (TCA) cycle, (iii) the electron transport system, and (iv) the pentose-phosphate pathway. The principal function of these catabolic pathways is the production of energy, usually in the form of ATP, to sustain mechanical, synthetic or osmotic work. In addition, the pathways produce intermediates for synthetic reactions, generate reducing power in the form of NADH or NADPH for synthetic reactions and provide a mechanism for macromolecular degradation. The evidence for the existence of these pathways in cestodes consists essentially of demonstration of the enzymic steps and identification of the various intermediates in the pathway.

Some useful, general studies on intermediary metabolism include those on: *Moniezia expansa* (*59–61, 664*); *H. diminuta* (*400, 531, 590, 612, 667*); *H. microstoma* (*665, 666*); *Echinococcus* spp. (*488, 498, 500*); *Mesocestoides corti* (*399*); *Cotugnia digonopora* (*618, 619*); *Schistocephalus solidus* (*406*); and *Ligula intestinalis* (*502*).

The adult: carbohydrate metabolism

Table 5.4. *End-products of carbohydrate breakdown in cestodes, besides* CO_2. *(Data from Barrett, 1981; Rahman & Mettrick, 1982; Pampori et al. 1984a,b; McManus & Sterry, 1982)*

Species	Aerobic	Anaerobic
Echinococcus granulosus (Adult)	Succinate, lactate, acetate	Similar to aerobic
(Protoscolex)	Lactate, pyruvate, acetate, succinate, ethanol	Similar to aerobic; larva with more succinate, no pyruvate
Hymenolepis diminuta (Adult)	Lactate, acetate, succinate	Similar to aerobic with more succinate
Hymenolepis microstoma (Adult)	Lactate, succinate, acetate, propionate	Similar to aerobic with more succinate
Moniezia expansa (Adult)	Lactate, succinate	Similar to aerobic with more succinate
Cotugnia digonopora (Adult)	Lactate, pyruvate, succinate, malate	—
Spirometra mansonoides (Adult)	—	Acetate, propionate, traces of lactate, succinate
(Larva)	—	Acetate, lactate, traces of propionate, succinate
Schistocephalus solidus (Plerocercoid)	Acetate, propionate	Propionate, acetate
Diphyllobothrium dendriticum (Adult and Plerocercoid)	Succinate, lactate	Similar to aerobic
Ligula intestinalis (Adult and Plerocercoid)	Lactate, succinate, acetate, propionate, pyruvate, malate	Similar to aerobic
Mesocestoides corti (Tetrathyridium)	Lactate, succinate, acetate	Similar to aerobic with more succinate

The characteristic feature of carbohydrate breakdown in cestodes is the production of a range of complex end-products, usually organic acids, even under aerobic conditions (Table 5.4). This contrasts with predominantly aerobic organisms, such as most free-living metazoa, where the end-product of glycolysis is almost exclusively lactic acid formed from pyruvic acid. Lactic acid is produced as a result of rapid muscular contraction carried out essentially under anaerobiosis and its production ensures a rapid expenditure of energy without the limitation due to the rate of diffusion of oxygen. The anaerobic phase is followed by an aerobic phase, where pyruvic acid is metabolised to acetyl-coenzyme A which is in turn oxidised completely to

Intermediary carbohydrate catabolism

CO_2 and water in the Krebs cycle. An 'oxygen debt' is built up if oxygen does not become available after the anaerobic phase.

Glycolysis

Cestodes, like most organisms, can oxidise carbohydrate via glycolysis and carry out substrate-linked phosphorylations, resulting in the formation of ATP, at the 3-phosphoglycerate kinase and pyruvate kinase steps. A highly active sequence of glycolytic enzymes has been demonstrated biochemically in a variety of cestodes (Table 5.5) and measurement of the steady-state levels of glycolytic intermediates in freeze-clamped specimens – i.e. parasites frozen rapidly in liquid nitrogen – confirms operation of the classical Embden–Meyerhof pathway (59, 63, 500, 502, 665). These metabolite measurements have also allowed the calculation of mass action ratios for the individual glycolytic enzymes (Table 5.6) with the subsequent identification of non-equilibrium reactions; these are likely to be regulatory sites (control points) in the pathway.

The theoretical background concerning regulation of fluxes through metabolic pathways, the identification of regulatory points and the mechanisms whereby different catabolic and synthetic pathways are integrated has been expertly covered by Barrett (39). Earlier reviews of the regulation of respiratory metabolism in cestodes are those of Bryant (99, 100).

KEY GLYCOLYTIC ENZYMES

In light of the obvious importance of the glycolytic pathway in cestode metabolism, it is not surprising that a number of kinetic studies have been carried out on individual glycolytic enzymes. It should be noted, however, that no biologically significant differences have yet come to light between these and the corresponding vertebrate enzymes. Some of the more important glycolytic enzymes have been purified and/or characterised in several species, as follows.

Hexokinase

This is an important regulatory enzyme as it initiates glucose catabolism via glycolysis and the pentose-phosphate pathway, and activates the formation of glycogen and complex carbohydrates from glucose. Hexokinase is characteristically inhibited by its product, glucose-6-phosphate. A hexokinase, with broad specificity, has been partially purified from *H. diminuta* (401), although a separate galactokinase is also present in this worm (402, 403). Similarly, the hexokinase from *Bothriocephalus scorpii* can phosphorylate glucose, galactose and fructose but not mannose or glucosamine (916). In contrast, four distinct hexokinases, catalysing specifically the phosphorylation of glucose, fructose, mannose and glucosamine were reported in early work on *E. granulosus* (3).

85

Table 5.5. *Specific activities of the glycolytic enzymes in a variety of cestodes*

	Specific activities								
Enzyme	Mesocestoides[a] corti (tetrathyridia)	Echinococcus[b] granulosus (protoscoleces, ovine strain)	Echinococcus[b] multilocularis (protoscoleces)	Schistocephalus[c] solidus (plerocercoids)	Ligula[d] intestinalis (plerocercoids)	Bothriocephalus[e] gowkonsensis (adults)	Khawia[e] sinensis (adults)	Triaenophorus[e] crassus (plerocercoids)	Triaenophorus[e] crassus (adults)
Hexokinase	2	3	4	10	6	39	44	0	10
Glucosephosphate isomerase	—	6964	3321	599	5218	260	360	111	254
Phosphofructokinase	8	394	347	89	156	32	0	0	0
Aldolase	—	125	171	198	636	60	225	26	186
Triosephosphate isomerase	—	17 170	10 306	4825	12 231	—	—	—	—
Glyceraldehyde-3-phosphate dehydrogenase	571	3370	3571	1661	3307	—	—	—	—
Phosphoglycerate kinase	—	1876	1619	1126	2734	—	—	—	—
Phosphoglycerate mutase	—	3897	1375	60 } combined	2168	—	—	—	—
Enolase	—	2407	1359	744	—	—	—	—
Pyruvate kinase	5	132	83	116	77	38	69	7	34
Lactate dehydrogenase	102	611	652	320	351	17	18	94	36

Notes: Specific activities are expressed as nmol/min per mg soluble protein at 25°C (*M. corti*) or 30°C.
Data from [a] Köhler & Hanselmann (1974); [b] McManus & Smyth (1982); [c] Körting & Barrett (1977); [d] McManus & Sterry (1982); [e] Körting (1976).

Table 5.6 Mass action ratios of reactions catalysed by glycolytic enzymes

Enzyme	Apparent equilibrium constant	Mass action ratios in					
		Echinococcus granulosus[a] (protoscoleces, ovine strain)	Echinococcus multilocularis[a] (protoscoleces)	Moniezia expansa[b] (adults)	Hymenolepis microstoma[c] (adults)	Schistocephalus solidus[d] (plerocercoids)	Ligula intestinalis[e] (plerocercoids)
Hexokinase	$1.5 \times 10^2 - 5.5 \times 10^3$	6.9×10^{-2}	7.7×10^{-3}	2.2×10^{-2}	8.3×10^{-2}	—	1×10^{-3}
Phosphoglucomutase	5.5×10^{-2}	15.2	2.2	—	—	2.4×10	11.7
Glucosephosphate isomerase	$2.8 \times 10^{-1} - 5 \times 10^{-1}$	1.5×10^{-1}	2.5×10^{-1}	4.4×10^{-1}	2.6×10^{-1}	1.0×10^{-1}	2.9×10^{-2}
Phosphofructokinase	1×10^3	5.8×10^{-1}	1.1	4.8×10^{-1}	26	3.2×10^{-1}	7.7
Aldolase	$6.8 \times 10^{-5} - 13 \times 10^{-5}$	2.2×10^{-2}	10.3×10^{-3}	3.3×10^{-6}	4.1×10^{-5}	$7 \times 10^{-5} - 13 \times 10^{-5}$	11.1×10^{-5}
Triosephosphate isomerase	$3.6 \times 10^{-2} - 6.5 \times 10^{-2}$	2	4.8	3.1×10^{-1}	3.6×10^{-1}	3.2×10^{-1}	4
Phosphoglycerate mutase	1.7×10^{-1}	6.3×10^{-1}	—	1.6×10^{-1}	1.0×10^{-1}	5×10^{-1} (combined)	—
Enolase	$5.3 \times 10^{-1} - 6.3$	2.3	2.8	1.2	3.4		1.6
Pyruvate kinase	$2.2 \times 10^3 - 15 \times 10^3$	8	10.8	1.7	2.6	2×10^{-1}	1.4

Notes:
When the apparent equilibrium constant exceeds the mass action ratio by more than 20 the reaction is considered to be catalysed far from equilibrium and hence is likely to play a regulatory role in glycolysis. Thus, in the species presented, hexokinase, phosphofructokinase and pyruvate kinase are regulatory enzymes. Data from [a] McManus & Smyth (1982); [b] Behm & Bryant (1975a); [c] Rahman & Mettrick (1982); [d] Beis & Barrett (1979); [e] McManus & Sterry (1982).

The adult: carbohydrate metabolism

Glycogen phosphorylase
This performs a role similar to that of hexokinase and controls the activation of glycogen for synthetic reactions and in catabolism (to give glucose-1-phosphate, which enters the glycolytic sequence). Despite the fact that glycogen is the main energy reserve in cestodes, relatively little is known of its properties, although its activity is stimulated by AMP in adult *H. diminuta* (*590*). In addition, the cysticercoids of *H. diminuta* have *a* and *b* forms of phosphorylase and their interconversion, similar to the situation in mammals, is regulated by a 3', 5'-cyclic-AMP-dependent protein kinase and a phosphorylase phosphatase (*544, 546*).

Phosphofructokinase
This enzyme catalyses the conversion of fructose-6-phosphate to fructose-1,6-bisphosphate (FBP) and is the key regulatory enzyme of glycolysis. The properties of phosphofructokinase (PFK) have been investigated in some detail in adult *Moniezia expansa* (*60*) and in plerocercoids of *S. solidus* (*65*), where its activity is modulated by a number of compounds, including ATP, AMP, fructose-6-phosphate, Mg^{2+}, Mn^{2+}, K^+ and NH_4^+. In general, the PFKs from both species exhibit properties similar to those of the enzymes from mammalian sources and they probably regulate glycolysis in the same manner as their mammalian counterparts. The inhibitory effects of ATP on the PFK from *Schistocephalus solidus* and the relief of this inhibition by AMP are shown in Fig. 5.2.

Pyruvate kinase
The conversion of phosphoenolpyruvate (PEP) to pyruvate is catalysed by pyruvate kinase, another potential regulatory enzyme. The properties of cestode pyruvate kinases and the effect of possible modulators have been investigated extensively. Most are activated by fructose-1,6-bisphosphate (FBP). ATP, lactate, malate, calcium and bicarbonate have all been described as additional modulators of pyruvate kinase in different cestodes although there is no consistent pattern, and precise modulation varies between species. Particular studies include those on: *Mesocestoides corti* (*399*); *Hymenolepis diminuta* (*117, 568*); *M. expansa* (*60, 98*); *Ligula intestinalis* (*487*); *Bothriocephalus gowkongensis*, *Khawia sinensis*, *Triaenophorus crassus*, *S. solidus* (*405*); *E. granulosus* and *E. multilocularis* (*500*); *Dipylidium cati*, *D. caninum*, *Taenia pisiformis*, *Moniezia benedeni*, *Anoplocephala perfoliata* (*286*); *Spirometra erinacei* (*240*); and *Bothriocephalus acheilognathi* (*927*).

The multienzymic nature of pyruvate kinase has been investigated in detail in *M. expansa* (*98*), *H. diminuta* (*117*) and *S. erinacei* (*240*). These worms possess FBP-sensitive and FBP-insensitive pyruvate kinase isoenzymes. In *H. diminuta*, as many as five pyruvate kinase isoenzymes (for definition and usefulness, see Chapter 6) occur during development (Fig. 5.3) and it seems likely that differential expression of these different forms of the enzyme may help to control the specific composition of excreted end-products by the various life cycle stages. The nature and regulation of the end-products secreted in *H. diminuta* are discussed further below.

Lactate dehydrogenase
Lactate dehydrogenase (LDH) catalyses the reversible terminal reaction in glycolysis which results in the formation of lactic acid. It is present in a number of cestodes (Table 5.5), which is not surprising as most species excrete lactate (Table 5.4). Even *S. solidus*, which produces mainly acetate and propionate, has an active

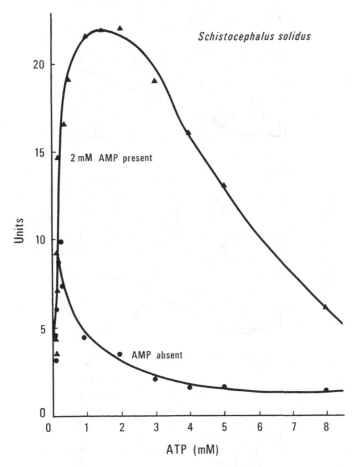

Fig. 5.2. Plot of the activity of phosphofructokinase from plerocercoids of *S. solidus* against the concentration of ATP in the presence and absence of 2 mM AMP. The concentration of fructose 6-phosphate was 1 mM. (Reprinted with permission from *International Journal for Parasitology*, 12, Beis, I. & Theophilidis, G. Phosphofructokinase in the plerocercoids of *Schistocephalus solidus* (Cestoda: Pseudophyllidea), © 1982; Pergamon Journals Ltd.)

LDH. Rather surprisingly, the properties of LDH have been investigated fully only in *Hymenolepis* spp. Burke *et al.* (*108*) purified 128-fold a single species of LDH from *H. diminuta* which resembled the H form of mammalian LDH. Other workers (*448, 567, 630, 926*) were able to distinguish electrophoretically at least two LDH isoenzymes in *H. diminuta* and *H. microstoma*. The kinetic parameters for LDH, similar in both species (Table 5.7), indicate that pyruvate reduction is favoured over lactate oxidation and that pyruvate, once formed, would be rapidly reduced to lactate, with the subsequent oxidation of NADH.

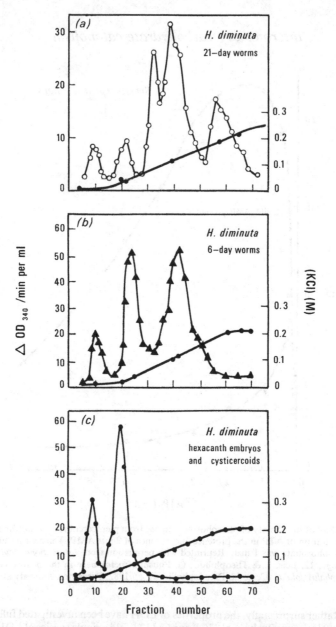

Fig. 5.3. Isozymes of pyruvate kinase during development of *Hymenolepis diminuta*. Enzyme precipitated by 70% $(NH_4)_2 SO_4$ was eluted from DEAE cellulose in a linear, 0.0–2.5 M KCl gradient. (*a*) Parasites from mature, 21-day infections in rats. The same pattern occurred in 10-day parasites in which reproductive organs are well developed but no gravid proglottides are present. (*b*) Parasites from 6-day infections, which lack reproductive organs. The same pattern was observed in 4-day infections and in the anterior proglottides of 21-day parasites which also lack reproductive organs. Eight-day parasites, in which reproductive organs are present, contained the 0.08 M isozyme and the activity of the 0.04 M isozyme was much reduced. (*c*) Hexacanth larvae from gravid proglottides of 21-day parasites, and cysticercoids developed from such larvae in the beetle, *Tenebrio molitor*. (After Carter & Fairbairn, 1975.)

90

Table 5.7. *Characteristics of LDH of* Hymenolepis microstoma *and* H. diminuta. *(Data from Pappas & Schroeder, 1979)*

Parameter	*Hymenolepis microstoma*	*Hymenolepis diminuta*
K_m (lactate; mM)	3.8	6.3–7.8
K_m (pyruvate; mM)	0.51	0.17–0.25
K_m (NADH; mM)	0.011	0.004–0.30
K_m (NAD; mM)	0.017	0.30
Substrate specificity		
D(−)-lactate	No reaction	No reaction
NADP	No reaction	No reaction
NADPH	No reaction	No reaction
Isoenzymes	At least two	At least two
pH optimum	6.6–6.8	5–7.4
Effect of NEM and pCMB	Strong inhibition	Strong inhibition
Molecular weight	160 000	141 000

Notes: NEM, *N*-ethylmaleimide; pCMB, *para*-chloromercuribenzoate.

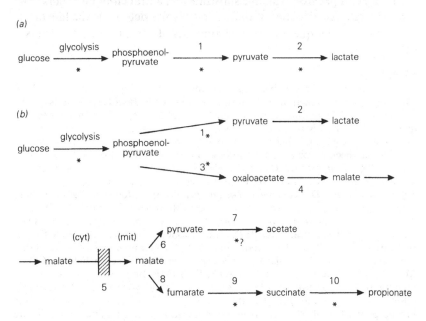

Fig. 5.4. Two types of energy metabolism in cestodes. (a) *Type 1*: homolactate fermentation. (b) *Type 2*: Malate dismutation. Reaction 3 involves a carboxylation step; decarboxylation occurs at 6, 7 and 10. Reducing equivalents are generated at reactions 6 and 7; one reducing equivalent is used at reaction 9. Thus, when the mitochondrial compartment is in redox balance and malate is the sole substrate, twice as much propionate as acetate is produced. Key: 1, pyruvate kinase; 2, lactate dehydrogenase; 3, phosphoenolpyruvate carboxykinase; 4, malate dehydrogenase; 5, mitochondrial membrane; 6 malic enzyme; 7, pyruvate dehydrogenase complex; 8, fumarase; 9, fumarate reductase; 10, succinate decarboxylase complex. * indicates reactions at which ATP is synthesised from ADP; cyt, cytosol; mit, mitochondrion. (After Bryant & Flockhart, 1986.)

91

The adult: carbohydrate metabolism

Respiratory end-products

Cestodes produce a range of end-products as a result of their respiratory metabolism (Table 5.4). Bryant & Flockhart (*104*) have usefully divided the patterns of respiratory metabolism among parasitic helminths into three types. The metabolism of larval and adult cestodes fits broadly into the first two categories of this biochemical classification and these are illustrated in Fig. 5.4. *Type 1* contains the homolactate fermenters in which carbohydrate is degraded, via glycolysis, to lactate and excreted. The ANU (Australian) strain of *H. diminuta* tends towards this type of metabolism (see below).

Most cestodes which have been investigated, however, conform to the second category, *type 2*, which is characterised by a CO_2-fixation step. Carbohydrate is degraded to the level of PEP by glycolysis, the steps involved being similar to those in mammalian tissue. At this point, the enzymes pyruvate kinase and phosphoenolpyruvate carboxykinase (PEPCK) compete for available substrate and a branch-point occurs (Fig. 5.4). The relative activities of these two enzymes determine the fate of the PEP and the subsequent types and amounts of end-products formed (see below).

PEPCK

This catalyses the reversible reaction involving carboxylation of PEP to oxaloacetate. The reaction is important energetically because it results in the production of ATP or another nucleoside triphosphate, depending on the species in question. The PEPCK reaction is also significant because it maintains the cytosolic pool of malate, which, in helminths, represents the major mitochondrial substrate. PEPCK has been reported from a number of cestodes including: *Schistocephalus solidus* (*406*); *L. intestinalis* (*502*); *Bothriocephalus gowkongensis, Khawia sinensis, Triaenophorus crassus* (*405*); *Mesocestoides corti* (*399*); *Echinococcus* spp. (*500*); *Dipylidium cati, D. caninum, Taenia pisiformis, Moniezia benedeni* and *Anoplocephala perfoliata* (*286*). More extensive studies are those on PEPCK from *M. expansa* (*61*), *H. diminuta* (*568, 680, 956*) and *Spirometra erinacei* (*240*). Data from these studies indicate that PEPCK in cestodes catalyses oxaloacetate formation rather than PEP production (in contrast to the mammalian enzyme) and that the enzyme is probably regulatory, with its possible modulators including HCO_3^-, Mg^{2+}, Mn^{2+}, lactate, GTP, GMP, ATP, ITP, IMP, fumarate, succinate and α-ketoglutarate. In addition, PEPCK is predominantly cytoplasmic, although two different isoenzymes may exist in the cytoplasm and mitochondria. Some of the kinetic properties of PEPCKs from cestodes and other parasitic helminths are compared in Table 5.8.

The fate of phosphoenolpyruvate

In the presence of pyruvate kinase, pyruvate and ATP may be formed with the subsequent conversion of pyruvate to lactate (or, exceptionally, other

Intermediary carbohydrate catabolism

Table 5.8. *PEPCK in cestodes and comparison with PEPCKs in other parasitic helminths. (Data from Fukumoto, 1985)*

Enzyme source	pH optimum	K_m (mM) for			
		PEP	HCO_3^-	GDP	IDP
Spirometra erinacei	5.6	0.25	4.5	0.025	0.085
Hymenolepis diminuta	5.6	0.039	3.3	0.021	0.078
Moniezia expansa	6.6	0.09	—	0.51	0.15
Moniliformis dubius	5.5	0.069	4.3	0.002	0.017
Fasciola hepatica	5.8–6.2	0.24	4.18	0.022	0.084
Ascaris suum	5.8	0.066	3.2	0.013	0.076

Notes: PEPCK, phosphoenolpyruvate carboxykinase.

products, including alanine by transamination and ethanol through acetaldehyde (*492, 498, 500, 918*). Alternatively, PEPCK catalyses the fixation of carbon dioxide into PEP, with the formation of oxaloacetate and a nucleotide triphosphate. Malate dehydrogenase then converts oxaloacetate to malate. Malate enters the mitochondrion and undergoes a dismutation to pyruvate (via an NAD or NADP-linked malic enzyme) and to fumarate (under the action of fumarase). Pyruvate is further oxidised by the pyruvate dehydrogenase complex to acetate, which is excreted.

Fumarate acts as substrate for an important enzyme complex, fumarate reductase, which occurs in a number of cestodes (*698*). Fumarate reductase accepts reducing equivalents (NADH or NADPH), generated by malic enzyme, and transfers the electrons to fumarate, which is thus reduced to succinate. The reduction of fumarate to succinate is accompanied by the phosphorylation of ADP, one ATP being produced for each fumarate reduced. This phosphorylation involves the cytochrome chain. Succinate may be excreted or is converted by another enzyme complex, succinate decarboxylase, to propionate and excreted.

Type 3 fermentation, characterised by a series of reactions between mitochondrial end-products that yield branched-chain fatty acids (e.g. 2-methylvalerate, 2-methylbutyrate), occurs in *Ascaris*, a number of other intestinal nematodes and in some trematodes (*490*). However, cestodes, with the exception of *Bothriocephalus scorpii*, which excretes methylbutyrate (*107*), have not been shown to produce branched-chain fatty acids as end-products of respiratory metabolism. Some or all of the enzymes of the TCA cycle may be present in cestodes in addition to the type 1 and type 2 fermentation pathways. The extent to which the cycle may contribute to carbohydrate metabolism in cestodes is considered below.

The adult: carbohydrate metabolism

The PEPCK/pyruvate kinase branchpoint

It is not possible to generalise on the control of the PEP branchpoint in cestodes because different metabolites can act as modulators of pyruvate kinase and PEPCK in different species. The branchpoint has been most studied in *Moniezia expansa* (*59, 61, 98, 103*) and in *H. diminuta* (*568, 612, 650*).

Moniezia expansa
The major end-products of carbohydrate metabolism in this species are lactate and succinate. The relative amount formed of each depends on the presence or absence of oxygen in the incubation medium, the presence or absence of glucose, and the presence or absence of fumarate. For example, anaerobiosis leads to an increase in lactate production, which is accompanied by a fall in the intracellular level of malate. Malate is an inhibitor of pyruvate kinase in *M. expansa*, so the fall in malate levels results in an increase in pyruvate kinase activity leading to a rise in lactate production. Conversely, under aerobic conditions, malate levels increase, pyruvate kinase activity is inhibited and lactate production is decreased (Fig. 5.5).

A model incorporating two different types of mitochondria has been postulated to explain these events (*103*). Thus, succinate (or NADH) is oxidised in 'aerobic' mitochondria, which possess a functional electron transport system linked to

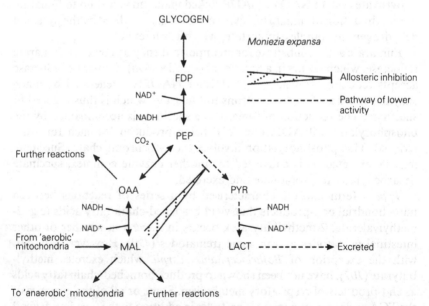

Fig. 5.5. Proposed pathways of metabolism in *Moniezia expansa* scoleces under aerobic conditions: cytosolic reactions. (After Bryant & Behm, 1976.) OAA, oxaloacetate; PYR, pyruvate, MAL, malate, LACT, lactate; for other abbreviations, see the text.

94

oxygen, to produce malate, which is translocated to the cytoplasm. Simultaneously, malate is metabolised to succinate in 'anaerobic' mitochondria, which contain no cytochrome oxidase, by the fumarate-reductase system. This is clearly an ingenious concept, which may also apply in other cestodes, although its experimental confirmation will provide a major challenge.

Hymenolepis diminuta

It is generally accepted that *H. diminuta* excretes lactate, succinate and acetate as its major respiratory end-products (apart from CO_2) and recent nuclear magnetic resonance (n.m.r.) analysis (Fig. 5.6) (*62, 83*) has confirmed earlier data obtained by more conventional approaches. There is, however, less agreement as to the precise quantity of each end-product excreted by this worm and how its respiratory metabolism is regulated.

High resolution [13]C n.m.r. spectroscopy is a non-invasive technique that can be used to study metabolic pathways in intact organisms, organs or cells. Metabolism of ^{13}C-enriched substrates may be followed spectroscopically for a number of hours. The rates of formation of metabolites can be monitored and end-products identified. This procedure provides the biochemist with a complete overview of the metabolism of a given substrate without destruction of the biological material and it has the additional advantage of enabling the observer to follow the metabolism of specific carbon nuclei and their metabolites.

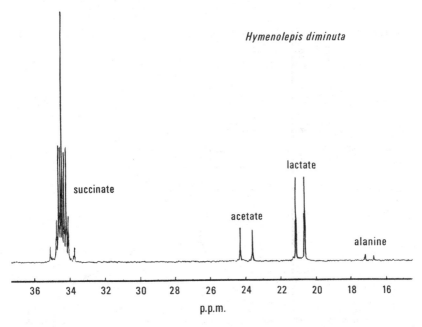

Fig. 5.6. Proton decoupled [13]C n.m.r. spectrum at 75.47 MHz of the excretory products of *Hymenolepis diminuta* fed D-[6-^{13}C]glucose. p.p.m., parts per million. (After Blackburn *et al.*, 1986.)

The adult: carbohydrate metabolism

Succinate accumulation in *Hymenolepis diminuta*

It has been suggested by Podesta *et al.* (*650*) that ambient CO_2 levels determine the amount of succinate and lactate excreted by intestinal helminths. A summary of the events which they propose lead to succinate accumulation in *H. diminuta* is provided in Fig. 5.7. There is considerable release of CO_2 in the post-prandial intestine of the rat host as the acidic chyme is passed into the duodenum and there is an acidification of worm tissues as the CO_2 diffuses in. The acidification is countered by worm secretion of H^+ and by mobilisation of Ca^{2+} from the calcareous corpuscles (Chapter 4), thus releasing carbonate. The increased osmolality and greater HCO_3^- concentration activate PEPCK, which favours succinate production over lactate. Succinate, being a dicarboxylic acid, is twice as effective as lactate in metabolic disposal of H^+. With a decrease in passage of stomach chyme into the intestine, there would be a decrease in acid stress on the worms, accompanied by a concomitant decrease in tissue HCO_3^- concentrations and osmolality, which would favour lactate production.

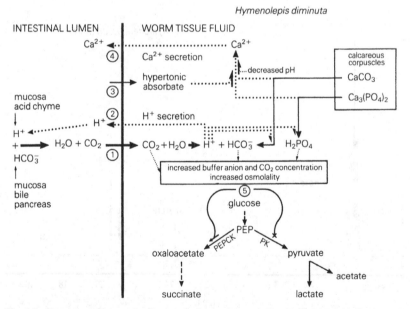

Fig. 5.7. Summary of events leading to succinate accumulating in *Hymenolepis diminuta*. 1, Tissue acidification by ambient CO_2. 2, H^+ secretion. 3, Tonicity of fluid absorbed by the worm. 4, Ca^{2+} secretion. 5, Metabolic consequences of tissue buffering problem. (After Podesta *et al.*, 1976.)

Intermediary carbohydrate catabolism

Table 5.9. *Values for acid excretion by* Hymenolepis diminuta. *(Data from Ovington & Bryant, 1981)*

Acid excretion (%)				
Succinic	Acetic	Lactic	Lactic/ succinic	Reference
		80–90	> 8–9	*673*
		38–98	> 4–10	*429*
80	13	7	0.09	*200*
62	25	13	0.21	*935*
30	50	17	0.57	*145*
5	11	84	16.8	*612*[a]

Note:
[a] ANU strain.

This hypothesis, despite its merits has, however, been questioned by Ovington & Bryant (*612*), who noted in their experiments with *H. diminuta* that succinate excretion was only a small fraction of total acid production, while acetate production did not vary over a wide range of CO_2 concentrations. In addition, whereas lactate excretion decreased with increasing HCO_3^-, succinate excretion increased only marginally. Whilst conceding that the excretion of organic acids may serve to maintain tissue pH in the presence of high pCO_2, they disagreed that high CO_2 levels cause a major shift from lactate to succinate production. In conclusion, they suggested that, as intestinal parasites have access to oxygen in their normal environment, the energy metabolism of *H. diminuta*, as well as that of other intestinal helminths, is adapted to fluctuating O_2 and CO_2 tensions and is not directed solely by the high ambient CO_2 levels of the gut. This important aspect of helminth biochemistry clearly warrants further investigation.

Strain variation and metabolism

Another interesting feature of carbohydrate metabolism in *H. diminuta* relates to the contradictory reports of its excretory end-products (Table 5.9). Many factors could have contributed to these inconsistencies, including the use of different definitive hosts (rats of various strains), different intermediate hosts (species or strain of beetle), different experimental and analytical protocols and ages of hosts and parasites (*104, 531*). A further complication relates to the work of Coles & Simpkin (*145*), who have shown

The adult: carbohydrate metabolism

differential quantitative excretion of end-products along the length of *H. diminuta*, with increasing emphasis on succinate production in more mature segments. Nevertheless, it is probable that the major reason for the apparent disparity in results from different groups is related to strain variation in the tapeworm itself. Subsequent work has substantiated the existence of strain variation in *H. diminuta* (*400, 627, 628*) and at least four 'strains' are now recognised.

The phenomenon of biochemical strain variation in helminths is proving of increasing interest to parasite biochemists and the area has been comprehensively reviewed by Bryant & Flockhart (*104*). Probably the most remarkable examples of strain variation in cestodes occur in the hydatid organisms, *Echinococcus* spp. Accounts of the biochemistry and physiology (*492*) and the extent and significance of strain variation (*501, 865*) are available for this group.

In addition to DNA and protein differences (Chapter 6) and differences in developmental characteristics *in vitro* (Chapter 10), marked interspecific and intraspecific (i.e. strain) differences in intermediary and energy metabolism have been recorded (*488, 498, 500*). Most work has been performed on protoscoleces of the UK sheep/dog and horse/dog strains of *E. granulosus*, and *E. multilocularis*, which possess a type 2 metabolism. A general scheme for their carbohydrate metabolism, is presented in Fig. 5.8, although the three forms consume different amounts of oxygen and endogenous glycogen and produce different concentrations of metabolic end-products *in vitro* (Table 5.10). Similar variations occur in Kenyan and Australian strains of *E. granulosus* (*492*). Limited work on the adult stage of *E. granulosus* (*105, 488*) suggests a type 2 metabolism, similar to the protoscolex.

Mitochondrial reactions

A number of important biochemical reactions occur intramitochondrially and some of these are described below. One major drawback to the study of energy metabolism in cestodes is the failure, due to technical difficulties, to produce isolated mitochondrial preparations of the purity which can be obtained from mammalian tissues. Generally, mitochondrial fractions of cestode origin are contaminated with other cellular components such as glycogen, endoplasmic reticulum and various membranes, and some of the data obtained with such mitochondria should, therefore, be treated with caution.

Malic enzyme
The key position of malic enzyme, which oxidatively decarboxylates malate to pyruvate inside the mitochondrion, in cestode respiratory metabolism has been discussed above. The malic enzyme of *H. diminuta* has been most investigated (*221*,

Intermediary carbohydrate catabolism

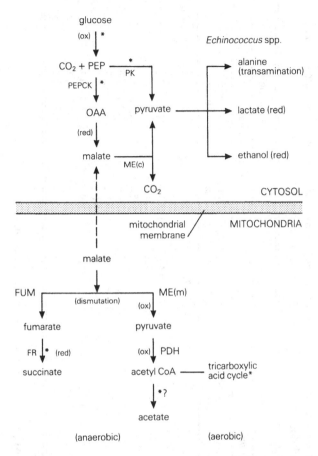

Fig. 5.8. Respiratory pathways in *Echinococcus* spp. *, sites of ATP synthesis; (ox) oxidative, and (red) reductive processes; PK, pyruvate kinase; OAA, oxaloacetate; ME(c), ME(m), 'malic' enzyme (cytosolic) or (mitochondrial); FR, fumarate reductase; PDH, pyruvate dehydrogenase complex. (After McManus & Bryant, 1986.)

698). It has been purified (445) and shares some properties in common with malic enzymes from mammals and birds in being NADP-dependent, heat-stable and able to decarboxylate oxaloacetate. The malic enzyme of *H. microstoma* also has a marked specificity for NADP (216), contrasting with that of *Spirometra mansonoides*, which appears to be both NAD- and NADP-linked (220). Malic enzyme has been demonstrated in a range of other cestodes including: *Mesocestoides corti* (399), *Schistocephalus solidus* (406), *Moniezia expansa* (60), *Echinococcus* spp. (500) and *L. intestinalis* (502).

Table 5.10. *The carbohydrate metabolism of the horse and sheep strains of* Echinococcus granulosus *and* E. multilocularis. *(Data from McManus & Smyth, 1978)*

	Aerobic			Anaerobic		
	E. granulosus		E. multi-locularis	E. granulosus		E. multi-locularis
	Horse strain	Sheep strain		Horse strain	Sheep strain	
Glucose taken up	0	0	43.7±7.2	0	0	30.0±7.3
Glycogen utilized (as glucose)	45.7±12.1	101.7±3.2	57.0±10.5	115.7±8.7	107.5±2.0	59.8±4.4
Oxygen consumed	84.2±6.0	119.5±6.2	248.0±13.7	—	—	—
Lactate	54.9±5.1	23.8±2.1	30.1±4.3	58.5±4.1	24.8±2.0	33.9±3.7
Succinate	7.3±0.5	14.6±3.5	33.5±2.5	61.0±9.9	42.4±2.6	97.9±9.7
Acetate	23.4±1.5	62.0±3.1	59.3±2.8	24.8±1.3	81.6±2.7	65.5±3.1
Propionate	1.6±0.1	0	1.0±0.1	3.0±0.2	0	6.7±0.3
Pyruvate	1.6±0.4	0.9±0.2	1.6±0.3	1.9±0.3	1.3±0.3	2.1±0.3
Malate	1.0±0.1	1.9±0.1	2.7±0.2	0.8±0.1	1.9±0.1	4.5±0.2
Ethanol	7.2±0.8	7.5±0.3	2.6±0.1	2.8±0.2	5.9±0.2	5.0±0.3

Note:
The results (expressed as nmol/mg dry wt per 3 h) are means ± s.e. of four determinations.

Intermediary carbohydrate catabolism

Table 5.11. *Mitochondrial fumarate reductase activity of adult*
Hymenolepis diminuta. *(Data from Fioravanti & Saz, 1980)*

Substrate	Addition	Activity (nmol/min per mg[a])
NADH	None	12.7
NADPH	None	3.4
NADH	Fumarate	68.2
NADPH	Fumarate	3.2

Notes:
[a] Activity is expressed per mg mitochondrial protein. Assays were performed in
air by following the disappearance of reduced pyridine nucleotide at 340 nm.

Transhydrogenases
Mitochondria from adult *H. diminuta* exhibit an NADH-coupled fumarate
reductase (Table 5.11). This presents a potential dilemma with respect to the
utilisation of intramitochondrial reducing equivalents by this worm. As reducing
equivalents are generated by the malic enzyme in the form of NADP, a mechanism
for the transfer of hydride ions from NADPH to NAD to produce NADH is
required so that electron-transport-associated activities can proceed and terminate
with the reduction of fumarate to succinate. Such a mechanism does exist in *H.
diminuta* as there is a non-energy-linked, membrane-associated transhydrogenase
(*214, 217, 221, 476*). This transhydrogenase, which also occurs in *H. microstoma*
(*216*) and *Spirometra mansonoides* (*220*) catalyses the reaction:

$$NADPH + NAD^+ + H^+ \rightleftharpoons NADP^+ + NADH + H^+.$$

The mitochondria of these three cestodes also contain an NADH/NAD
transhydrogenase (*215, 219, 220*), although its physiological role remains obscure.
A proposed scheme (*477*) for the coupling of malic enzyme with fumarate reductase
and the intramitochondrial localisation of the individual reactions is shown for *H.
diminuta* in Fig. 5.9.

The tricarboxylic acid cycle

In mammalian tissues, complete oxidation of carbohydrate is carried out
aerobically via the TCA cycle. The cycle functions catabolically, in conjunc-
tion with the electron transport system, to yield ATP, and biosynthetically
as a fundamental source of intermediates for other pathways. Although
restricted to a small number of species, those cestodes that have been
investigated in detail exhibit a complete, or near complete, sequence of

The adult: carbohydrate metabolism

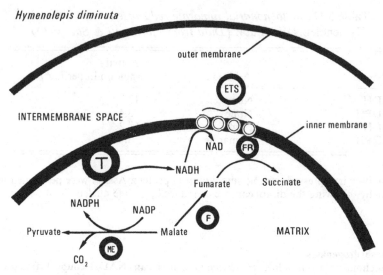

Hymenolepis diminuta

Fig. 5.9. Proposed scheme for the intramitochondrial metabolism of malate by *Hymenolepis diminuta*. Abbreviations: ME, 'malic' enzyme; F, fumarase; T, transhydrogenase; FR, fumarate reductase; ETS, electron transport system. Once within the matrix compartment, malate oxidation, as catalysed by 'malic' enzyme, results in NADPH formation. Via the activity of fumarase, malate also is converted to fumarate in the matrix compartment. NADPH then serves as a substrate for the inner-membrane-associated transhydrogenase and transhydrogenation between NADPH and matrix NAD is a scalar reaction associated with the matrix side of the inner membrane. Matrix NADH so formed reduces the electron transport system via a site on the matrix side of the inner membrane permitting fumarate reductase activity. The reduction of fumarate to succinate results in succinate accumulation within the matrix compartment. (After McKelvey & Fioravanti, 1985.)

Krebs' cycle intermediates (Table 5.12) and/or enzymes (Table 5.13). Nevertheless, certain key enzymes, especially aconitase and isocitrate dehydrogenase, are very low in activity or are undetectable in species such as *H. diminuta*, whereas only very small amounts of $^{14}CO_2$, a characteristic end-product of the TCA cycle, were liberated *in vitro* from [^{14}C]glucose by the tetrathyridia of *Mesocestoides corti* (*399*) and adults of *Cotugnia digonopora* (*618*). The classical TCA cycle is, therefore, unlikely to function to any significant extent in these cestodes.

In general it is difficult to assess experimentally the importance of the TCA cycle to the energy budget of cestodes, since, as we have seen, partly oxidised metabolites are excreted as end-products even under aerobic conditions. Notwithstanding, there is increasing evidence that, under aerobic conditions, certain cestodes such as *Schistocephalus solidus* (*406*) and *Echinococcus* spp. (*500*) are capable of catabolising substantial

102

Table 5.12. Steady-state content of the tricarboxylic acid cycle intermediates in several cestodes

Metabolite	Content (nmol/g fresh wt)				
	Echinococcus[a] granulosus (ovine strain, protoscoleces)	Echinococcus[a] multilocularis (protoscoleces)	Ligula[b] intestinalis (plerocercoids)	Ligula[b] intestinalis (adults)	Schistocephalus[c] solidus (plerocercoids)
Citrate	1360	910	127	167	221
Isocitrate	77	—	—	—	12
2-Oxoglutarate	14	21	—	—	71
Succinate	500	1141	716	675	25
Fumarate	10	8	—	—	—
Malate	786	846	187	80	134
Oxaloacetate	8	10	50	84	8

Notes:
Data from [a] McManus & Smyth (1982); [b] McManus & Sterry (1982); [c] Beis & Barrett (1979).

103

Table 5.13. *Specific activities of the tricarboxylic acid cycle in a variety of cestodes*

Enzyme	Hymenolepis diminuta[a] (adults)	Bothriocephalus gowkonensis[b] (adults)	Khawia sinensis[b] (adults)	Triaenophorus crassus[b] (adults)	Ligula intestinalis[c] (plerocercoids)	Schistocephalus solidus[d] (plerocercoids)	Echinococcus multilocularis[e] (protoscoleces)	Echinococcus granulosus[e] (protoscoleces, ovine strain)	Mesocestoides corti[f] (tetrathyridia)
Citrate synthase	4	8	10	0	17†	59	23	36	1512†
Aconitase	0	2	0	0	8†	3	43	62	56†
Isocitrate dehydrogenase (NAD)	—	6†	0†	0†	0†	1	16	29	0†
Isocitrate dehydrogenase (NADP)	0	3	2	4	21†	5	193	73	682†
Oxoglutarate dehydrogenase	—	—	—	—	420†	22	46	20	30†
Succinate dehydrogenase	65	100	20	42	111†	119	11	7	1535†
[Fumarate reductase]	[45]	[7]†	[0]†	[0]†	[25]†	[5]	[1]	[2]	[363]†
Fumarase	45	157	112	188	101	59	39	39	468†
Malate dehydrogenase	2803	4360	9861	6313	1596	2316	7023	6809	114050

Notes:
Specific activities are expressed as nmol/min per mg protein at 30°C[a–e] or nmol/min per mg wet wt at 25°C[f].
†, mitochondrial fraction.
Data from [a] Ward & Fairbairn (1970); [b] Körting (1976); [c] McManus (1975a); [d] Körting & Barrett (1977); [e] McManus & Smyth (1982); [f] Köhler & Hanselmann (1974).

amounts of carbohydrate via a functional TCA cycle. In addition, the work on *Moniezia expansa* (*103*) described above has raised the interesting possibility that there could be separate aerobic and anaerobic mitochondria in cestodes, the former being concerned with fumarate reduction and the latter possessing a classical TCA cycle. Whether or not the two sorts of mitochondria are present in the same cell or tissue remains to be established.

At least four TCA cycle enzymes are under regulatory control in mammals: citrate synthase, NAD-linked isocitrate dehydrogenase, 2-oxoglutarate dehydrogenase and succinate dehydrogenase. Possible modulators of these enzymes have not yet been studied in cestodes although knowledge of the regulation of citrate synthase, in particular, might give a better indication as to the metabolic fate of citrate *in vivo*. Regardless of the extent of the TCA cycle operating classically, it is clear that at least part of the cycle, operating in reverse from succinate to oxaloacetate, occurs in all cestodes examined to date.

Acetate and propionate production

A number of cestodes produce propionate and/or acetate as respiratory end-products (Table 5.4). In mammals, propionate and acetate are formed by the decarboxylation of succinate and pyruvate, respectively, but the corresponding pathways in cestodes have been little studied.

Acetate formation

It is likely that pyruvate, the product of the oxidative branch of the mitochondrial dismutation reaction, is further metabolised in cestodes to acetyl-CoA by oxidation with NAD^+, as catalysed by the lipoamide-dependent mitochondrial pyruvate dehydrogenase complex. This enzyme has been reported in *H. diminuta* (*935*) and *S. solidus* (*406*). The acetyl-CoA is then hydrolysed to acetate. During this step, ATP synthesis may occur through the conservation of the acetyl-CoA energy-rich thioester bond by the combined action of an acyl-CoA transferase and a thiokinase (*398*) as follows:

$$\text{acetyl-CoA} + \text{succinate} \rightleftharpoons \text{acetate} + \text{succinyl-CoA}. \quad (1)$$
$$\text{succinyl-CoA} + \text{ADP} + P_i \rightleftharpoons \text{succinate} + \text{CoA} + \text{ATP}. \quad (2).$$

An acyl-CoA transferase has been detected in sonicated mitochondrial preparations from *Spirometra mansonoides* (*643*).

Propionate

Propionate formation in cestodes probably proceeds essentially via reversal of the reactions required for the conversion of propionate to succinate in animal tissues (Fig. 5.10). Two of the enzymes involved, propionyl-CoA carboxylase and methylmalonyl-CoA mutase have been demonstrated in the mitochondria of *S. mansonoides* (*643*, *884*). An acyl-CoA carboxylase, which can catalyse the carboxylation of propionyl-CoA, has also been isolated from this worm (*533*). The proposed pathway of propionate formation is associated with net ATP synthesis

The adult: carbohydrate metabolism

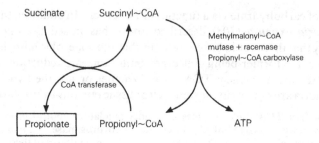

Fig. 5.10. A proposed pathway for propionate formation in cestodes (After Barrett, 1981.)

(*643*) and this may result from the coupling of ADP phosphorylation to the lysis of the carboxyl function of methylmalonyl-CoA. Net ATP production can only be achieved if the decarboxylase system is functionally coupled to acyl-CoA transferase (reaction (1), p. 105), which allows the recycling of the energy-rich thioester bond of its CoA intermediates (*398*).

Electron transport

General comments

In mammals and other aerobic organisms, the Embden–Meyerhof pathway and TCA cycle result in oxidation of all the carbon atoms of the carbohydrate molecule to CO_2, and for each glucose molecule a total of 12 pairs of hydrogen atoms (two from the Embden–Meyerhof cycle and 10 from the TCA cycle) are produced. Ten of the pairs of hydrogen atoms are transferred to the carrier molecules NAD^+ or $NADP^+$, while one pair (produced at the succinate $\rightarrow$ fumarate step) is transferred to FAD. Electrons from the reduced carrier molecules are transferred in turn to flavoproteins, thereby regenerating $NAD(P)^+$ or FAD, to other components including coenzyme Q or ubiquinone, to a complex cytochrome system, comprising at least five cytochromes b, c_1, c, a and a_3 (cytochrome oxidase) and finally to molecular oxygen to form water. This respiratory chain, whose components are embedded in the mitochondrial inner membrane, is known as the *electron transport system*. During the transfer of electrons along the electron transport chain, ATP is synthesised by a mechanism which is not fully understood. This aerobic production of ATP is called *oxidative phosphorylation*.

Detection of the components of an electron transport system, involving cytochromes, is based on (*a*) identification of the various cytochromes by absorption and difference spectroscopy at room and liquid-nitrogen temperatures, and (*b*) the use of 'specific' inhibitors, such as antimycin A and cyanide, for the various

106

electron transport carriers to demonstrate their inhibitory effect on the oxidation of various substrates in isolated mitochondrial preparations. Although the basic principles of the mammalian electron transport system are well established, there is considerable disagreement on many important details (*9*). It is not surprising, therefore, to find that the electron transport system in cestodes – which is technically a much more difficult system to investigate – is very poorly known.

In cestodes the major pathways of carbohydrate metabolism, described above, are anaerobic, although a number of the key reactions involved occur inside the mitochondria. Nevertheless, all species examined so far appear to utilise oxygen, at least *in vitro*. This raises two important questions: is oxygen uptake accompanied by ATP synthesis, and, if oxidative phosphorylation occurs, what is its overall contribution to the energy balance of the cestode?

Electron transport components

The electron transport system of cestodes has been comprehensively reviewed by Cheah (*129*); earlier reviews are those by Smyth (*796*) and Bryant (*97*). To date, only the electron transport systems of large intestinal cestodes – *Moniezia expansa, Taenia hydatigena, T. taeniaeformis, H. diminuta* – have been investigated in detail. The mitochondria of such cestodes have structural characteristics similar to those present in mammalian tissues (*25, 298*), although variation in number, form and size has been noted in different strobilar regions such as the tegument, parenchyma and reproductive tissue (*455*).

Extensive work by Cheah (*121, 122, 123, 128, 130*), mainly with *M. expansa*, has shown that large cestodes possess a cytochrome chain which differs from the mammalian system in being branched and possessing multiple terminal oxidases (Fig. 5.11). One branch resembles the classical chain with cytochrome a_3 as its terminal oxidase. The terminal oxidase of the alternative pathway, which branches at the level of rhodoquinone or vitamin K, is an *o*-type cytochrome. Cytochrome *o* is an autoxidisable *b*-type cytochrome which is commonly found in micro-organisms, parasitic protozoa and plants. The classical chain constitutes about 20% of the oxidase capacity in cestodes and cytochrome *o* is quantitatively the major oxidase. Cyanide-insensitive respiration – i.e. where oxygen uptake occurs in the presence of cyanide – is characteristic of most helminths (*39*). Cytochrome *o* binds cyanide much less strongly than cytochrome a_3, and it seems reasonable, therefore, to equate cyanide-insensitive respiration with the non-classical pathway.

In support of Cheah (*121*), Allen (*11*) demonstrated for the first time the

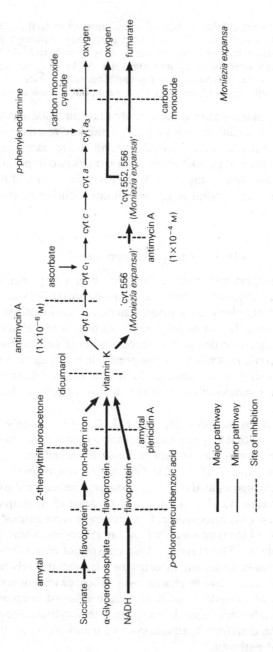

Fig. 5.11. Proposed electron transport system of *Moniezia expansa*. (After Cheah, 1983.)

Table 5.14. *Respiratory control exhibited by isolated* Hymenolepis diminuta *mitochondria with a range of substrates. (Data from Yorke & Turton, 1974)*

Substrate	O$_2$ uptake at 37°C (ng-atom/min per mg protein)		Respiratory control ratio
	− ADP	+ ADP	
DL-Glycerol-3-phosphate	47.3	60.5	1.3
Succinate	15.5	21.4	1.4
DL-Isocitrate	9.5	16.0	1.7
L-Glutamate	5.9	9.0	1.5

Note: The respiratory control ratio was calculated as the ratio of oxygen uptake following, and prior to, the addition of ADP. O$_2$ uptake was measured polarographically.

presence of rhodoquinone, but not ubiquinone, in an adult cestode, *M. expansa*. Similarly, adults of *S. mansonoides* contain rhodoquinone but not ubiquinone, although both quinones occur in the eggs and coracidia of this species (*50*). In contrast, ubiquinone and vitamin K have been detected in adults of the pseudophyllid *Penetrocephalus ganapatii* (*177*). Why these quinones should vary qualitatively between species and even between life cycle stages is unclear.

Oxidative phosphorylation

Now that it is established that cestodes possess all the components of a electron transport system, is the latter functional? Weinbach & von Brand (*952*) failed to demonstrate either respiratory control or oxidative phosphorylation in *T. taeniaeformis*, although they regarded this as a technical rather than a physiological problem. However, there is good evidence that isolated mitochondria from *M. expansa* (*124–127*) and *H. diminuta* (*663, 978*) are capable of oxidative phosphorylation and respiratory control. The demonstration that a preparation of *H. diminuta* mitochondria will oxidise a range of substrates, exhibiting respiratory control, is shown in Table 5.14. Similarly, mitochondria from *Diphyllobothrium latum* can oxidise NADH (*728*) and succinate (*729*). It is likely that the classical mammalian-type part of the cytochrome chain in cestodes is capable of oxidative phosphorylation, but there is no evidence for ATP synthesis occurring on the alternative branch from the quinone or vitamin K/cytochrome *b* complex to cytochrome *o*.

The adult: carbohydrate metabolism

Cestodes, as we have seen, fix CO_2, and the electron transport system plays a key role in the reduction of fumarate to succinate (*129, 220, 221*). Anaerobically, NADH is reoxidised by the reduction of fumarate to succinate. The NADH-linked fumarate reductase reaction results in a site 1 electron-transport-associated phosphorylation of ADP (Fig. 5.12). So there is good evidence that oxygen or fumarate can act as an electron acceptor in cestodes. In *M. expansa*, for example, it has been suggested (*121, 122*) that oxygen is the sole terminal electron acceptor for the minor classical pathway involving cytochrome a_3 as its terminal oxidase. The alternative branch, with cytochrome *o* as terminal oxidase, could interact directly with either oxygen or fumarate (see Fig. 5.11). It is clear, therefore, that cestodes have mechanisms for performing respiratory metabolism under both aerobic and anaerobic conditions. However, the relative contribution of aerobic oxidative phosphorylation to their overall energy budget is, because of numerous technical problems, difficult to assess. This area is rather controversial and the controversy may well have arisen from the misconception that the mammalian intestinal lumen, which large adult cestodes occupy, is anoxic. We now know that, in the rat, this environment has a pO_2 of 40–50 mmHg (*530*), which is sufficient to drive aerobic metabolic pathways. Nevertheless, the fact that cestodes produce reduced end-products under aerobic conditions indicates that oxidative phosphorylation cannot supply all their energy requirements. Furthermore, some species (*H. diminuta, H. nana, S. mansonoides, Mesocestoides corti*) can be cultured *in vitro* for long periods under totally anaerobic conditions, whereas others (*H. microstoma, T. crassiceps, Echinococcus* spp.) survive and develop better in the presence of oxygen (*796*; Chapter 10). Energy production in large cestodes such as *H. diminuta* is, therefore,

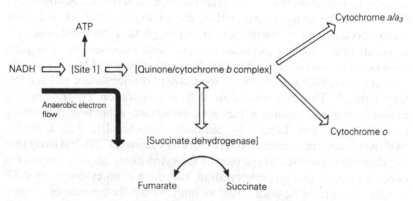

Fig. 5.12. Anaerobic electron flow in cestodes. (Modified after Barrett, 1981.)

110

regarded by some workers (*221*) as being essentially anaerobic, with fumarate rather than oxygen acting as terminal electron acceptor. Others (*129, 663*) regard cestodes as having a substantial aerobic component to their energy metabolism. It is possible that both capacities co-exist within the same organism, which is thus adapted to possible fluctuations of oxygen tension within its environment. What actually applies *in vivo*, however, remains an open question.

Hydrogen peroxide formation

When *Moniezia expansa* and several other helminths are incubated *in vitro* under aerobic conditions, measurable quantities of the highly toxic hydrogen peroxide are formed. Hydrogen peroxide may arise as an end-product of cytochrome *o* reacting with oxygen (*121*) or through the action of superoxide dismutase. The physiological function of this enzyme, which has been reported in *H. diminuta*, *M. expansa*, *Schistocephalus solidus* and *T. taeniaeformis* (*64, 437, 637*), may be to protect cells against the potentially harmful effects of the superoxide free radical ($O_2^-\cdot$) by catalysing the reaction:

$$2O_2^-\cdot + 2H^+ \rightleftharpoons O_2 + H_2O_2.$$

Whether cestodes actually produce significant quantities of hydrogen peroxide *in vivo* is debatable. It may well be that its formation *in vitro* is artifactual and the result of unphysiological (hyperbaric) oxygen tensions in the incubation medium. In mammals, hydrogen peroxide is destroyed by the enzyme catalase but as cestodes lack appreciable catalase activity (*637*), this role may be taken over by the peroxidase reported in the mitochondria of several species (*64, 396, 478, 637, 703, 705, 714*).

Other pathways

The pentose-phosphate pathway

An alternative pathway to the Embden–Meyerhof pathway of glycolysis for conversion of carbohydrates to pyruvate is the pentose-phosphate pathway (Fig. 5.13). Its main role is not ATP production but to provide NADPH for fat synthesis, and pentoses (in particular, D-ribose-5-phosphate) for nucleic acid synthesis. The pathway can also convert pentoses to hexoses, which can then be further metabolised by glycolysis. With regard to cestodes, a

111

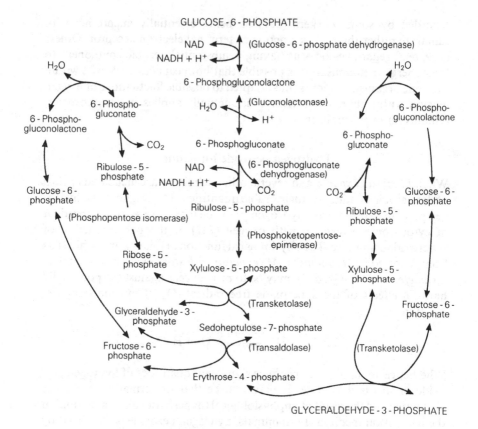

Fig. 5.13. The pentose phosphate pathway. (After Smyth, 1969.)

complete sequence of enzymes has been demonstrated only in larval *E. granulosus*, where up to 20% of glucose utilised *in vitro* may be via this pathway (*4, 6*). Evidence for the existence of the pentose-phosphate pathway is very fragmentary in other cestodes and, despite the fact that the first two enzymes of the cycle, glucose-6-phosphate dehydrogenase and 6-phosphogluconate dehydrogenase, have been reported in *Anoplocephala perfoliata, M. benedini, T. saginata, T. pisiformis, D. caninum* (*698*), *E. multilocularis* (*500*), *S. solidus* (*406*) and *L. intestinalis* (*502*), its biological significance in these species is uncertain.

The glyoxylate pathway

The glyoxylate cycle has been demonstrated in micro-organisms, germinating seeds and also in *Ascaris* eggs (*39*). It effectively 'short circuits' the TCA

Other pathways

cycle by bypassing the two decarboxylation steps involving isocitrate dehydrogenase and oxoglutarate dehydrogenase (decarboxylase). Two additional enzymes, isocitrate lyase (which cleaves isocitrate to succinate and glyoxylate) and malate synthase (which condenses a second acetyl-CoA with glyoxylate to give malate) are involved. There have been few attempts to demonstrate the glyoxylate pathway in cestodes and its potential role in carbohydrate metabolism, therefore, remains unproven. Nevertheless, n.m.r. studies (83) have shown that *H. diminuta* produces some [4-^{13}C]succinate when incubated in D-[6-^{13}C]glucose and its formation, which cannot be explained on the basis of the commonly accepted pathways for glucose metabolism in *H. diminuta*, may be via the glyoxylate cycle and/or the TCA cycle operating in the forward direction. Histochemical evidence supports the existence of a functional glyoxylate cycle in *H. diminuta* and *Triaenophorus nodulosus* (542, 553, 554, 556).

6

The adult: proteins and nucleic acids

Proteins

General considerations

Proteins are involved in a variety of cellular functions, including structural support, the formation of contractile systems and molecular transport. In addition, proteins can act as hormones, some are toxins, many are antigenic and the largest group, the enzymes, catalyse fundamental synthetic and degradation reactions in cells. Another very important group of proteins is the immunoglobulins (antibodies) (Table 11.1, p. 285); these combine specifically with antigens and are key components of the host immune response.

The total protein content of cestodes is lower than that of most invertebrates being generally between 20% and 40% of the dry weight (796), although values of over 60% have been reported for Echinococcus spp. (488, 498) (Table 4.1).

STRUCTURAL PROTEINS

Structural proteins are key components of cells but, despite the fact that they provide potentially important targets for chemotherapy, they have been little studied in cestodes. The reported structural proteins include sulphur-rich keratin (Fig. 7.7, p. 173), which makes up the hooks and embryophores in taeniid cestodes, and sclerotin (a quinone tanned protein) (Fig. 7.8, p. 173), a principal component of the egg shells of Pseudophyllidea and possibly other groups (Chapter 7). The most abundant structural protein in animals, collagen, has been characterised from larvae of *Bothriocephalus scorpii* and *Nybelinia* spp. (977) and from the cysticerus of *Taenia solium* (*Cysticercus cellulosae*) (885, 886). Collagen from C. cellulosae is similar to that from vertebrates, being composed of a molecule 280 nm in length with a molecular weight of 300 000, comprising three subunits of M_r 100 000 each. The amino acid composition (Table 6.1) is also

Proteins

Table 6.1. *Amino acid composition of soluble collagen of* Cysticercus cellulosae. *(Data from Torre-Blanco, 1982)*

Amino acid	Residues/1000 total amino acid residues		
	20 h hydrolysate	48 h hydrolysate	Average
Hydroxyproline	0	0	0
Lysine	14.64	15.36	15
Histidine	4.85	4.73	5
Hydroxylysine	18.80	19.20	19
Arginine	54.96	55.21	55
Aspartic acid	54.85	54.65	55
Threonine	28.35	21.03	28
Serine	27.04	22.38	27
Glutamic acid	97.50	97.68	98
Proline	216.00	218.60	217
Glycine	313.80	310.80	312
Alanine	55.21	55.81	56
Half-cystine	0	0	0
Valine	27.95	28.32	28
Methionine	23.50	22.60	23
Isoleucine	21.28	21.88	22
Leucine	25.66	25.91	26
Tyrosine	6.28	6.34	6
Phenylalanine	8.22	8.33	8

typical of vertebrate collagens, although one major difference is the apparent absence of hydroxyproline. Whether this affects its biophysical properties has yet to be evaluated. The contractile protein myosin has not been studied in cestodes, although paramyosin from *H. diminuta* has, and appears from its amino acid composition, tactoid banding and molecular weight to be similar to paramyosins from other sources (*967*). Actin, a key component of the cell cytoskeleton has, like the contractile proteins, a highly conserved structure throughout the animal kingdom, and it is unlikely that actin from cestode sources, surprisingly yet to be studied biochemically, will differ substantially. In addition to actin filaments, the other important components of the cell's cytoskeleton are microtubules. These are believed to play a key role in the movement of secretory products in various animal cells and consist of tubulin, a globular polypeptide of M_r 50 000. The primary mode of action of certain benzimidazole compounds (e.g. mebendazole) against cestodes and nematodes is thought to be due to their binding to parasite tubulin (*190*). The distribution of microtubules and actin filaments of cells cultured from the germinal layer of *E. granulosus*

The adult: proteins and nucleic acids

(Chapter 10) has been shown to be remarkably similar to that of mammalian cells during interphase and mitosis (719).

The alkaloid colchicine binds tightly to tubulin and this characteristic has been used (Fig. 6.1) to isolate a tubulin-like fraction from *H. diminuta*, with properties similar to tubulin from other organisms. Furthermore, colchicine affects the qualitative distribution of [^{3}H]proline-incorporated protein in this worm, with label accumulating in the parenchyma (195). This suggests that colchicine inhibits translocation in the tegument and provides evidence that microtubules within the internuncial processes facilitate movement of cell products from tegumentary cytons (Chapter 2) to the body surface for subsequent release.

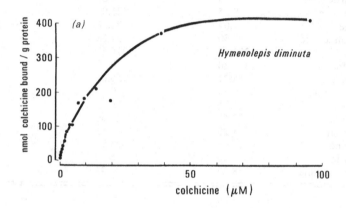

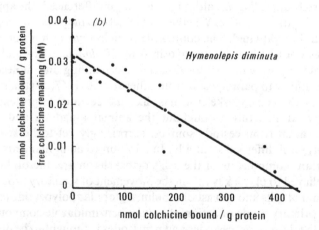

Fig. 6.1. (*a*) Colchicine binding in *Hymenolepis diminuta*. (*b*) Scatchard plot of data taken from (*a*). (After Watts, 1981.)

116

Proteins

Fractionation of soluble proteins from cestodes has revealed, in addition to other components, a variety of lipoprotein and glycoprotein conjugates. Many of these complexes are antigenic, and much effort, particularly in taeniid cestodes, has focussed on trying to isolate and characterise these antigens with a view to using them in specific immunodiagnosis and possibly as vaccines (*690, 962*; Chapter 11).

Unquestionably the most highly characterised soluble antigenic component in taeniids is a lipoprotein designated as antigen 5. Early work indicated that antigen 5 was specific to *E. granulosus* but the presence of anti-antigen 5 antibodies in sheep and humans infected with other larval cestodes has now been demonstrated (*754*). It is an iodinatable, concanavalin-A-binding molecule of variable, high M_r of approximately 400 000, and comprising subunits of 60 000–70 000 M_r which themselves dissociate on disulphide bond reduction to two subunits of approximately 20 000 and 40 000 M_r. The two subunits of antigen 5 have recently been identified putatively in *E. granulosus* hydatid cyst fluid (HF) using radioiodination, immunoprecipitation, and SDS/polyacrylamide gel electrophoresis and immunoblotting (*754*). Such immunochemical techniques are proving invaluable for the physico-chemical characterisation of proteins and polypeptide antigens of a variety of different parasites. The prominent labelled antigens in HF, which included both parasite and host components, are shown in Fig. 6.2. The two subunits of antigen 5 are arrowed at 20 000 and 38 000 M_r (lane 3); the two other major parasite bands arrowed at 12 000 and 16 000 M_r are probably two of the subunits of antigen B, the other major lipoprotein antigen of HF. Of particular physiological interest is the fact that the larger subunit of antigen 5 bears the phosphorylcholine hapten, which appears to be responsible, at least in part, for the majority of cross-reactions with other helminth infections during hydatid serology. A phosphorylcholine-containing component has also been demonstrated in the turbot tapeworm *Bothriocephalus scorpii* and this has been shown to bind host C-reactive protein (CRP) (*223*). By binding CRP, it is possible that the parasite masquerades as 'self' and eludes destruction by the immune defences of the host. This is a derivative of the molecular mimicry hypothesis, a contentious topic in parasite immunology, which will be elaborated on later (Chapter 11).

Proteins of host origin, including immunoglobulins and bovine serum albumin, have been identified in cyst fluids of several taeniid cestodes including *E. granulosus* (Fig. 6.2) and *E. multilocularis* (*492, 754*), *T. taeniaeformis* and *T. crassiceps* (*624*) and *T. saginata* (*470*). How these proteins enter the cysts is not known but it may be

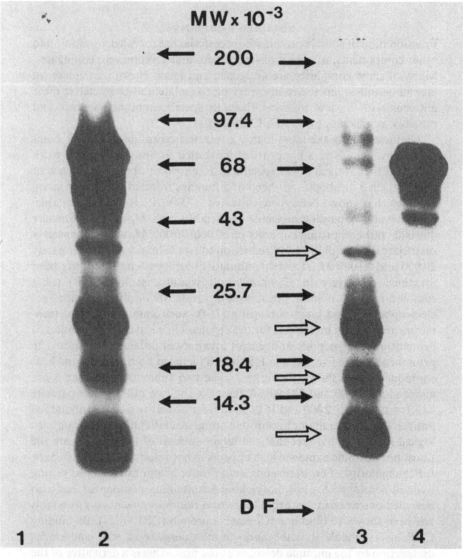

Fig. 6.2. Prominent parasite antigens and host components in *Echinococcus granulosus* sheep cyst fluid (SCF). Labelled fluid was immunoprecipitated with: lane 1, normal rabbit serum; lane 2, rabbit anti-SCF; lane 3, preabsorbed rabbit anti-SCF; lane 4, rabbit anti-sheep whole serum. Antigens were reduced prior to electrophoresis. The major parasite antigens are shown arrowed at M_r 38 000, 20 000, 16 000 and 12 000. D.F., diffusion front. (Reprinted with permission from *Molecular and Biochemical Parasitology*, 25, Shepherd, J. C. & McManus, D. P., Specific and cross-reactive antigens of *Echinococcus granulosus* hydatid cyst fluid, © 1987, Pergamon Journals Ltd.)

Proteins

by diffusion, through fissures in the cyst membranes (*624*) or by endocytosis, for which there is evidence in the bladder of *T. crassiceps* (*880, 881*). The ionic nature of proteins may also play an important role in their own absorption (*344*).

Tegumental proteins

The complexity of the cestode tegument was discussed in Chapter 2. The brush border-plasma membrane subtending the tegument is a 'fluid-mosaic', consisting of proteins, glycoproteins and lipoproteins, embedded in a lipid bilayer (Fig. 2.3, p. 10). An array of membrane-bound proteins, many of which exhibit enzyme activity, is associated with the tegumental brush border plasma membrane. This important area has been comprehensively reviewed by Pappas (*624*).

One very useful approach for investigating these and other proteins is by SDS/polyacrylamide gel electrophoresis (SDS/PAGE) and a number of studies have investigated the protein composition of the brush border plasma membrane in cestodes using this technique (*30, 237, 397, 464, 491, 538, 624*). SDS/PAGE protein profiles (*491*) for isolated brush border fractions (enriched with microtriches and brush border membrane) from protoscoleces of *E. granulosus* are shown in Fig. 6.3. A complex of at least 50 polypeptides, ranging in M_r from more than 200 000 to less than 14 000 can be identified (Fig. 6.3 (*a*)). The two major periodic acid–Schiff (PAS)-staining bands of M_r 110 000 and >200 000 probably represent glycopeptides and/or glycoproteins (Fig. 6.3 (*b*)); the other PAS-positive components running close to or with the tracking dye are probably glycolipids (*b*). These PAS-positive components probably originate from the glycocalyx, the properties of which were described in Chapter 2.

TEGUMENTAL ENZYMES

Alkaline phosphatase, acid phosphatase, 5'-nucleotidase, monoacyl hydrolase, ribonuclease, type 1 phosphodiesterase, adenosine triphosphatase, adenyl cyclase, glycosyl transferase, esterases and disaccharidase have been biochemically or cytochemically demonstrated in the tegument of various cestodes (*152, 210, 250, 374, 491, 620, 624–626, 651, 718, 763, 776, 898*). Several of these enzymes – phosphatases, 5'-nucleotidase and phosphodiesterase – probably have a digestive and/or absorptive function but the role of the others is uncertain.

Ca^{2+}-ATPase has been used as an enzymic marker for the brush border of *H. diminuta*, although its precise function is open to question. Two forms of this enzyme occur within the tegument of the parasite. One appears to be calmodulin dependent and is active at a higher pH than the other (Table 6.2). Calmodulin appears to be a ubiquitous Ca^{2+} receptor which plays a part in most of the Ca^{2+}-regulated processes that have been studied in eukaryotic cells. It is a highly conserved protein, so it is not surprising that

119

The adult: proteins and nucleic acids

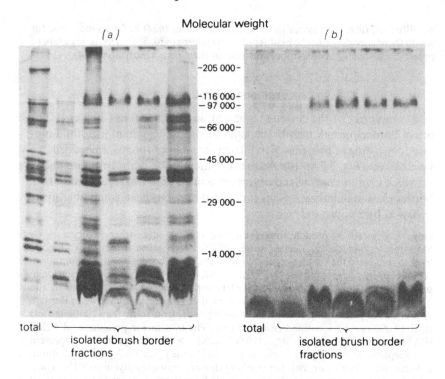

Fig. 6.3. SDS/PAGE (5–15% linear gradient gel) separated proteins of the total worm homogenate and isolated brush border fractions from protoscoleces of *Echinococcus granulosus* (horse strain). (*a*) Coomassie blue staining; (*b*) Periodic acid–Schiff (PAS) staining. (After McManus & Barrett, 1985.)

calmodulin isolated from *H. diminuta* has chemical and activatory properties similar to those of ovine brain calmodulin (*94*).

Little information is available concerning receptor sites in the brush border plasma membrane of cestodes, which is rather surprising in light of their absolute reliance on the tegument for nutrient uptake. However, a specific, high-affinity receptor for cyanocobalamin (an analogue of cobalamin, vitamin B_{12}), which may contain a relatively exposed membrane protein, has been isolated from spargana of *Spirometra mansonoides* (*238*). A time course of radiolabelled cyanocobalamin binding to isolated plasma membranes is shown in Fig. 6.4. *S. mansonoides*, in a manner similar

Table 6.2. *The effect of dialysis, Ca² ⁺, EDTA and calmodulin on ATPase activity of* Hymenolepis diminuta. *(Data from Hipkiss et al., 1987)*

Treatment	Ca^{2+}-ATPase activity	
	pH 5.5	pH 7.5
Undialysed control	2.00 ± 0.10	3.00 ± 0.20
EDTA dialysed	0.06 ± 0.01	0.09 ± 0.02
+ 10 mM Ca^{2+}	0.43 ± 0.01	0.61 ± 0.01
+ 10 mM Ca^{2+} + calmodulin	0.44 ± 0.01	0.72 ± 0.02

Note: Mean values, $\pm$ S.D., are activities expressed in milliunits (nmol phosphate released/min per mg protein); number of determinations $= 4$.

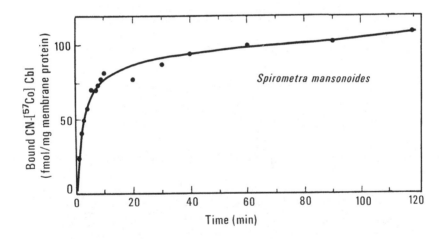

Fig. 6.4. Time course of CN-[^{57}Co]Cbl (cyanocobalamin) binding to microtriches membranes of *Spirometra mansonoides*. (After Friedman *et al.*, 1983.)

The adult: proteins and nucleic acids

to that of some other cestodes (e.g. *Diphyllobothrium latum*), takes up large amounts of cobalamin. This it converts to adenosyl-cobalamin, which acts as a coenzyme for methylmalonyl-CoA mutase. This is an integral enzyme in the formation of propionate, a major end-product of energy metabolism in some cestodes including *S. mansonoides* (Table 5.4).

ORIGIN OF TEGUMENTAL PROTEINS

Most assessments of proteins and/or enzymic activities associated with the cestode surface have not attempted to establish whether the proteins under study originate from the host or parasite. One possible approach for identifying proteins synthesised specifically by parasites is outlined in Fig. 6.5. It involves radiolabelling surface components (*494*) followed by their fractionation by immunoprecipitation, SDS/PAGE and autoradiography. An alternative approach is that of Atkinson & Podesta (*30*), who radiolabelled adults of *H. diminuta* with [^{14}C]leucine *in vitro*. Subsequent one- and two-dimensional gel electrophoretic separation and staining of the

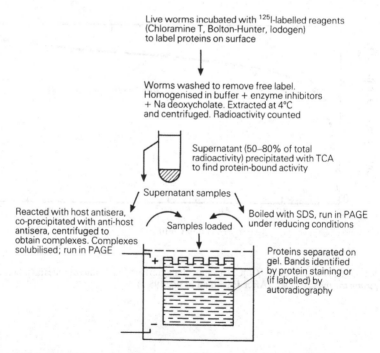

Fig. 6.5. A protocol for labelling and analysing cestode surface components. TCA, trichloroacetic acid. (After Wakelin, 1984*a*.)

122

protein constituents from lysates of whole worms, brush border denuded worms and brush border fractions demonstrated the diversity of the proteins in *H. diminuta*. Fluorographic analysis of the same polyacrylamide gels identified the proteins synthesised by the cestode; 17 of these polypeptides were associated with the brush border and these are depicted in the fluorograph shown in Fig. 6.6. This work has been extended to investigate synthesis of proteins in different regions of the strobila and in oncospheres of *H. diminuta* (*764, 765*).

It is well recognised that host immunoglobulins are adsorbed on to the tapeworm surface (*52, 57, 879, 882*) and these may act as 'blocking' antibody as a defence against cell-mediated host immune mechanisms (*686*). Other host proteins including lipase (*720*) and trypsin (*743*) are also absorbed. The radioactivity recovered from adult *H. diminuta* incubated in [³H] trypsin for increasing time periods is shown in Fig. 6.7. Several species (*H. diminuta, H. microstoma, Moniezia expansa, Ligula intestinalis*) have also been shown to absorb α-amylase, and adsorption appears to lead to an increase in amylolytic activity (*39*). This is a possible manifestation of the phenomenon of membrane (=contact) digestion (Chapter 2), whereby membrane-bound molecules are believed to be capable of digesting large molecules with which they come in contact. This hypothesis remains to be proved, since the validity of the experiments purporting to demonstrate membrane digestion in cestodes has been questioned (*861*). The uptake of proteins across the cestode tegument by endocytosis is discussed in Chapter 2.

Protein polymorphism

GENERAL CONSIDERATIONS

It is now recognised that a substantial number of proteins, especially enzymes, are polymorphic in that they exist in the cell as multiple molecular forms differing in certain of their physico-chemical properties. Each form of a polymorphic enzyme is called an isoenzyme or isozyme. Electrophoretic techniques provide convenient methods whereby this protein heterogeneity can be investigated and the approach has been widely exploited to characterise parasites. In short, aqueous parasite extracts are electrophoresed, or focused isoelectrically, and separated proteins are stained generally (usually with Coomassie Blue) or more specifically with a histochemical (enzyme) stain (the zymogram technique). Further details of individual procedures and the use of the approach in parasite identification are to be found in a number of recent reviews (*104, 258, 413, 536, 615, 856*).

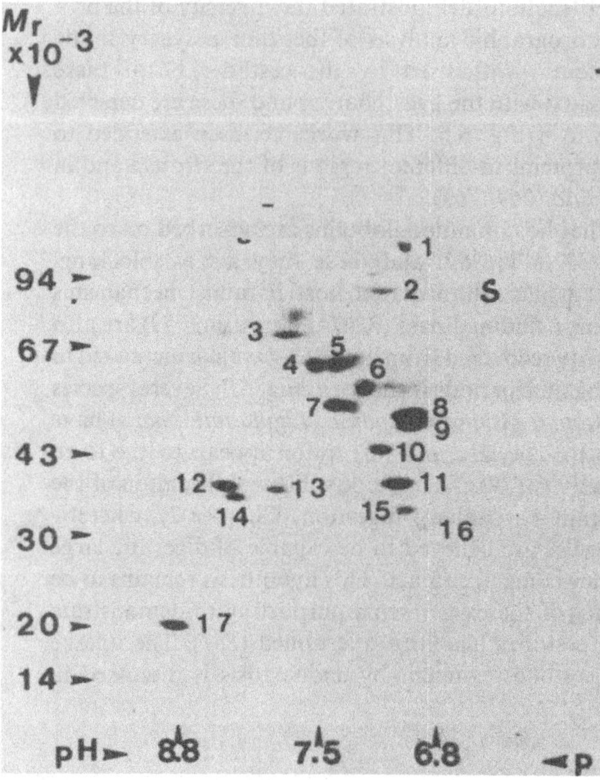

No.	$M_r \times 10^{-3}$	pI
1	108	7.00
2	90	7.25
3	73	7.75
4	65	7.50
5	65	7.40
6	59	7.25
7	54	7.45
8	54	6.90
9	51	6.90
10	46	7.10
11	38	7.05
12	38	8.20
13	37	7.80
14	36	8.10
15	35	6.95
16	32	6.85
17	20	8.80

Fig. 6.6. A fluorograph of the major polypeptides synthesised in the brush border of adult *Hymenolepis diminuta*, following their separation by two-dimensional PAGE; the apparent molecular weight (M_r) and isoelectric point (pI) of each protein are also shown. (After Atkinson & Podesta, 1982.)

Proteins

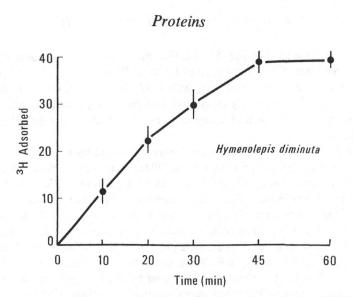

Fig. 6.7. Radioactivity (^{3}H absorbed) recovered from *Hymenolepis diminuta* incubated in [^{3}H]trypsin (6.6 μg/ml) for increasing periods (min). (After Schroeder & Pappas, 1980.)

Several studies have distinguished cestodes by differences in their total protein profiles following electrophoresis or isoelectric focusing and the following have been examined recently: *Taenia* spp. (*109*), *Bothriocephallus scorpii* (*679*), *Hymenolepis* spp. (*179, 893, 894*), *Eubothrium crassum* and *Diphyllobothrium crassum* (*783*). McManus (*489*) reported enzyme analyses, following isoelectric focusing of extracts from *L. intestinalis* and *Schistocephalus solidus*. Of several tested, the only enzyme to exhibit polymorphism was phosphoglucomutase in *L. intestinalis*. Similarly, four enzymes were tested but only malate dehydrogenase varied in its mobility and number of isoenzymes between *M. expansa* of cattle and sheep origin (*321*). Other work on enzyme electrophoresis in cestodes includes that on *Taenia* spp. (*432–434*), *Raillietina tetragona* (*34*), and *B. scorpii* (*677, 678*).

IDENTIFICATION OF *ECHINOCOCCUS* SPECIES AND STRAINS

Most work on protein polymorphism in cestodes has undoubtedly been performed with the hydatid organisms *Echinococcus* spp., because their identification presents a number of challenging problems. The life cycles involve various hosts while much inter- and intra-specific variation occurs and this has important implications for the epidemiology of hydatid disease (*495, 501, 865*).

Protein profile differences following isoelectric focusing (*414*), in combination with other criteria, have convincingly shown that at least three strains of *E. granulosus* occur in Australia. Protein patterns of two of these

The adult: proteins and nucleic acids

strains are illustrated in Fig. 6.8. Furthermore, the enzyme glucose phosphate isomerase (GPI) distinguishes between *E. granulosus* isolated from camels and those from sheep and cattle (*431, 433, 434*). A more extensive survey with 10 enzymes by isoelectric focusing (*499*) confirmed the distinction between horse and sheep isolates of *E. granulosis* referred to in Chapter 5.

The distinctiveness of camel isolates is supported by a comprehensive survey on hydatid material in Kenya (*504*). Two enzyme patterns after isoelectric focusing suggest that most cysts of sheep, goat, cattle and human origin are of one kind, which, experimental work indicates, also infects many dogs in one area. However, some material from goats and cattle appears to be distinct, with banding patterns in the former being similar to those of camel isolates. The enzyme patterns obtained with the Kenyan isolates are shown in Fig. 6.9. Variation between different geographical isolates of bovine *E. granulosus*, identified by enzyme electrophoresis of GPI, suggest that cattle can harbour more than one strain of *E. granulosus* (*301*). This variation is illustrated in Table 6.3.

Amino acids

GENERAL COMMENTS

All tissues contain free amino acids and a much larger fraction of bound (i.e. protein) amino acids. In cestodes, the free amino acid pool ranges from 100–400 mg/100 g fresh weight. Table 6.4 illustrates the qualitative differences that have been reported for the amino acids of some hymenolepidid cestodes (*79*). In general, however, there is no major departure in the amino acid content of cestodes from that of other organisms.

The amino acids occurring in a number of other adult and larval cestodes have been documented and recent data are available for *Parionella* spp., *Skrjabinia* spp., *Raillietina* spp. (*77, 346*), *Bothriocephalus scorpii* and *Nybelinia* spp. (*977*); *Taenia hydatigena* (*636*); *Diphyllobothrium* spp. (*236, 271, 275*); *Amoebotaenia cuneata* (*78*); *Ligula intestinalis* (*164*); *E. granulosus* (*29*); *Gangesia* spp. (*587*); *Lytocestus indicus*, *Introvertus raipurensis* and *Lucknowia indica* (*598*). Some earlier data have been provided by Smyth (*796*) and von Brand (*911*).

AMINO ACID UPTAKE

Amino acid uptake, as with other nutrients, has been studied more extensively in cestodes than in any other helminth group. This is because cestodes have no gut, and nutrient uptake must occur across the surface epithelial syncytium. Cestodes generally, and *H. diminuta* in particular, thus provide excellent models for the study of a number of membrane functions of broad

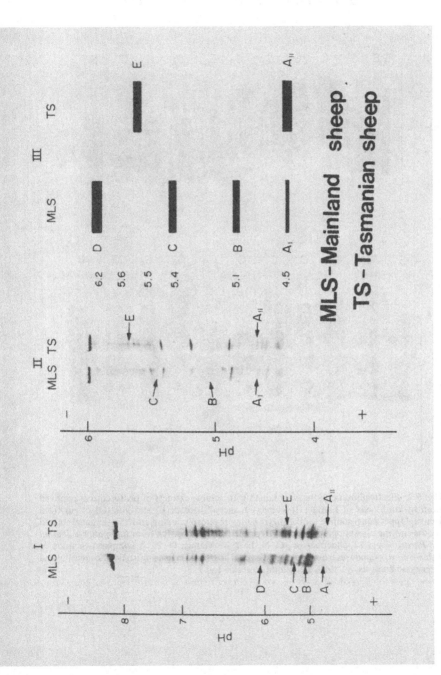

Fig. 6.8. A comparison of protein banding patterns of *Echinococcus granulosus* (protoscoleces) of sheep origin from the mainland of Australia (MLS) and Tasmania (TS). I, pH range 3.5–9.5; II, pH range 4.0–6.5; III, diagram showing the differences between material from the mainland and Tasmania after common protein bands have been removed. Numbers and letters denote different proteins. (Reprinted with permission from *International Journal for Parasitology*, **14**, Kumaratilake, L. M. & Thompson, R. C. A., Biochemical characterization of Australian strains of *Echinococcus granulosus* by isoelectric focusing of soluble proteins, © 1984, Pergamon Journals Ltd.)

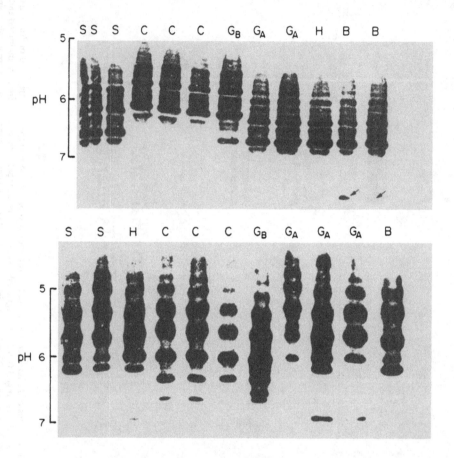

Fig. 6.9. Electrophoretic patterns obtained with soluble extracts of protoscoleces removed from hydatid cysts of human(H), camel(C), sheep(S), cattle(B) and goat(G) origin from Kenya. Upper panel stained for the enzyme glucose phosphate isomerase; lower panel stained for the enzyme phosphoglucomutase. (Reprinted with permission from *International Journal for Parasitology*, **12**, Macpherson, C. N. L. & McManus, D. P., A comparative study of *Echinococcus granulosus* from human and animal hosts in Kenya using isoelectric focusing and isoenzyme analysis, © 1982, Pergamon Journals Ltd.)

Proteins

Table 6.3. *Species of origin, geographical location, condition and glucose phosphate isomerase zymogram patterns of* E. granulosus *cyst extracts. (Data from Harrison et al., 1986)*

Host species	No. of samples	Geographical location	Protoscoleces present	Strain
Bovine	2	Eire	+	Equine
Bovine	3	Switzerland	+	Equine
Bovine	2	Kenya	+	Ovine
Bovine	3	Scotland	−	Ovine
Porcine	1	Switzerland	+	Equine
Equine	1	Switzerland	+	Equine
Ovine	1	Kenya	+	Ovine
Caprine	1	Kenya	+	Ovine
Human	1	Cyprus	+	Ovine
Human	1	Yugoslavia	+	Ovine
Gerbil	1	Human origin	+	Ovine

application to membrane biology. Discussions of membrane transport mechanisms, some of the limiting features of the cestode tegument in relation to nutrient uptake, and the apparent homeostatic control of the amino acid composition in the intestinal lumen are in Chapters 3 and 5. It is well established that dipeptides and tripeptides are actively transported by the intestinal mucosa (Chapter 3) but, surprisingly, uptake of these molecules has yet to be investigated in cestodes.

A series of recent publications covering both theoretical and experimental aspects of cestode membrane biology and the uptake of nutrients including amino acids is available (*23, 57, 579, 580, 622–624, 647–649*). A valuable earlier review is that by Pappas & Read (*629*). Recent investigations on amino acid uptake include those on *H. diminuta* (*364, 365, 466, 939*) and *E. granulosus* (*366*).

Studies, predominantly with *H. diminuta*, indicate that amino acid uptake in larval and adult cestodes occurs via specific, mediated systems, although diffusion may contribute at high substrate concentrations. Most of the well-recognised features of mediated transport have been demonstrated in that amino acid uptake obeys Michaelis–Menten kinetics, is stereospecific (for the α-amino group), can be inhibited both competitively and non-competitively, shows a distinct pH optimum and has a large temperature coefficient. Exceptionally, *T. crassiceps* cysticerci appear to take up proline and glutamate solely by diffusion, although other amino acids enter this worm by active mediated transport (*632*). In contrast to the situation in mammals, amino acid uptake in cestodes does not appear to be

129

The adult: proteins and nucleic acids

Table 6.4. *Qualitative differences in amino acids of hymenolepidid cestodes. (Data from Bhalya et al., 1985)*

	H. diminuta	H. microstoma	H. palmarum	Staphylepis rustica
Alanine	+	+	−	−
β-Alanine	−	+	−	−
β-Aminoisobutyric acid	+	+	+	+
Arginine	+	+	+	−
Aspartic acid	+	+	+	−
Citruline	+	+	−	−
Cysteine	+	−	+	+
Cystine	+	−	+	−
3, 4-Dihydroxyphenylalanine	−	−	+	−
Glycine: alanine	+	−	−	−
Glutamic acid	+	+	−	+
Glycine	+	+	−	+
Histidine	+	+	+	−
Hydroxyproline	+	−	−	−
Isoleucine	+	+	+	−
Leucine	+	+	−	−
Lysine	+	+	−	+
Methionine	+	+	−	+
Norleucine	−	−	+	−
Phenylalanine	+	+	+	+
Proline	+	+	+	−
Serine	+	+	−	+
Threonine	+	+	−	+
Tryptophan	+	−	+	−
Tyrosine	+	+	+	+
Ornithine	−	+	−	−
Taurine	+	+	−	−
Urea	−	+	−	−
Valine	+	+	−	−

Note: +, present; −, absent.

coupled to ion transport and shows virtually equal affinity for D- and L-amino acids. The two systems are, nevertheless, similar in that amino acid uptake in both involves multiple uptake sites with some overlap in their specificities; at least six transport loci occur in *H. diminuta* (Table 6.5). It has been suggested that the γ-glutamyl cycle may be involved in the transport of amino acids into *Moniezia benedini*, as three of the key enzymes are present (423). Whether this cycle occurs in other cestodes has not been investigated.

Table 6.5. *A summary of the major amino acid transport systems in* Hymenolepis diminuta. *(Data from Pappas, 1983a)*

	Transport system					
	Dicarboxylic	Glycine	Serine	Leucine	Phenylalanine	Dibasic
Major amino acids interacting	Aspartic Glutamic Methionine	Glycine Methionine	Serine Alanine Threonine Methionine Valine Proline	Leucine Isoleucine Methionine	Phenylalanine Tyrosine Histidine Methionine	Arginine Lysine
Overlapping amino acids	Serine Alanine Glycine	Serine Theonine Alanine	Glycine	Glycine Serine Threonine Alanine Valine	Leucine Isoleucine	Histidine

131

The adult: proteins and nucleic acids

Protein and amino acid catabolism

PROTEASES

The breakdown of proteins to give amino acids requires the action of proteolytic (protease) enzymes. However, amino acids do not provide an important energy source in cestodes and, although proteases have been described in several species, their physiological function generally remains obscure. The function of protease(s) (Fig. 8.12, p. 213) released from plerocercoids of *Spirometra erinacei* is, however, likely to be that of host tissue penetration and muscle digestion (*239, 425*). A cysteine protease has been purified from this worm by DEAE-cellulose adsorption and elution, gel filtration, ion-exchange and covalent chromatography and SDS/PAGE (*239*); it has a molecular weight of about 2.0×10^4 and exists as multiple forms detected by isoelectric focusing (Fig. 6.10).

Proteolytic activity has also been detected in *Schistocephalus solidus* (*777*), *Ligula intestinalis* (*514, 515, 516*), *Taenia saginata* (*287, 288*), *Moniezia expansa* (*182*) and *E. granulosus* (*491*).

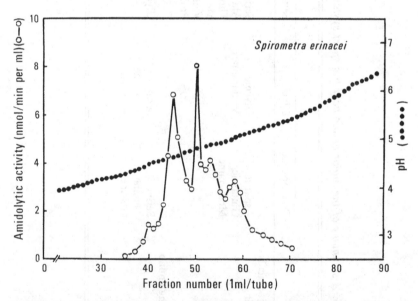

Fig. 6.10. Multiple forms of cysteine protease from *Spirometra erinacei* plerocercoids detected by isoelectric focusing and amidolytic activity with the synthetic substrate Bz-Arg-*p*NA. (After Fukase *et al.*, 1985.)

Proteins

Table 6.6. *Production of* $^{14}CO_2$ *from* Hymenolepis diminuta *incubated with different* ^{14}C-labelled amino acids. (Data from Wack *et al.*, 1983)

Amino acid[a]	Wet weight (mg) ± S.E.	μmol $^{14}CO_2$/g wet wt ± S.E.
Aspartate	520 ± 11	0.64 ± 0.22
Alanine	498 ± 7	0.15 ± 0.12

Notes:
[a] Glutamate, proline, serine, leucine, isoleucine, lysine, glycine and tryptophan did not produce significant amounts of $^{14}CO_2$.

AMINO ACID INTERCONVERSIONS

Little is known of the precise amino acid requirements of cestodes. Presumably, most, if not all, their required amino acids can be provided by the host and, as discussed above, cestodes have developed numerous transport systems to acquire these compounds. However, cestodes appear to have a limited ability to catabolise amino acids. This is exemplified by *H. diminuta*, which has been shown (*918*) to generate significant $^{14}CO_2$ only from ^{14}C-labelled aspartate and, to a lesser extent, alanine during incubation *in vitro* with 10 different amino acids (Table 6.6). The few amino acids that are catabolised in cestodes participate in two main pathways, namely transamination and oxidative deamination.

Transamination

In transamination, the amino group is transferred, by means of specific enzymes, directly to a keto acid (usually 2-oxoglutarate), which forms a substrate for the formation of the new acid. The most studied systems are:

(a) 2-oxoglutarate $\xrightleftharpoons{\text{transaminase}}$ glutamate,

(b) pyruvate $\xrightleftharpoons{\text{transaminase}}$ alanine.

Transaminases have been demonstrated in *Hymenolepis* spp., *Anoplocephala magna*, *Raillietina cesticillus*, *T. taeniaeformis*, *M. expansa* and *Lytocestus indicus* (*39, 669, 796*). Some 17 amino acids failed to act as amino acid donors in transamination reactions with *Hymenolepis* (*796*) and it appears that, compared with vertebrates, cestodes may have an extremely limited capacity for performing transaminations. This may reflect the fact that cestodes live in an environment rich in amino acids, in which case synthesis may play only a minor role in satisfying essential amino acid requirements.

133

The adult: proteins and nucleic acids

Oxidative deamination

Glutamate dehydrogenase, which catalyses the oxidative deamination of glutamate,

$$\text{glutamate} + \text{NAD}^+(\text{NADP}^+) + \text{H}_2\text{O} \rightleftharpoons 2\text{-oxoglutarate} + \text{NADH(NADPH)} + \text{NH}_4$$

has been partially purified and characterised from the cytosol of *H. diminuta* (*582*). The enzyme is specific for NAD$^+$ and it may play a role in the maintenance of redox potential in this worm. A flavine-linked L-amino acid oxidase, active with glutamate, has also been reported in *H. diminuta* (*796*) and this may provide an additional, minor pathway for oxidative deamination of amino acids in cestodes. Ammonia is produced during these reactions and is excreted. Glutamate can also be metabolised via the 4-aminobutyrate pathway to succinate. There are conflicting reports that this pathway, which bypasses the 2-oxoglutarate dehydrogenase reaction of the tricarboxylic acid (TCA) cycle, occurs in *M. expansa* (*153, 668*), but no other species have been investigated.

END-PRODUCTS OF NITROGEN METABOLISM

General considerations

The breakdown of proteins, purines and pyrimidines (see below) results in nitrogenous excretory products, but knowledge of intermediary nitrogen metabolism in cestodes is fragmentary. The nitrogenous compounds excreted are known for some species but the metabolic pathways whereby these products are produced are poorly understood. Ammonia is the major excretory product, but urea, uric acid and numerous other nitrogen-containing compounds, including amino acids and amines, have also been detected in cestodes (Table 6.7). The major source of ammonia is probably the oxidative deamination of glutamate, described above. Another possible source of ammonia is via the cleavage of urea by urease, although this enzyme has been detected only in two species, the trypanorhynchs, *Lacistorhynchus tenuis* and *Pterobothrium lintoni* (*796*). Interestingly, these cestodes live in the intestine of sharks, where they are exposed to a high level of urea, and the manufacture and utilisation of urease may be an adaptation to their environment. Whether the ammonia is used by these cestodes in synthetic reactions is not known, although the other product of urease activity, CO_2, appears not to be utilised (*621*).

134

Table 6.7. Nitrogen excretory products of cestodes. (Data from Smyth, 1969)

Excretory product	Hymenolepis diminuta	Echinococcus granulosus[a]	Cysticercus tenuicollis[a]	Lacistorhynchus tenuis	Taenia taeniaeformis[a]
Ammonia	+	−	−	+	+
Urea	+	+	+	−	+
Uric acid	−	+	+	−	−
Methylamine	−	−	−	−	+
Ethylamine	−	−	−	−	+
Propylamine	−	−	−	−	−
Butylamine	−	−	−	−	+
Amylamine	−	−	−	−	−
Heptylamine	−	−	−	−	+
Ethylene diamine	−	−	−	−	+
Cadaverine	−	−	−	−	−
Ethanolamine	−	−	−	−	+
1-Amino-2-propanol	−	−	−	−	+
Amino acids	?	+	−	−	+
Creatinine	−	+	+	−	+
Betaine	−	+	−	−	−

Note: +, present; −, absent.
[a] Larva.

135

The adult: proteins and nucleic acids

Urea excretion

Production of urea by cestodes suggests the existence of the urea (Krebs–Henseleit) cycle, which is shown in Fig. 6.11. One of the key enzymes, arginase, has been widely reported in cestodes (796, 185–187). However, some of the other enzymes, notably carbamyl phosphate synthetase and ornithine transcarbamyl, are either absent or present in only low amounts (39) and it is doubtful if a complete cycle operates in cestodes. It is likely that the urea excreted by tapeworms comes from the activity of arginase alone. The uric acid produced and excreted by cestodes probably arises from the breakdown of purines (39).

Amines

The excretion of amines is unusual in animals. Amines are highly toxic and one method employed by vertebrates to detoxify them is via monoamine oxidase, an enzyme which has been detected in *H. diminuta* (569). Amines can arise from the decarboxylation of the appropriate amino acid, e.g. glycine and alanine can give rise to methylamine and ethylamine, respectively. Another possible source of amines may be the reduction of azo or nitro compounds (39) and azo- and nitro-reductase activity has been reported from *M. expansa* (180, 181). Furthermore, the physiologically active amines octopamine, dopamine, adrenalin and serotonin (5-hydroxytryptamine) have been demonstrated in cestodes (283, 296, 435, 681, 682, 758, 859), where they probably function predominantly as neurotransmitters (see Chapter 2).

Serotonin is tentatively regarded as an excitatory neurotransmitter in

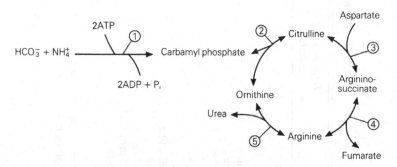

Fig. 6.11. The (Krebs–Henseleit) ornithine cycle. Numbers refer to enzymes as follows. (1) Carbamyl phosphate synthetase (E.C.2.7.2.a). (2) Ornithine transcarbamylase (E.C.2.1.3.3). (3) Arginino-succinate synthetase (E.C.6.3.4.5). (4) Arginino-succinate lyase (E.C.4.3.2.1). (5) Arginase (E.C.3.5.3.1). (After Smyth, 1969.)

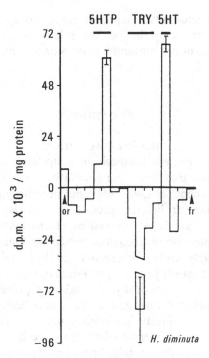

Fig. 6.12. Chromatography of ³H-labelled extracted products following incubation of homogenates of *Hymenolepis diminuta* in [³H]tryptophan: 5HT, 5-hydroxytryptamine; 5HTP, 5-hydroxytryptophan; TRY, tryptophan; or, origin; fr, solvent front. The data were obtained by subtracting the radioactivity in boiled homogenates (blanks) from that in unboiled homogenates. Radiolabelled tryptophan was metabolised in the latter to [³H]5HTP and [³H]5HT. (After Ribeiro & Webb, 1983a.)

Diphyllobothrium dendriticum and has been shown to be a stimulatory neurotransmitter affecting muscle contraction in *H. diminuta* (*278*) and in larvae of *Grillotia erinaceus* (*932*). Serotonin may also have a hormonal role in metabolic regulation in cestodes, as it has been shown to have a considerable effect on glucose uptake and intermediary metabolism in *H. diminuta* when supplied exogenously (*163, 268, 531*). In addition to taking up large amounts of exogenous serotonin from the rat gut via specific serotonin receptors (*510*), *H. diminuta* has the enzymic capacity to synthesise serotonin from both tryptophan and 5-hydroxytryptophan (*681, 683*). This is illustrated by results of thin-layer chromatography of the ³H-labelled extracted products, following incubation of homogenates of *H. diminuta* in [³H]tryptophan (Fig. 6.12). Another important amino acid derivative which has been detected in cestodes (*295*) is choline. This is a major

137

The adult: proteins and nucleic acids

constituent of phospholipids (Chapter 4) and is a component of the inhibitory neurotransmitter acetylcholine. The role of acetylcholine and other potential neurotransmitter molecules in cestode neurotransmission is discussed in Chapter 2.

Protein synthesis

GENERAL ACCOUNT

Our knowledge of protein synthesis in cestodes is somewhat limited, although it is known that, like other helminths, they are capable of rapid production of polypeptides (*39*). This rapid protein synthesis is probably associated with their prolific egg production and their high tegumental turnover. The area has been reviewed by Harris (*300*), who noted that studies with cestodes have proceeded on two fronts. One involves incorporation of radioactively labelled precursors into the protein of intact worms, an approach used to study protein synthesis in cysticercoids of *H. diminuta* (*363*). Numerous biosynthetically radiolabelled protein antigens of *T. taeniaeformis* larvae and oncospheres have also been identified by this method (*90*). The previously discussed studies (p. 122) (*30, 764, 765*) involving one- and two-dimensional electrophoresis and fluorography of proteins synthesised in the strobila, brush border and oncospheres of *H. diminuta*, further demonstrate the value of the approach. The other method utilises cell-free protein-synthesising systems and has been applied in both homologous and heterologous assays with cestodes.

Homologous systems

Cell-free systems capable of synthesising polypeptides have been prepared from protoscoleces of *E. granulosus* (*7*), larval *T. crassiceps* (*588*) and *H. diminuta* (*633*). In general, these studies have demonstrated that protein synthesis in cestodes, although showing some specificity, is similar to that in mammals in that it requires polysomes, amino acid adenylates, aminoacyl-tRNAs, pH 5 fraction, ATP, GTP, magnesium and either sodium or potassium ions.

Heterologous systems

Several studies have used a cell-free system prepared from rabbit reticulocytes to translate cestode messenger RNA (mRNA) *in vitro*. Thus, a small portion of the proteins synthesised in this heterologous system, using mRNA from *T. crassiceps*, were shown to be antigenic (*5*). The molecular weights of the translated polypeptides were generally low (13 000–22 000),

138

Nucleic acids

Table 6.8. *The incorporation of [^{35}S] methionine into protein in a rabbit reticulocyte lysate system as directed by eight different RNA preparations from* E. granulosus *(horse strain). (Data from McManus et al., 1985)*

RNA preparation	[^{35}S] Methionine incorporation c.p.m./μl
1	48 145
2	75 215
3	65 185
4	44 035
5	36 190
6	56 830
7	68 805
8	71 890
No RNA	7480

Note: Total RNA (3 μg) was added in each case and translation mixtures (18 μl) were incubated for 90 min at 30°C.

but one protein of 260 000 was also synthesised. In a similar study (*493*), translation of the total RNA from *E. granulosus* (Table 6.8), *E. multilocularis* and *T. crassiceps* gave significant incorporation of [^{35}S]methionine into synthesised proteins. Many polypeptides, ranging in molecular weights from 10 000 to 205 000, were produced, a number of which proved to be antigenic. Furthermore, *in vitro* translation of RNA derived from *Cysticercus cellulosae* yielded a protein that could be immunoprecipitated by rabbit anti-pig immunoglobulin (*173*). The synthesis of a host-like determinant may be of some survival value to the parasite by enabling it to evade the host immune response (Chapter 11). Whether other cestodes can similarly manufacture host-like components has not been explored.

Nucleic acids

General considerations

Our knowledge of nucleic acids and nucleic acid metabolism in cestodes is very limited when compared with many other organisms. Nevertheless, some recent advances have been made and it is clear that new approaches in

The adult: proteins and nucleic acids

molecular biology are providing highly specific methods for identifying cestodes and for studying their interrelationships at the genetic level. Moreover, these techniques will make it possible to understand more fully the cestode genome and its expression. This will facilitate the characterisation and synthesis of cestode gene products with immunodiagnostic and/or vaccine potential and will allow the analysis of differential gene expression during cestode development and differentiation. The value of such information in relation to the study of the biology of cestodes and other parasitic organisms has been highlighted (*472, 473, 492, 493, 767*).

As much of the terminology used in molecular biology may be unfamiliar to some readers, it is appropriate to define some of the vocabulary and this is given in an appendix to this chapter. There are two types of nucleic acids, the ribonucleic acids (RNA) and the deoxyribonucleic acids (DNA). Genetic information is carried in the linear sequence of nucleotides in DNA. Each molecule of DNA contains two complementary strands of deoxyribonucleotides which contain the purine bases, adenine and guanine and the pyrimidines, cytosine and thymine. RNA is single-stranded, being composed of a linear sequence of ribonucleotides; the bases are the same as in DNA with the exception that thymine is replaced by the closely related base uracil. DNA replication occurs by the polymerisation of a new complementary strand on to each of the old strands.

Expression of the genetic information occurs when a specific segment of DNA (called a coding region or gene) is copied in the nucleus into RNA by a process called DNA *transcription*. The RNA is modified before it leaves the nucleus as a mRNA molecule to direct the synthesis of a specific protein on a ribosome by *translation*.

PURINES, PYRIMIDINES AND NUCLEOSIDES

Nucleic acid synthesis, a basic feature of all living organisms, has been shown experimentally in cestodes (*349, 979, 980*). There is no clear evidence that cestodes have the capacity to synthesise purines *de novo*, although some indirect evidence suggests that this may occur in regenerating tetrathyridia of *Mesocestoides corti* (*315*). Cestodes can incorporate free purines and purine nucleosides, obtained from their hosts, into nucleic acids by salvage pathways (*39*). The enzymes involved in these salvage pathways have not generally been studied in cestodes, although the key enzymes adenosine deaminase, adenosine kinase and AMP deaminase have been demonstrated in *H. diminuta* (*246, 248*).

As with purines, there is indirect evidence from studies *in vitro* that regenerating tetrathyridia of *M. corti* can synthesise pyrimidines *de novo* (*315*). Furthermore, aspartate transcarbamylase, the first enzyme in the pathway, has been demonstrated in *Moniezia benedini* (*39*), while five of the six pathway enzymes have been measured in *H. diminuta* (*326*). It appears, therefore, that at least some cestodes have the capacity to synthesise pyrimidines by the biosynthetic route. Little is known of pyrimidine salvage pathways in cestodes, although the key enzyme thymidine kinase has been

Nucleic acids

Table 6.9. *Purine and pyrimidine transport systems in* Hymenolepis diminuta. *(Data from Barrett, 1981)*

Thymine-uracil carrier		Hypoxanthine carrier I		
Transport site	Activator site	Transport site	Activator site	Hypoxanthine carrier II
Thymine	Thymine	Adenine	Adenine	Hypoxanthine
Uracil	Uracil	Guanine		Purine?
5-Bromouracil	5-Bromouracil	Hypoxanthine		(Adenine)
5-Aminouracil	5-Aminouracil	6-Methyluracil?		
		Uracil?		
(Adenine)	5-Bromouracil?			
(Hypoxanthine)				
(6-Methyluracil?)				
(Purine?)				

Note: Compounds in parentheses bind non-productively (i.e. they are not transported).

detected in *H. diminuta* (*39, 208*). Cestodes possess specific pyrimidine transport systems (see below) and it is likely that they can incorporate pyrimidines into nucleic acids by salvage pathways of the type found in other helminths (*490*).

Absorption of purines and pyrimidines by cestodes occurs by a combination of passive diffusion and mediated transport. In *H. diminuta*, purine and pyrimidine uptake is very complex and seems to involve at least three carrier systems (Table 6.9), two of which appear to bind several substrate molecules simultaneously (*631*). Pyrimidine transport was thought to involve allosteric regulation because the relation between initial uptake and substrate concentration was sigmoidal. However, more recent work (*890*) has indicated that the sigmoidal kinetics of pyrimidine transport in *H. diminuta* is an isotope effect, obtained only when $2\text{-}^{14}\text{C}$-labelled pyrimidines were used; absorption kinetics of methyl-^{14}C- and ^3H-labelled pyrimidines were hyperbolic. Nucleosides (thymidine, uridine, adenosine and guanosine) are absorbed by *H. diminuta, H. citelli* and *H. microstoma* via a specific sodium-dependent, mediated system involving at least two carriers (*347*). Interestingly, the mechanism displays a diurnal periodicity in *H. diminuta* (*616, 617*).

NUCLEIC ACID COMPOSITION
Nucleic acids normally constitute about 5–15% (by weight) of tissues and there is usually two to eight times as much RNA as DNA. Cestodes appear to conform to this plan as, for example, DNA and RNA represent 0.8% and

141

The adult: proteins and nucleic acids

7.3% of the dry weight, respectively, in *E. multilocularis*; the values for *E. granulosus* depend on the strain analysed and vary between 0.4% and 0.8% for DNA and between 5.2% and 8.9% for RNA (*488, 498*) (Table 4.1).

DNA
The base composition of nuclear DNA, usually expressed as the molar percentage of guanine plus cytosine (G + C), is characteristic of an organism. Even within a species, wide variations in G + C percentage may occur (e.g. the G + C percentage varies between 30% and 71% for different strains of the bacterium *Staphylococcus aureus*) (*600*), although the precise cause of this variation is not known. With the exception of *Orygmatobothrium* spp., the G + C content of nuclear DNA from a variety of cestodes (Table 6.10) has been shown to be in the range of most other invertebrate groups (*394, 600*). Whether these base ratios vary within individual cestode species has not been investigated.

When total DNA is subjected to high speed density gradient centrifugation in caesium chloride, a large band of nuclear DNA and one or more small satellite bands are usually observed. Up to four satellite DNA bands have been detected in *H. citelli*, *H. diminuta* and *H. microstoma*, one of which is mitochondrial DNA (mtDNA) (*394*). The mtDNA of *H. diminuta* has been isolated (*118*) and has been shown to be a typical circular molecule. The characteristics of *H. diminuta* DNA are shown in Table 6.11. In contrast, *E. multilocularis* and *E. granulosus* produced two distinct DNA bands after fractionation in caesium chloride, but there was no evidence that the DNA from either band represented mtDNA (*493*). There is presumably so little mtDNA in comparison to nuclear DNA in these organisms that it is completely masked in preparations of total DNA by this method. That this is the case has been shown by a recent study (*976*), where a different procedure, based on the selective precipitation of nucleic acids by cetyltrimethylammonium bromide (CTAB), was employed to extract mtDNA from isolated mitochondria. Some 300 g and 50 g, respectively, of *Taenia* spp. and *Echinococcus* sp. tissue yielded approximately only 1 ng mtDNA.

In mammals, some 10% of the DNA is highly repetitive, a further 20% is moderately reiterated and the remainder represents non-repetitive or single copy sequences. The rate of renaturation of DNA has been used to estimate that 16% of the DNA in *H. diminuta* is repetitive (*746*). Furthermore, the rate of reassociation of the single copy component in this species has allowed calculation of its haploid genome size, which is approximately 1.5×10^8 base-pairs (*746*). This is similar to that reported for other helminths (*767*) and there is no reason to believe that the genome size of other cestodes will vary significantly from this value. The renaturation of DNA from *H. diminuta* compared with that of the planarian *Dugesia tigrina* is shown in Fig. 6.13. The free-living flatworm DNA renatured faster, thereby indicating that the tapeworm is the more complex organism.

Table 6.10. *The guanine + cytosine content of cestode DNAs compared with other groups (Data from Normore, 1976; Barrett, 1981)*

	G + C content (%)
Cestodes	
Orygmatobothrium sp.	61
Schistocephalus sp.	47
Lacistorhynchus tenuis	45
Phyllobothrium sp.	44
Echinococcus multilocularis	44
Hymenolepis microstoma	39
H. citelli	39
H. diminuta	38
Gyrocotilideans	36–45
Trematodes	34–47
Nematodes	41
Acanthocephalans	39
Bacterial phages	28–70
Viruses of chordates	35–74
Bacteria	23–80
Fungi	22–70
Algae	37–68
Bryophytes	41–57
Gymnospermae	35–40
Angiospermae	34–49
Protozoa	19–66
Porifera	35–60
Coelenterates and Aschelminthes	37–41
Mollusca	31–47
Arthropoda	32–46
Echinodermata	35–46
Chordata	36–47

RNA

There are three different types of RNA in cells, namely ribosomal RNA (rRNA), transfer RNA (tRNA) and messenger RNA (mRNA). *E. granulosus* is the only cestode in which RNA has been investigated in depth (*8*) and the evidence indicates that the RNA species in this worm and their formation conform with those of other eukaryotes. The rRNA had sedimentation coefficients of 29.4 and 19.6, was rich in guanosine-5'-monophosphate and had a G + C content of around 50%. In addition, a light RNA fraction, with a sedimentation coefficient of 6.3, was isolated

143

Table 6.11. *Characteristics of* Hymenolepis diminuta *DNA. (Data from Carter et al., 1972)*

	Buoyant density in CsCl (g/ml)	% G+C from buoyant density	Thermal denaturation (°C)	% G+C	Mean length (μm)
(a) Adult worm, nuclear DNA	1.696	36.7	83.6	34.9	
(b) Adult worm, mitochondrial DNA	1.691	31.6	82.2	32.6	4.76

144

Nucleic acids

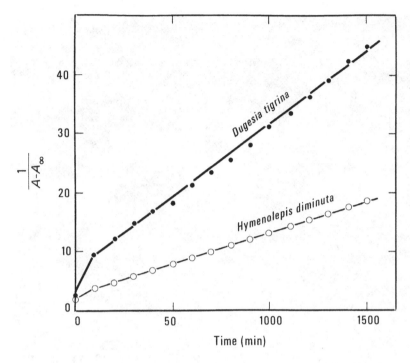

Fig. 6.13. Renaturation of *Hymenolepis diminuta* and *Dugesia tigrina* DNAs. Absorbance at 260 nm. (After Searcy & MacInnis, 1970.)

which corresponded to tRNA (G + C content of 59%). Two further heavier fractions (sedimentation coefficients of 35 and 40–42.5) were also apparent and these were probably the precursors of the rRNA. Pulse labelling of *E. granulosus* RNA also demonstrated the presence of short-lived mRNA in these experiments. The two rRNA species in *E. granulosus* can be demonstrated by agarose gel electrophoresis under non-denaturing conditions (*493*). A typical electrophoretic separation of *E. granulosus* RNA, together with RNAs from *Escherichia coli* and *Schistosoma mansoni* for comparison, is shown in Fig. 6.14.

ISOLATION OF NUCLEIC ACIDS

If recombinant DNA and related techniques in molecular biology are to be applied meaningfully to cestodes, functional nucleic acids have to be prepared such that the DNA can be cleaved by restriction endonuclease enzymes (which digest the double-stranded DNA at highly specific sites, each recognizing a precise nucleotide sequence) and the RNA can be used to

145

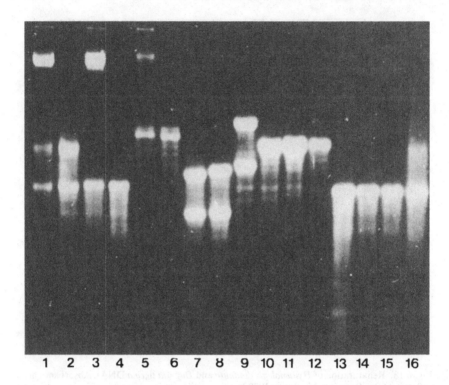

Fig. 6.14. Agarose gel electrophoresis of *Echinococcus granulosus* (horse strain) RNA. The samples in lanes 1, 2, 7 and 16 were prepared in 10 mM phosphate buffer, pH 6.8, and were not denatured. The samples in lanes 3, 4, 8 and 13–15 were denatured in 50% (w/v) dimethylsulphoxide in 10 mM phosphate buffer, pH 6.8. The samples in lanes 5, 6, 9 and 10–12 were denatured in 1 M glyoxal/3 M urea in 10 mM phosphate buffer, pH 6.8. Lanes 1, 3 and 5, *E. granulosus* total nucleic acids (10 μg); lanes 2, 4 and 6, *E. granulosus* RNA (5 μg); lanes 7, 8 and 9, *Escherichia coli* RNA (10 μg); lanes 10–16, *Schistosoma mansoni* RNA (10 μg). The *E. coli* RNA was included as a molecular weight marker; the larger subunit is approximately 1 000 000 and the smaller subunit is 500 000. The RNA was visualised by staining with ethidium bromide and ultraviolet illumination. (After McManus *et al.*, 1985.)

prepare complementary DNA (cDNA) and can programme protein synthesis *in vitro*. A useful technique has been devised for the isolation of DNA and RNA from *Echinococcus* and other cestodes which satisfies these requirements (*493*). The method combines the simultaneous isolation of both nucleic acids and is illustrated in Fig. 6.15. If DNA is not required, RNA can be prepared separately by lysis of parasites in guanidine

Frozen parasites (1 ml packed worms) crushed in liquid nitrogen.

|

Powder thawed into 4 ml extraction buffer.

|

Equal volume extraction buffer containing 1% SDS then added,
followed by 1 mg Proteinase K.

|

Incubate for 3 h.

|

Phenol extraction, back extraction and re-extraction.

|

Nucleic acids in supernatant precipitated in 0.3 M sodium acetate and
2.5 vol. ethanol overnight at $-20°C$ and pelleted at $10\,000 \times g$, 10 min, 0°C.

|

Total nucleic acids dried and dissolved in
4 ml 10 mM Hepes, pH 7.5.

|

3 vol. of 4 M LiCl added and left overnight.

/\

| RNA precipitate pelleted at $10\,000 \times g$, 10 min. Washed in 70% ethanol, dried and dissolved in 0.1 ml 10 mM Hepes, pH 7.5. | DNA isolated from supernatant by addition of 2.5 vol. ethanol and subsequent spooling. Washed in 70% ethanol, dried and dissolved in 0.5 ml TE buffer. |

Fig. 6.15. A method for the combined isolation of RNA and DNA from cestodes. TE buffer is 10 mM Tris·HCl (pH 8.0), 1mM EDTA. (After McManus *et al.*, 1985.)

The adult: proteins and nucleic acids

thiocyanate followed by centrifugation of the extract through a CsCl cushion. Total RNA is pelleted at the bottom of the centrifuge tube and can be used directly for *in vitro* translation assays and cDNA synthesis, subsequent to cDNA cloning. Alternatively, polyadenylated messenger RNA can be isolated from the total RNA by passage through a column of either oligo(dT)-cellulose or poly(U)-Sepharose. Numerous other techniques are available for the isolation of nucleic acids, including the CTAB method (*976*) referred to above. Details of these and other general procedures in molecular biology have been described by Maniatis *et al.* (*509*).

MANIPULATION OF NUCLEIC ACIDS

Once cestode DNA or RNA has been purified, it can be subjected to various procedures of genetic manipulation, which are summarised in Fig. 6.16. Except for *H. diminuta* and other large and readily available species (*Taenia hydatigena, T. taeniaeformis, Moniezia expansa*), physiological and biochemical studies on cestodes are hindered by the problem of insufficient supply of parasite material. Cloning and expressing cestode genes in bacteria can help to circumvent this difficulty and currently there is great interest in constructing recombinant DNA libraries for this purpose. In addition,

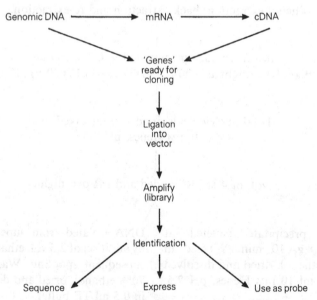

Fig. 6.16. A general scheme of procedures for genetic manipulation. (After Simpson *et al.*, 1986.)

Nucleic acids

expression libraries can be used to clone genes coding for immunologically important cestode antigens, using serum from infected individuals (475) or from animals immunised with parasite extracts as probes. Various strategies for library construction are currently being tried with cestodes, ranging from the cloning of fragments of genomic DNA to using cDNA synthesised from mRNA isolated from specific life cycle stages.

Cloning genomic DNA

An example of the former cloning strategy, outlined in Fig. 6.17, is provided by Rishi & McManus (695), who have constructed a small, size-selected genomic DNA library in a plasmid vector of *Escherichia coli* using total DNA isolated from *Echinococcus granulosus*. Subsequent differential

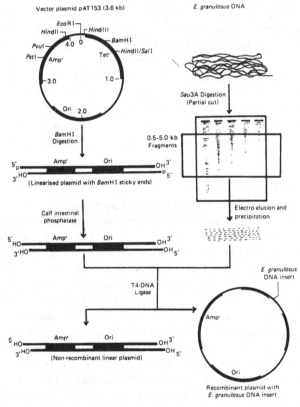

Fig. 6.17. Genomic DNA cloning using the plasmid vector, pAT153. kb, 10^3 base-pairs. (After Rishi & McManus, 1987a.)

149

screening of the library with DNAs from a panel of taeniid cestodes has identified recombinant plasmids containing *Echinococcus-* and *E. granulosus*-specific inserts. These recombinant plasmids can be used as genetic markers in strain characterisation (see below) and have potential as DNA probes for application in a field microscope assay for distinguishing eggs of *Echinococcus* from those of other taeiniids. A similar procedure has been used to clone specific DNA fragments from *T. solium* (*695*) and the approach is appropriate for cloning genes or parts of genes from other cestodes. That this is the case is shown by the fact that the entire mitochondrial genome of *T. hydatigena* has been cloned into a plasmid (*976*), the first time that this has been achieved for any helminth parasite.

Cloning cDNA

Another cloning strategy was provided by the pioneering work of Bowtell *et al.* (*91*), who were the first to construct an expression library using cDNA synthesised from a parasitic helminth. They isolated mRNA from strobilocerci of *T. taeniaeformis* and used it to make a cDNA expression library in the bacteriophage vector, λgt11amp3, a derivative of the more commonly used λgt11. Immunological screening of this library with sera from mice that had been hyperimmunised with eggs identified a number of distinct parasite antigens (*91, 92*). Whether any of these antigens are host-protective (see Chapter 11) remains unproven. A simplified flow chart for constructing libraries using a bacteriophage vector such as λgt11 is given in Fig. 6.18.

DNA PROBES AND CESTODE IDENTIFICATION

One of the exciting opportunities provided by molecular technology is the possibility of examining genetic variation in cestodes directly and in detail at the DNA level. The DNA sequence of an organism's genome is the ultimate basis of all biological classifications, and direct analysis at this level overcomes an inherent problem of other identification procedures such as isoenzyme analysis or the use of monoclonal antibodies. Such techniques analyse the expressed products of the genome and can show life-stage or environmentally mediated, including host-induced, variation, whereas DNA procedures do not. The most practical method for determining interspecific or intraspecific DNA sequence variation in a cestode is to take sequences from a genomic DNA library (not necessarily produced from the cestode under study) and to test its specificity. The genetic function of the sequence may be unknown, but this is generally unimportant for its taxonomic value. A simple way to test cloned sequences for specificity is by Southern blot hybridisation (*818*), illustrated in Fig. 6.19.

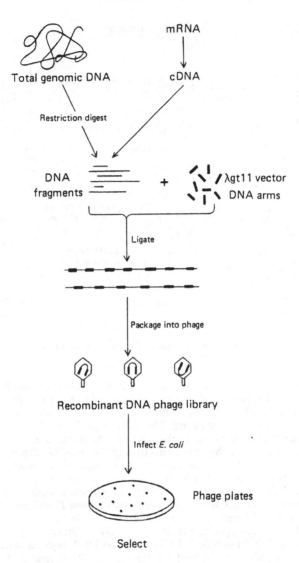

Fig. 6.18. A flow chart for constructing genetic libraries using the bacteriophage expression vector λgt11 (After Simpson *et al.*, 1986.)

The adult: proteins and nucleic acids

DNA Probing

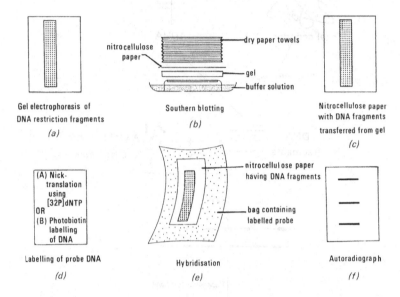

Fig. 6.19. The technique of Southern hybridisation.

Briefly, cestode DNA is digested with a restriction endonuclease and the resultant DNA fragments are transferred (blotted) from the gel to a nitrocellulose filter, and probed with a labelled (usually with ^{32}P or with non-radioactive biotin or photobiotin) cloned DNA sequence. The labelled probe will hybridise with the complementary sequence bound to the filter and the position of hybridisation is revealed by autoradiography or colorimetrically. The position depends upon the size of the genomic fragment, subsequent to endonuclease digestion, and differences in fragment size are normally a reflection of exact base sequence. The approach has proved useful for discriminating a variety of taeniid cestodes, including *E. granulosus* from *E. multilocularis* (*496*) and *T. solium* from *T. saginata* (*695*). It can also readily distinguish between strains of *E. granulosus* (*162, 496, 497*) and provides a powerful technique for speciation studies on cestodes in general. Figure 6.20 shows the hybridisation of ^{32}P-labelled pEG18 (a cloned DNA segment, specific for *E. granulosus*) and pSM889 (a cloned fragment of the RNA gene of *Schistosoma mansoni*) to Southern blots of genomic DNA extracted from United Kingdom isolates of *E. granulosus* of sheep, cattle and horse origin. The patterns of hybridisation clearly distinguish between the horse and sheep strains and indicate a very close affinity between *E. granulosus* from cattle and sheep.

Nucleic acids

(a)

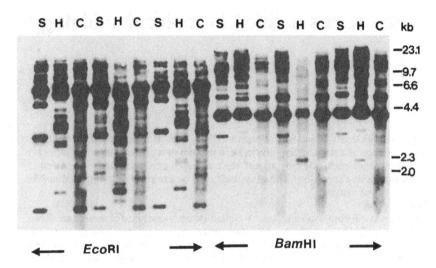

(b)

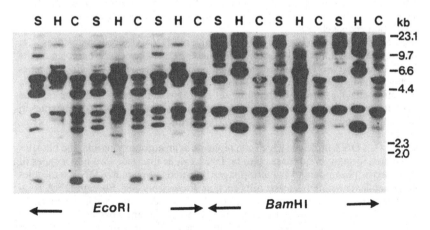

Fig. 6.20. The hybridisation of ^{32}P-labelled pEG18 (a) and ^{32}P-labelled pSM889 (b) to *Echinococcus granulosus* restriction fragments. S, sheep origin; H, horse origin; C, cattle origin; kb, 10^3 base-pairs.

APPENDIX

Some terms defined in molecular biology

Gene
DNA is divided into functional segments called genes. The gene specifies the amino acid sequence of a single polypeptide, i.e. a protein or a protein subunit. Genes are separated from each other by spacer regions which contain signals important in regulating gene expression. The gene itself may be interrupted by internal spacer regions called introns.

Transcription
The genetic information in a gene is copied (transcribed) into a messenger RNA molecule (mRNA), preserving the sequence by complementary base-pairing. The introns are cut and the mRNA molecule is transported into the cytoplasm where it directs the synthesis of protein at the ribosomes. The sequence of bases is translated into a sequence of amino acid residues by a triplet code wherein three bases specify one amino acid.

Complementary DNA, or cDNA
cDNA is DNA which has been synthesised enzymically from mRNA. This procedure takes advantage of the ability of the enzyme reverse transcriptase to synthesise double-stranded DNA from a single-stranded RNA template.

Restriction endonucleases
These are enzymes which occur in bacteria and cut DNA at specific sequences (e.g. the enzyme EcoRI, derived from *E. coli* (thus the derivation of the name), cuts the DNA chain between guanine and adenine in the sequence guanine-adenine-adenine-thymine-thymine-cytosine). Each time a particular DNA is cleaved by a restriction enzyme, precisely the same set of fragments is generated. The enzyme DNA ligase effectively works in a reverse manner to that of restriction enzymes in that it splices pieces of DNA back together again.

Vector
A vector is a DNA molecule which can replicate as an autonomous unit and has sites into which foreign, in our case cestode, DNA can be inserted. Two major types of vector exist, plasmids and bacteriophages. Plasmids are circular DNA molecules which replicate inside bacterial cells such as *Escherichia coli*. Foreign DNA fragments with a maximum size of 10 kb (1 kb = 10^3 bases) can be cloned in plasmids. As bacteria divide, the plasmids are transmitted to all the daughter bacteria, thus giving rise to many identical copies of the one plasmid. Commonly used plasmids in recombinant DNA work are pBR322 and its analogue, pAT153. Both plasmids have resistance to the antibiotics amipicillin and tetracycline, and several useful sites for restriction enzymes. Bacteriophage lambda is an important model for molecular biologists and has become a useful cloning vector in that fragment sizes of up to

15–20 kb can be inserted. Lambda is a bacterial virus with two different cycles. It can grow lytically, as do most viruses, by simply replicating within the bacterium. When mature viral particles have been assembled, the bacterium bursts and releases large numbers of infective virus, which infect and lyse other bacteria. Lambda can also grow lysogenically. In this case, the phage genome is propagated along with the bacterial DNA. The genes controlling this behaviour are located in the middle of the phage genome, while the genes required to propagate lambda as a functional virus are found in the two arms. The genes which control the lysogenic life cycle of lambda are not required for lytic gowth and can be replaced by DNA from another source. In this way, phage lambda can be used as a vector for pieces of foreign DNA. An expression vector is a vector in which the site of cloning is situated next to a bacterial promotor which directs synthesis of mRNA. The principle is to insert a piece of cDNA containing a protein-coding sequence adjacent to a promotor. The bacterium will then synthesise a mRNA containing the foreign cDNA sequence and this will be expressed as a hybrid protein. Recombinant clones are then screened by antibody directed against the protein of interest. Expression of foreign genes has been achieved in *E. coli* using both plasmid (e.g. pUC18) and bacteriophage vectors (e.g. λgt11).

Recombinant DNA libraries

Insert DNA can be extracted directly from a parasite, cut into fragments using restriction enzymes and cloned into a vector. This is called genomic DNA. Alternatively, mRNA is isolated and converted into cDNA. Mixtures of clones containing a variety of insert molecules are called libraries (i.e. genomic DNA or cDNA libraries) and interesting clones are sought in the libraries by screening with probes. The probes may be nucleic acid, in which case the DNA (with a sequence complementary to that of nucleic acid probe) of the clone is allowed to hybridise to the probe. Otherwise, if the library is an expression library, the probe is an antibody and we rely on the clone making a protein and having the protein recognised by a defined antibody.

7

The biology of the egg

General account

The egg is essentially the only developmental stage (except a coracidium in certain orders) which is exposed to the external environment. A number of morphological and physiological adaptations allow the egg to survive in this hazardous environment – and at the same time respond sensitively to those stimuli which enable it to hatch in the appropriate place in the appropriate host.

Eggs can be divided into two groups: group I (p. 170), those from cestodes whose hosts have aquatic associations; and group II (p. 170), those whose hosts have terrestrial associations. The protective membranes developed in these two groups tend to be strikingly different. In group I, for example, the pseudophyllidean egg has a sclerotin (p. 172) capsule (Fig. 7.1(b)) and a ciliated embryophore, whereas in group II, the cyclophyllidean egg usually (Fig. 7.1(a)) has a thick keratin embryophore and a thin capsule. In the Taeniidae the embryophore is made of closely fitting keratin blocks (Fig. 7.1(c)) which separate under the action of gut enzymes and allow the oncosphere to hatch (p. 191).

The egg and reproductive system thus present an elegant model for the study of a number of areas of physiological interest, especially the ultrastructure and cytochemistry of the spermatozoa, ova and embryonic envelopes (see p. 166), as well as the physiological processes involved in egg hatching and the subsequent 'activation' of the released oncosphere.

Morphology of the reproductive system

Male

SPERMATOGENESIS

With rare exceptions (e.g. *Dioecocestus*), cestodes are hermaphrodite and the male system follows the typical platyhelminth plan.

156

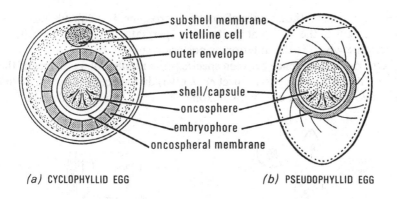

(a) CYCLOPHYLLID EGG   *(b)* PSEUDOPHYLLID EGG

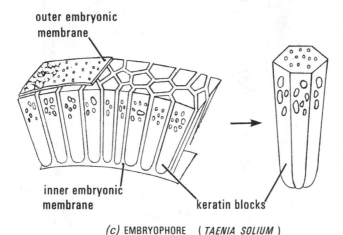

(c) EMBRYOPHORE (*TAENIA SOLIUM*)

Fig. 7.1. Morphology of the cyclophyllid and pseudophyllid egg. (*a*) A typical cyclophyllid egg, based on *Taenia* sp. (*b*) A typical pseudophyllid egg, based on *Diphyllobothrium* sp. (After Smyth, 1963.) (*c*) Embryophore of the egg of *Taenia solium* showing presence of keratin blocks held together by a cementing substance; under enzyme action, these blocks become separated during the hatching process. (After Wang *et al.*, 1981.)

Early work on spermatogenesis has been reviewed in the first edition (*796*). More recent work has been reviewed by Euzet *et al.* (*196*), Davis & Roberts (*171*), Ubelaker (*889*) and Lumsden & Specian (*462*). Ultra-structure studies have shown that cestode spermatozoa generally (but see below) have the same pattern of organisation in all orders. The chief features are: (*a*) a long thread-like body; (*b*) an elongated nucleus; (*c*) cortical microtubules underlying the plasma membrane; (*d*) the absence of

157

mitochondria; (*e*) the absence of a typical acrosome. It is difficult, however, to distinguish the different parts of a spermatozoon – or, indeed, to determine which is the anterior or posterior end!

One major difference between species is the number of axonemes of the 9 + 1 pattern. On this basis, Euzet *et al.* (*196*) divided cestode spermatozoa into two types, with one or two axonemes (Fig. 7.2 and Table 7.1):

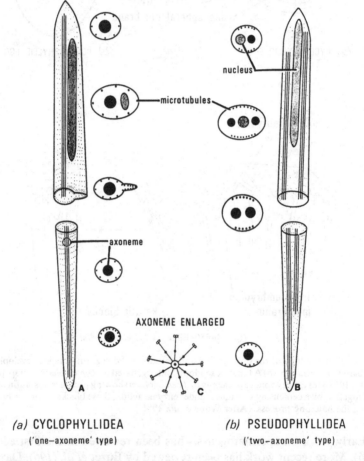

(a) CYCLOPHYLLIDEA
('one–axoneme' type)

(b) PSEUDOPHYLLIDEA
('two–axoneme' type)

Fig. 7.2. The two types of spermatozoa found in cestodes; note the absence of an acrosome and mitochondria. (*a*) The 'one-axoneme' type: found typically in the Cyclophyllidea but also reported in the Caryophyllidea, Diphyllidea and Tetraphyllidea and (more rarely) in the Pseudophyllidea. (*b*) The 'two-axoneme' type: found typically in the Pseudophyllidea, but also reported in the Proteocephalidea, Tetrarhynchidea and the Tetraphyllidea. (Modified from Ubelaker, 1983; after Euzet *et al.*, 1981.)

Morphology of the reproductive system

Table 7.1. *Cestode spermatozoa: species with one or two axonemes*

Order	Species	Reference
Type A. With two *axonemes*		
Proteocephalidea	*Proteocephalus longicollis*	*840*
Pseudophyllidea	*Diphyllobothrium latum*	*458*
	Bothriocephalus clavibothrium	*846*
Tetrarhynchidea	*Lacistorhynchus tenuis*	*835*
[Trypanorhyncha]		
Tetraphyllidea	*Onchobothrium uncinatum*	*564*
	Acanthobothrium filicolle	*563, 564*
Type B. With one *axoneme*		
Pseudophyllidea	*Bothrimonas sturionis*	*482*
Cyclophyllidea	*Hymenolepis nana*	*796*
	H. diminuta	*387, 831*
	H. microstoma	*833*
	Ophryocotyloides corvorum	*851*
	Catenotaenia pusilla	*833*
	Taenia hydatigena	*212*
	Echinococcus granulosus	*41, 571, 833*
	Moniezia expansa	*833*
	Raillietina carmeostrobilata	*653*
	Inermicapsifer madagascarensis	*833*
	Monoecocestus americanus	*482*
Diphyllidea	*Echinobothrium affine*	*196*
	E. typus	*196*
Tetraphyllidea	*Echeneibothrium beauchampi*	*566*
	Phyllobothrium gracile	*561*
	Pseudanthobothrium sp.	*482*
Caryophyllidea	*Glaridacris catostomi*	*842*

Type A. The 'one-axoneme' type: typical of the Cyclophyllidea but also reported from the Caryophyllidea, Diphyllidea and Tetraphyllidea.

Type B. The 'two-axoneme type': typical of the Pseudophyllidea, but found (more rarely) in the Proteocephalidea, Tetrarhynchidea and Tetraphyllidea. This is considered to be the primitive type, as it has been found in the free-living platyhelminths.

The one-axoneme type has been shown to develop from the two-axoneme type. Thus, in *Phyllobothrium gracile*, although during spermatogenesis two basal bodies are formed, one of these is aborted (*561*). In *Echinococcus multilocularis*, both basal bodies initiate the formation of an axoneme, but one of these is subsequently aborted during elongation of the other (*45, 46*).

159

The biology of the egg

Microtubules occur in a single row below the plasma membrane of spermatozoa, although their distribution varies in different species (*171*). In *H. diminuta* and the Cyclophyllidea (Fig. 7.2), in general, they completely encircle the spermatozoon (*171, 212, 835*), often exhibiting helical periodicity. In *Phyllobothrium gracile* (Fig. 7.2(*b*)) the cortical microtubules may be confined to two demi-circles, lateral or external to the axonemes (*561, 564*). The significance of these different distributions of microtubules in relation to potential sperm activity is not known.

Although most authors report the absence of mitochondria, there is a possibility that these may be present in a modified form. Thus, in *Taenia hydatigena*, Featherston (*212*) described the presence of 'septate' structures similar to those found in insect spermatozoa, where they are known as '*neberkern*' – bodies which have been shown to be derived from mitochondria. In *Ophryocotyloides corvorum*, by light microscopy, 'granular' mitochondria have been described, which lie scattered during the early stages of spermatogenesis but later pass into the tail region (*851*).

PHYSIOLOGY AND BIOCHEMISTRY OF SPERMATOZOA

Very little is known regarding the physiology or biochemistry of spermatozoa. The significance of the absence of mitochondria and an acrosome is difficult to determine. In vertebrate and most invertebrate spermatozoa, mitochondria are abundant and form a substantial 'middle piece' and the acrosome secretes a lytic enzyme which assists in the penetration of the egg. That cestode spermatozoa can function satisfactorily without these organelles suggests they may have metabolic requirements different from those of other groups and that penetration of the cestode egg is, in some way not understood, easier than that of other invertebrate eggs.

Some evidence of the biochemical activities of cestode spermatozoa perhaps can be deduced from their cytochemistry, discussed below.

CYTOCHEMISTRY OF SPERMATOZOA

Early studies have been reviewed in the first edition (*796*). The development of new cytochemical techniques, especially at the electron microscopical level, has meant that more comprehensive investigations on the biochemical activities of spermatozoa – which cannot be undertaken by normal biochemical methods – are possible.

Cytochemical studies which throw some light on sperm metabolism are those on: *Acanthobothrium filicolle* (*564*), *Avitellina lahorea* (*209*), *Duthiersia fimbriata* (*769*), *Diphyllobothrium latum* (*796*), *Echeneibothrium beauchampi* (*566*), *Hymenolepis diminuta* (*542–560, 704*), *H. microstoma* (*134*), *Lacistorhynchus tenuis* (*835*), *Lytocestus indicus* (*769*), *Ophyrocotyloides corvorum* (*851*), *Phyllobothrium gracile*

160

(*561*), *Proteocephalus longicollis* (*840*), *Raillietina echinobothridia* (*769*), *R. johri* (*715, 716*), *R. tetragona* (*35*).

The significance of many of the above results has been reviewed by Davis & Roberts (*171*) and the following general conclusions can be drawn.

Cestode spermatozoa are rich in glycogen, large quantities being readily detectable at the light and electron microscopical levels. In the majority of species it is found in the form of beta particles (M_r about 900×10^6), in which form it is readily utilisable (*543*), an exception being *Diphyllobothrium latum* (see below). The reported distribution of glycogen within the sperm varies with species, but it does not appear to accumulate in the axoneme. Studies have shown that in *Hymenolepis diminuta* [³H]glucose is rapidly incorporated into the mature sperm (*454, 543*). The absence of mitochondria is reflected in the negative reactions for mitochondrial enzymes such as succinic dehydrogenase and α-glycerophosphate oxidase (*704*) and it is likely that spermatozoa rely chiefly on glycolysis for their energy requirements. In this respect, it may be significant that several enzymes of the phosphorylase system have been identified in the sperm of *H. diminuta* (*542, 544*).

In *H. diminuta*, glycogen occurs as alpha particles (*796*) and it has been suggested that these might be incorporated into the oocyte and used as an energy source during embryogenesis *ex utero* (*171*). Spermatozoa of *H. diminuta* have also been shown to be weakly positive for one oxidoreductase, D-glyceraldehyde-3-phosphate: NAD oxidoreductase (*554*) and three oligosaccharide hydrolases, β-D-glucosidase, α-D-galactosidase and β-glucuronidase (*555*).

Female

OOGENESIS

In comparison with spermatogenesis, there are comparatively few studies on the ultrastructure and cytochemistry of oogenesis. Work previous to 1970 has been reviewed by Rybicka (*721*) and Smyth (*796*). Much of the recent basic data on cestodes in general have been reviewed by Davis & Roberts (*170*), and Lumsden & Specian (*462*) and, in the Caryophyllidea, by Mackiewicz (*479*). As well as containing numerous mitochondria, polyribosomes, centrioles and Golgi complexes, oocytes characteristically possess RNA-containing granules variously known as 'vitelline granules', 'yolk substance', 'ribonuclear clusters' and similar terms, but their composition and function remain largely unknown. Their primary function is likely to be nutritive but they may also be implicated in the formation of the

The biology of the egg

proteinaceous capsule/shell in cyclophyllideans (see p. 179). Rather surprisingly, most oocytes appear to be almost free of glycogen, suggesting that the oocyte itself provides only minimal food reserves, which must therefore be provided by the vitelline cells, which also serve as a source of shell material (see below).

Insemination

The relevant literature has been reviewed by Williams & McVicar (*959*), Nollen (*599*), Oshmarin & Prokhova (*609*) and Smyth (*802*). Although the mechanical processes involved in insemination are well known, almost nothing is known regarding the physiology of this process or the fertilisation which results. There is no information as to whether in cestodes sexual maturation is co-ordinated by an internal, sexual endocrine system or whether external chemotactic attraction between worms takes place. Three types of sperm transfer are known: (*a*) self-insemination (*b*) cross-insemination and (*c*) hypodermic insemination.

SELF-INSEMINATION

General account

In self-insemination, the cirrus is inserted into the vagina in the same proglottid. The process has been well documented from *in vivo* specimens of the Tetraphyllidea (*Orygmatobothrium musteli, Dinobothrium septaria, Monorygma macquariae, Phyllobothrium* spp. and *Crossobothrium squali* (*959*)) and in the Cyclophyllidea (*Echinococcus granulosus* (*802, 815*), *E. multilocularis, E. oligarthrus* (*416*), *Ophryocotyle insignis, Raillientina* sp., *Choanotaenia exigua, Anoplotaenia dasyuri* (*959*)). Further evidence has come from single worm infections *in vivo* in *H. diminuta* (Table 7.2) and from experiments *in vitro* with the pseudophyllideans *Ligula intestinalis* (*227*), *Schistocephalus solidus* (*788, 802*), and *Spirometra* spp. (*73*).

In some species in which self-insemination occurs, e.g. *E. granulosus* (*815*) and *Phyllobothrium sinuosiceps, O. musteli, A. dasyuri* (*959*), a tegumental membrane sometimes occludes the genital pore in mature proglottides, effectively preventing the cirrus everting through the genital pore. This may assist self-insemination as well as preventing cross-fertilisation. Since the membrane may not occur in every proglottis, this does not necessarily preclude cross-insemination between other proglottides (see below).

In *Schistocephalus*, worms cultured free in liquid media fail to inseminate and the cirrus can often be observed projecting from the genital pore and the

Table 7.2. Hymenolepis diminuta:
*insemination by adult worms removed
from their rat hosts, exposed for 3 h to
[³H] thymidine and returned singly with
or without unlabelled worms to rats for 3
days. (Data from Nollen, 1975)*

Unlabelled worms per experiment	Number of experiments	Insemination by labelled worms	
		Self	Cross
0	8	8	—
1	5	5	4
2	4	4	6
3	4	4	12

receptaculum seminis is devoid of sperm. Insemination *in vitro* can, how-ever, be achieved by compressing the worms during culture within a cellulose tube (*788, 802*) (Fig. 10.2, p. 263); the same technique is also effective in inducing self-insemination in *Ligula in vitro*. By use of an appropriate culture tube, both inseminated and sterile (non-inseminated) worms can be produced in the same culture tube (Fig. 10.2). Compression alone is not, however, always effective in inducing insemination *in vitro* and difficulties have been encountered in obtaining fertile eggs of *Mesocestoides corti, E. granulosus* and *E. multilocularis* under such conditions (see Chapter 10).

Experiments with labelled spermatozoa
That both self-insemination and cross-fertilisation can occur in the same worm has been demonstrated unequivocally by the elegant experiments of Nollen (*599*) with *H. diminuta*. He used sperm labelled with [³H]thymidine and transplanted worms back to rats either singly or in combinations of labelled and unlabelled worms. In eight single-worm transplants (Table 7.2), all worms inseminated themselves, as demonstrated by the presence of labelled sperm in the receptaculum. In mixed labelled and unlabelled worm transplants, the labelled worms cross-inseminated with 92% of the unlabelled worms present.

CROSS-INSEMINATION
As already pointed out above, cross-insemination has been demon-strated in *H. diminuta* (*599*). This result has been confirmed by the elegant

163

experiments of Schiller (*737*), who showed that crosses between normal worms and variant worms resulted in an increase in the percentage of the variants in the offspring of the normal worms (Fig. 7.3). In other species, direct evidence of one-sided copulation or mutual copulation between proglottides of the same strobila, or detached proglottides or those of other strobila have been described for *O. musteli*, *Rhinebothrium flexile*, *Acanthobothrium quadripartum* and *Phyllobothrium* spp. (*959*).

The occurrence of cross-fertilisation in helminths can also be investigated by the examination of enzyme polymorphism and this technique has been elegantly applied by McManus (*489*) to populations of the pseudophyllidean cestodes, *Schistocephalus solidus* and *Ligula intestinalis*. It has already been pointed out that self-fertilisation has been demonstrated in these species cultured to maturity *in vitro*. In *S. solidus*, no polymorphic variants were detectable for the four enzymes investigated – lactate dehydrogenase, malate dehydrogenase, glucoseophosphate isomerase and phosphoglucomutase (PGM) – suggesting that cross-fertilisation does not occur in this species (at least in the population examined). In *L. intestinalis*, in contrast, one of the enzymes investigated, PGM, proved to be polymorphic. This enzyme appeared to be controlled by three loci, and one of these loci, designated PGM-2, is polymorphic, with three recognisable phenotypes readily identifiable by isoenzyme analysis.

This polymorphism, which was not related to the geographical origin of the infected fish hosts, is typical of a genetic polymorphism under the control of two co-dominant alleles. This type of balanced polymorphism indicates that cross-fertilisation must also occur, at least transiently, in *L. intestinalis*. The use of enzymes as markers in genetic studies in cestodes is clearly an important method for the detection of cross-fertilisation in this group.

In *Monoecocestus thomasi*, the vagina is functional only in immature proglottides in which neither male reproductive organs nor the ovary are mature. Cross-insemination therefore takes place between mature proglottides (in which the cirrus is developed) and immature proglottides in which the vagina is developed (*671*). The biological advantage of this arrangement is not clear.

HYPODERMIC INSEMINATION

Some cestodes (e.g. the Acoelidae) lack a vaginal opening and insemination apparently occurs by the well-armed cirrus being thrust into the tissues of the body wall from which spermatozoa migrate to the large seminal vesicle (*889*). The same mechanism occurs in some monogeneans, e.g. *Diclidophora*.

Morphology of the reproductive system

Hymenolepis diminuta

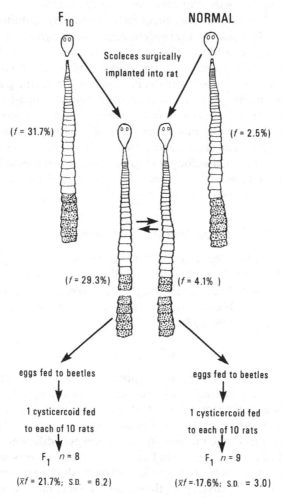

Fig. 7.3. Experiment designed to test if cross-fertilisation occurred with *Hymenolepis diminuta*. An F_{10} variant (with a testicular variation frequency of 31.7%) and a normal worm (variation frequency 2.5%) were simultaneously implanted into the duodenum of a rat. After worms became patent at 16 days post-implantation, the terminal gravid proglottides were removed and fed to separate groups of beetles. The resultant cysticercoids were fed to groups of 10 rats (one cysticercoid per rat). Analysis of the F_1 worms revealed a statistically significant *increase* in the variation of frequency in the progeny of the normal worm and a statistically significant *decrease* in the variation frequency of the variant (F_{10}) individual. (Compare $\bar{x}f = 17.6\%$ with 4.1% in the implanted worms and with 2.5% in the original worms before implantation. (After Schiller, 1974.)

The biology of the egg

Fertilisation

There are a number of descriptive accounts of spermatozoa–oocyte interactions in cestodes and the subsequent fertilisation (889) but little is known regarding the physiology of the process. All accounts agree that fertilisation takes place in the oviduct. The most recent accounts are those of *H. diminuta* (836) and *Acanthobothrium filicolle* (562). In the latter species, after the coiling of the spermatozoa around the oocyte, the plasma membrane of the sperm fuses with that of the ovum and, in addition to the nucleus, the axonemes, microtubules and the crested body also pass into the oocyte cytoplasm (562). Following fertilisation, these bodies (possibly superfluous spermatozoa, resulting from polyspermy) appear to be rejected. In *H. diminuta*, the elongated sperm nucleus becomes spherical after penetration and fuses with the oocyte nucleus, after which cleavage begins (836).

Egg: formation and structure

Physiological studies have centred chiefly on the pseudophyllidean and cyclophyllidean egg (Fig. 7.1). The formation of the capsule or egg shell (p. 171) has been studied extensively and these processes are dealt with in detail below. The field has been the subject of a number of reviews (138, 170, 451–453, 721, 796, 888, 889, 953).

Formation of embryonic envelopes

A fully formed egg is enclosed in a number of layers, and it is important for the study of the penetration of substances into the egg, as well as for understanding the physiology of hatching, that the nature of these layers be understood. Most workers have followed the terminology used by Rybicka (721), but there is some confusion in its use, as eggs of different species have been described as having three to five layers depending on whether or not their derivation is taken into account. Although all these layers essentially represent embryonic 'envelopes', those which are very thin are referred to as 'membranes', e.g. the *subshell membrane* and the *oncospheral membrane*. Some workers use the term membrane for all the layers, following the terminology used in vertebrate embryology.

The three basic embryonic envelopes are as follows (Figs. 7.4 and 7.11):

I. The *capsule* (shell/capsule = egg shell): the outermost covering layer, regarded by most investigators as being equivalent to the *egg shell* and is so regarded here. Some workers, however, speak of shell material fusing with the capsule and becoming transformed into the egg shell.

166

Egg: formation and structure

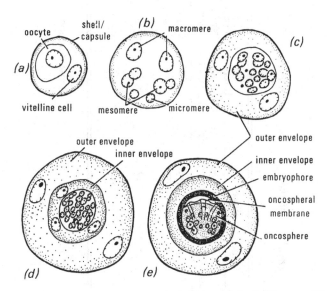

Fig. 7.4. Embryonic development in the Cyclophyllidea: (*a*) fertilised ovum; (*b*) cleaving embryo; (*c*) early preoncosphere; (*d*) late preoncosphere; (*e*) oncosphere. (After Rybicka, 1966.)

In *H. nana*, the shell material is described as being deposited against the inside of the capsule (*204*). The shell/capsule is well developed in the Pseudophyllidea, Tetraphyllidea and Trypanorhyncha. It is often poorly developed or absent in the Cyclophyllidea (Fig. 7.1(*a*)), in which case (especially in the Taeniidae) the embryophore (see below) is thickened and essentially functions as a protective 'shell'. A subshell membrane has been described in some eggs (Fig. 7.1(*a*)).

II. The *outer envelope*: a complex layer filling the space between the capsule and the inner envelope (Fig. 7.11).

III. The *inner envelope*: a syncytial layer showing much variation. Some workers divide this layer into two zones – zone I, a cytoplasmic layer and zone II, a gelatinous layer (Fig. 7.14). Part of this embryonic layer gives rise to the *embryophore* (Fig. 7.4) and also to the *oncospheral membrane* (Figs. 7.4, 7.11 and 7.14) (a very thin layer surrounding the oncosphere), which is often counted as a fourth layer. Additional layers, which may be further derived from the above basic envelopes have been reported in some species (e.g. *H. nana*; *204*), but it is beyond the scope of this text to discuss all the various modifications which can occur. Only those features which have a special physiological significance are discussed below.

Table 7.3. Hymenolepis nana: *histochemical reactions of the embryonic envelopes of the egg.* (Data from Fairweather & Threadgold, 1981a)

Test	Indicates presence of	Shell	Outer envelope	Inner envelope			Inter-filamentous material	Polar filament layer
				Embryophore	Cytoplasm	'Oncospheral membrane'		
PAS	Carbohydrates	++ᵃ	+	-	+++	ND	++++	+/++
PAS+amylase control		+	-/+	-	+	ND	-	-/+
Coomassie brilliant blue / Mercuric bromophenol blue	Proteins	++++	+	-?	+++	ND	-/+	+
Alcian blue	Acid mucopolysaccharides	++	+/++?	-	-	ND	++	+++

Notes:

ᵃ −, negative reaction; +, slight positive reaction; ++, moderate reaction; +++, strong reaction; ++++, intense reaction; ?, uncertain result; ND, not determined; PAS, periodic acid–Schiff.

Egg: formation and structure

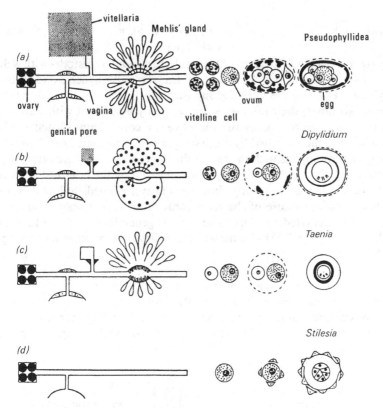

Fig. 7.5. The four types of egg-forming systems in cestodes. (a) Cestodaria, Pseudophyllidea, Trypanorhyncha, Tetraphyllidea. (b) Cyclophyllidea, e.g. *Dipylidium, Hymenolepis*. (c) Taeniidae, e.g. *Taenia*. (d) Thysanosominae, e.g. *Stilesia*. See the text. (After Löser, 1965b.)

The formation of these membranes from the fertilised egg in the Cyclophyllidea is shown in Fig. 7.4. Their structure and formation in *Hymenolepis* is considered in further detail on p. 179; their histochemistry is shown in Tables 7.3 and 7.6. In considering the functional morphology of eggs, it must be borne in mind that some eggs, such as those of the Pseudophyllidea (e.g. *D. latum*), hatch in water; others hatch in the alimentary canal of invertebrates or vertebrates (e.g. *H. diminuta, T. saginata*) and one species (*H. nana*), uniquely, can hatch in both its intermediate (beetle) and definitive (rat) host.

Types of eggs

There are four main types of cestode eggs (*451*), which may be divided broadly into two groups:

GROUP 1. THE PSEUDOPHYLLIDEAN-TYPE EGG (FIGS. 7.1, 7.5
AND 7.6)

Eggs with a thick, sclerotin (p. 172) capsule, produced by cestodes with well-developed vitellaria, i.e. Pseudophyllidea, Trypanorhyncha and Tetra-phyllidea. The cestodes of this group generally infect aquatic (rather than terrestrial) hosts; their life cycles are normally associated with water, in which the egg embryonates and the larva – a coracidium – hatches. The coracidium is then ingested by the first aquatic intermediate host. This type of egg is homologous with the egg of the Digenea (*810*). More rarely (e.g. *Archigetes*), the egg hatches only when ingested by the intermediate host The coracidium is frequently, but not always, ciliated, although some authors restrict the use of the term coracidium to a swimming embryo enclosed in a ciliated envelope. The latter is generally considered to be an embryophore (Fig. 7.1) but some workers believe it to be the inner envelope (*889*).

GROUP II. REMAINING EGG TYPES

These are produced by cestodes with either 'condensed' vitellaria, as in most cyclophyllideans, or lacking vitellaria, as in the Avitellina. Ubelaker (*889*) suggests that the latter could be considered as a separate group (group III).

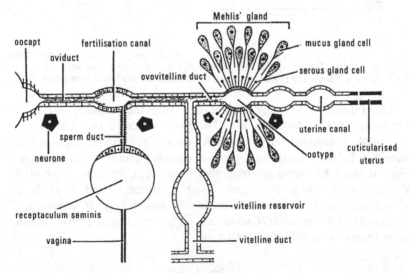

Fig. 7.6. Schematic representation of the egg-forming system of a pseudophyllidean cestode. (After Löser, 1965*a*.)

Egg: formation and structure

The eggs in this group are generally fully embryonated when laid and do not require a free-living aquatic stage in the life cycle. Although most of the well-known species have a terresterial intermediate host, usually an arthropod or a vertebrate, some species do have aquatic intermediate hosts, e.g. *Aploparaksis furcigera* (in oligochaetes), *Dicranotaenia coronula* (in copepods), *Diorchis ransomi* (in ostracods) and *Echinocotyle rosseteri* (in molluscs).

Three types of eggs are found in group II.

(i) The *Hymenolepis-type egg (Fig. 7.5(b))* with a thin capsule and a relatively thin embryophore; embryonated when laid, e.g. *Hymenolepis* and *Dipylidium*.

(ii) The *Taenia*-type egg (Fig. 7.5(c)) lacking an outer sclerotin capsule but often covered by a delicate capsule (which is normally lost in faecal eggs); the embryophore is very thick and striated and consists of closely packed keratin blocks (Fig. 7.1(c)), e.g. *Taenia* and *Echinococcus*.

(iii) The *Stilesia*-type egg (Fig. 7.5(d)) formed from genitalia lacking vitellaria, so that the egg is formed only from the components of the ovum and sperm. The uterus wall, however, lays down a thick, cellular covering, e.g. *Stilesia* and *Avitellina*.

Formation of pseudophyllid-type eggs (group I)

GENERAL MECHANISM

The shell/capsule

The formation of the capsule has been most frequently studied in the Pseudophyllidea where it forms the basis for a hard, stabilised, sclerotin egg shell. Investigations have largely centred around the biochemistry and cytology of the processes of egg shell formation. Ubelaker (*889*) further divides group I eggs into: subgroup I(i) eggs with thick operculate capsules which mature outside the uterus, e.g. some Pseudophyllidea; and subgroup I(ii) eggs with thin operculate capsules, e.g. Tetraphyllidea and Proteocephalidea.

Most work on egg formation has been carried out on *Diphyllobothrium latum* (*451, 796, 973*) and *Schistocephalus* (*796*) but other species examined more recently include: *Diphylobothrium* spp., *Diplogonoporus tetrapterus* (*327*); *Amphiptyches urna* (Gyrocotylidea), *Bothriocephalus scorpii* (*372*); *B. clavibothrium* (*844, 845*); *Bothridium pithionis*, *Khawia iowensis*, *Caryophyllaeus laticeps*, *Eutetrarhynchus ruficollis*, *Triaenophorus nodulosus* (*452*); *Glaridacris catostomi* (*843*);

171

The biology of the egg

Echeneibothrium maculatum (*451, 796*); *E. beauchampi* (*565*); *Proteocephalus longicollis* (*841*).

The mechanism (Figs. 7.5 and 7.6) whereby this type of egg, with its sclerotin egg shell, is formed, appears to be similar in all the above species and may be summarised as follows.

Ova released from the ovary pass down the oviduct where fertilisation takes place; a sphincter muscle controls the release of ova. 'Mature' vitelline cells, containing both shell and yolk precursors, pass from the vitelline glands to the vitelline reservoir where they may be temporarily stored. Vitelline cells pass from the reservoir to the ootype and here release the globules of shell precursors, which coalesce to form the egg shell, which may be deposited on, and combined with, a thin membrane probably representing the capsule. The egg shell, which is unhardened at this stage, thus encloses both the ovum and the remains of the vitelline cells – the latter serving as yolk reserves (= 'true' yolk?) for the developing embryo. As the eggs pass along the uterus and out through the uterine pore, the protein of the egg shell becomes 'tanned' and hardened by the quinone formed during the process (Figs. 7.7 and 7.8) and the eggs may visibly darken in colour. Scanning electron microscope studies (*973*) show that in *Diphyllobothrium latum* the shell displays a surface polygonal pattern with fine furrows and shallow pits. X-ray micro-analysis of the shell material reveals high peaks of phosphorus and sulphur, with low levels of Mg^{2+}, Ca^{2+} and K^+.

Chemistry of quinone tanning in the pseudophyllid-type egg

The egg shell in a number of pseudophyllid-type eggs has been shown to consist of a highly resistant protein, *sclerotin*. This is essentially a 'tanned' protein and one which occurs widely in the animal kingdom (*810*). 'Tanning' is accomplished by the action of an *o*-quinone derived enzymically from an *o*-phenol in the presence of oxygen (Fig. 7.8). The quinone probably reacts with free NH_2 groups in protein chains to form strong covalent cross-links, which result in the formation of the highly stable protein sclerotin. The structure may, however, be more complex than this, for in trematodes such as *Fasciola* there is evidence that the egg shell may be additionally stabilised by di-tyrosine as well as disulphide bonds, as in keratin (Fig. 7.7) (*810*).

The shell precursors in the vitelline cells – proteins, phenols and phenol oxidase (EC 1.14.18.1, monophenol *o*-diphenol:oxygen oxidoreductase) – can all be stained specifically with cytochemical reagents although the reactions are not as intense as in trematodes (*810*). The most useful of these reagents are probably (*a*) Fast Red Salt B, which stains phenolic materials orange/purple, and (*b*) catechol, which can be used for detecting the phenol oxidase. Details of these techniques are given by Smyth (*789*).

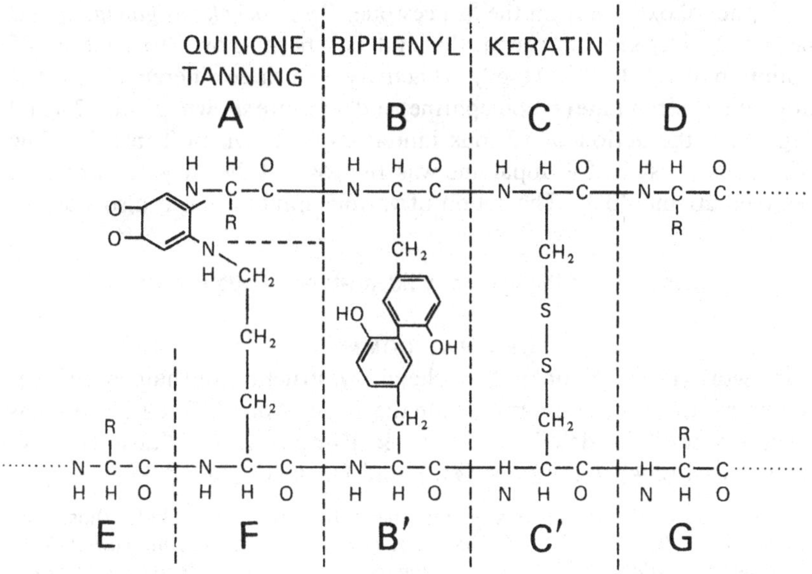

Fig. 7.7. Some of the covalent links which occur in structural proteins. ABCD represents one protein, EFB'C'G another. The N-terminal amino acid A is linked to an *o*-quinone which is also linked to a lysyl residue F(=quinone tanning); BB' represents two tyrosyl residues coupled by a biphenyl linkage; CC' represents two cysteinyl residues coupled by a cystine linkage (as in keratin); D, E and G are not cross-linked. (After Brunet, 1967.)

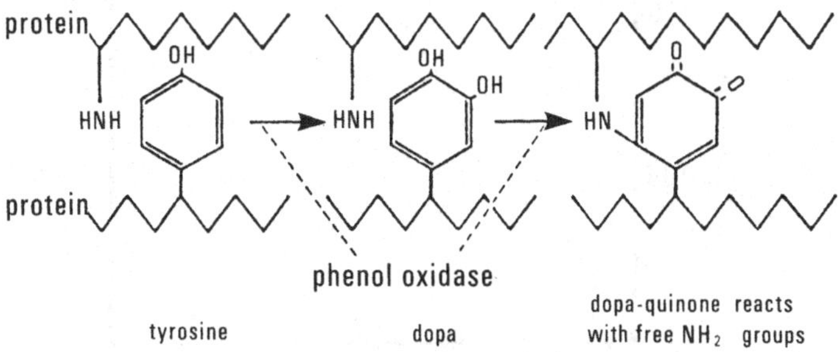

Fig. 7.8. Chemical reactions involved in quinone tanning. (After Smyth, 1969.)

173

The biology of the egg

A phenol oxidase from the fish cestode, *Penetrocephalus ganapatii*, has been characterised bio-chemically and has been shown to have a pH optimum of 7.4 (Fig. 7.9) (*362*). Its activity with three different substrates, dopamine, adrenaline (= epinephrine) and dopa are shown in Table 7.4 and Fig. 7.10; the action of various inhibitors is shown in Table 7.5. The Michaelis constant for dopamine was 189 μM and the enzyme was stable between 20 and 40°C. The action of various inhibitors was also studied.

Formation of the cyclophyllidean-type of egg (group II)

GENERAL COMMENTS

Early work (pre-1970) on the histochemistry, structure and hatching of eggs in this group has been summarised in the first edition (*796*). Later work has been reviewed in detail by Davis & Roberts (*170*), Fairweather & Threadgold (*204*), Lethbridge (*442*), and Ubelaker (*888, 889*).

Recent investigations, especially on the ultrastructure, include those on: *Anoplotaenia dasyuri* (*76*), *Avitellina lahorea* (*21, 209*), *Catenotaenia pusilla* (*834*), *Cittotaenia variabilis* (*143*), *Cotugnia digonopora* (*774*), *Dioecocestus acotylus* (*144*),

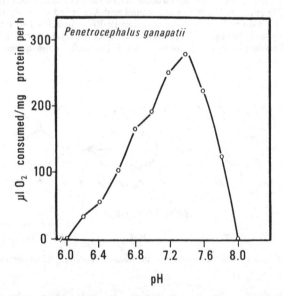

Fig. 7.9. The effect of pH on phenol oxidase activity in the pseudophyllidean cestode *Penetrocephalus ganapatii*, from the marine fish *Saurida* sp. The activity was measured in 10 mM sodium phosphate buffer. Optimum activity was reached at pH 7.4. (After Jayabaskaran & Ramalingam, 1985.)

Egg: formation and structure

Table 7.4. *Phenol oxidase in the soluble and particulate fractions from* Penetrocephalus ganapatii. *(Data from Jayabaskaran & Ramalingam, 1985)*

Substrate	Rate of oxidation (μl O_2/mg protein per h)	
	Soluble enzyme	Particulate enzyme
Epinephrine	33	352
Dopamine	26	300
Catechol	9	80
Dopa	12	80
Tyrosine	0	297[a]

Notes:
[a] The experiment was carried out in the absence of ascorbic acid-EDTA solution from the reaction mixtures. The activity was measured at 3–4 h of the reaction period, since the enzymes showed an initial lag period in the presence of the substrate tyrosine.

Echinococcus granulosus (837–839), E. multilocularis (725), Fimbriaria fasciolaris (380), Hymenolepis diminuta (331, 440, 442, 541, 559, 560, 640, 722), H. microstoma (134), H. nana (204), Moniezia expansa (21), Mesocestoides lineatus (151), Monoecocestus americanus (149), Oochoristica anolis (148, 150), Raillietina tetragona (35), R. echinobothridia (774), Stilesia globipunctata (21), Shipleya inermis (140–142), Taenia crassiceps (131), T. saginata (741), T. solium (428, 929).

Although the embryonic envelopes of all cyclophyllideans appear to be formed in a similar manner, this may not always be evident from the final form of the egg. This divergence undoubtedly reflects specific functional demands placed on cestode species by the habitats of eggs outside the host or by the ecology of their intermediate hosts, both of which may have important implications in the transmission strategies of various species (*148*).

In group II eggs, in contrast to those in group I, only a few vitelline cells (often one) interact with the oocyte (Fig. 7.4), with the result that, in some species, little true shell/capsule material is formed. In some species, e.g. *Avitellina lahorea*, the vitellaria are entirely lacking. This lack of protection by a shell is, however, compensated for in most cases by the presence of a

175

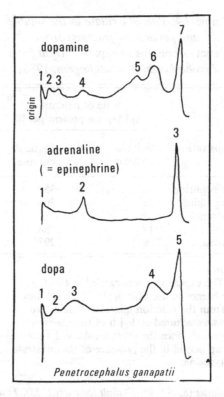

Fig. 7.10. Scanned zymogram pattern of phenol oxidase from the pseudophyllidean, *Penetrocephalus ganapatii*, with three different substrates. The gels were equilibrated in 10 mM sodium phosphate buffer, pH 7.4, stained for activity with 0.5 mM substrates in phosphate buffer for 8–10 h and scanned with an extinction recording densitometer. (After Jayabaskaran & Ramalingam, 1985.)

greatly thickened embryophore (Figs. 7.1(*a*),(*c*), 7.4 and 7.11). It is beyond the scope of this review to discuss all the variations in the form of the egg and its embryonic envelopes in this group and only the eggs of those species commonly used as experimental models are discussed in detail below.

SUBGROUP II.i: FORMATION OF THE *HYMENOLEPIS*-TYPE OF EGG

This type of egg is characteristic of the Cyclophyllidea, with the exception of the Taeniidae and the (subfamily) Thysanosominae.

The 'type' egg in this group was formerly that of *Dipylidium caninum*, whose egg was studied by Pence (*639, 640*). More recently, however, the eggs of several species of *Hymenolepis* have been examined in great detail

Egg: formation and structure

Table 7.5. *Effect of various inhibitors on the phenol oxidase of*
Penetrocephalus ganapatii. *(Data from Jayabaskaran & Ramalingam,*
1985*)*

Inhibitors	Concentration $\times 10^2$ (mM)	Phenol oxidase activity (μl O_2/mg protein per h)
Potassium cyanide	—	258
	0.1	204
	0.5	167
	1.0	142
	5.0	55
	10.0	45
Phenylthiourea	—	258
	0.5	211
	1.0	202
	5.0	186
	10.0	167
	50.0	135
Diethyldithiocarbamate	—	258
	5	238
	10	227
	50	153
	100	111

and the *Hymenolepis*-type egg is perhaps more appropriate as a basic
reference type. The structure is best known in *H. diminuta* (*331, 888, 889*),
H. nana (*204*) and *H. microstoma* (*134*). Because the eggs of these species are
so widely used as experimental models, their structure is discussed in some
detail below.

Eggs of *Hymenolepis diminuta* and *H. nana*
The structure, ultrastructure and formation of the hymenolepidid egg has
been reviewed in detail by Ubelaker (*888*). Its general morphology is shown
in Fig. 7.14). Although there are only the usual three basic embryonic
membranes (p. 179) in the developing egg – shell/capsule, outer envelope,
and inner envelope – the fully formed egg often appears to be more complex
due to further differentiation of these layers. The following structures can
be recognised (Fig. 7.11).

1. Shell/capsule
This forms the tough, outer layer of the egg. It consists largely of protein, the
precise composition of which is uncertain. The reported absence of phenol
or phenolase indicates that it is not the tanned protein sclerotin (*440, 640*),

177

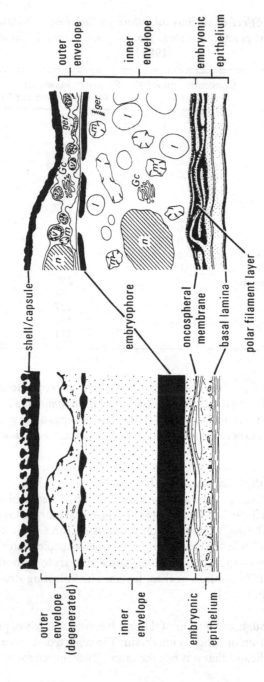

Fig. 7.11. Comparison of the fine structure of the membranes of the eggs of *Hymenolepis diminuta* and *H. nana*. Note: in *H. nana* (in contrast to *H. diminuta*): (a) the embryophore is thin and incomplete – a feature which may facilitate hatching in the gut of the definitive host; (b) there is an additional polar filament layer between the oncospheral membrane and the basal lamina. *Gc*, Golgi complex; *ger*, granular endoplasmic reticulum; *l*, lipid bodies; *m*, mitochondria; *n*, nucleus; *v*, vacuoles. (After Fairweather & Threadgold, 1981*a*; Holmes & Fairweather, 1982.)

178

as in the Pseudophyllidea (p. 172). Likewise, the evidence for -SS- or -SH-bonds, which would indicate the presence of 'keratin', is somewhat conflicting (*331, 439, 541*). The histochemical reactions of the egg shell are shown in Tables 7.3 and 7.6. Moreover, different 'isolates' of *H. diminuta* give different cytochemical reactions (Table 7.6) – a result in keeping with the differences found in the biochemistry (Chapter 5).

Three components appear to be involved in the formation of the egg shell: (*a*) a vitelline cell, (*b*) the embryonic capsule and (*c*) a uterine secretion. The mechanism appears to be as follows. A single vitelline cell, which together with the fertilised ovum, will form the 'egg' (Fig. 7.4), releases secretory granules which form a membrane-like layer, the embryonic capsule, which then encloses them. As embryonation proceeds, 'shell' material is produced by the outer envelope and, as it polymerises, accumulates on the inner side of the capsule to form the 'egg shell'. In *H. diminuta*, the original capsule normally becomes obscured but capsule material can be detected in the eggs of some other species (e.g. *Catenotaenia pusilla*). In addition to these two components, the shell may be built up further by secretions from the uterus. This process may be more important in *H. nana* where the uterus is more than just a simple sac enclosing the eggs, but is subdivided into partitions, which ramify between the oncospheres (*204*).

2. The subshell membrane

The presence of a subshell membrane has been noted by a number of workers (*439, 440, 541*), but it is often difficult to see and its embryonic origin is unknown. In *H. diminuta*, it is probably represented by the cytoplasmic layer (zone I; see below). This membrane appears to be relatively impermeable to many substances and is unaffected by proteases, carbohydrases or lipases. It appears to be a mucopolysaccharide–protein complex and may be important in preventing premature hatching, as well as providing 'back up' protection for the egg shell against a hostile external environment.

3. The inner envelope

This envelope is a syncytial layer involving two cells filling the space between the embryo and the egg shell; it is responsible for the formation of the embryophore and oncospheral membrane (*204*). After the latter is formed, what is left of the inner envelope becomes the cytoplasmic and gelatinous layers – zones I and II, respectively, in *H. diminuta* (Fig. 7.14). The latter has the ability to swell readily in dilute salt solutions, a property which may facilitate the escape of the oncosphere during hatching (p. 191) (*442*).

4. The embryophore

The proteinaceous embryophore is synthesised by the inner envelope and in *H. nana* this process is highlighted by the presence of granular endoplasmic

Table 7.6. Hymenolepis diminuta: histochemical reactions of the embryonic envelopes of the egg. (Data from Holmes & Fairweather, 1982)

Test	Indicates presence of	Variety[a]	Shell	Inner envelope			Penetration gland	Hooks
				Cytoplasmic layer	Gelatinous layer	Embryophore		
Coomassie brilliant blue	Protein	R	+[b]	+++	++	++	++	−
		I	++	+++	++	+++	+++	−
Mercuric bromophenol blue (alcoholic)	Protein	R	+++	+++	++	++++	+++	−
		I	+++	+++	++	++++	+++	+
Alkaline fast green	Basic protein	R	+++	+++	++	++++	++	+
		I	+++	+++	++	++++	++	++
Paraldehyde-fuchsin	Disulphide and sulphydryl groups	R	−	−	−	+++	+++	++
		I	−	−	−	+++	++	++
DMAB method	Tryptophan	R	+	−	+	−	++	++
		I	+	+	+	−	++	++
Sakaguchi reaction	Arginine	R	−	−	−	−	−	−
		I	−	−	−	−	+	−
Millon reaction	Tyrosine	R	++	−	−	−	++	−
		I	++	++	+	+	++	−
PAS	Carbohydrate	R	−	++	−	−	−	−
		I	−	++	−	−	−	−
PAS + amylase control	Carbohydrate	R	−	+++	−	−	+	−
		I	−	+++	−	−	+	−
Alcian blue	Acid mucopolysaccharides	R	−	+++	−	−	+	−
		I	+	+++	−	−	+++	−
Sudan black	Lipid	R	−	+++	−	−	++	−
		I	−	+++	−	−	++	−

180

Toluidine blue								
α-Metachromasia	R	+	−	−	−	−	−
	I	+ + +	−	−	−	−	−
β-Metachromasia	R	−	+ + +	+ + +	+ + +	+ + +	−
	I	−	+ +	+ +	+ +	+ +	−
γ-Metachromasia	R	−	−	−	−	+ +	−
	I	−	−	−	−	+ +	−
Azure A								
α-Metachromasia	R	−	−	−	−	−	−
	I	+ +	+ +	+ +	+ + +	+ + +	−
β-Metachromasia	R	−	+ +	+ +	+ + +	+ + +	−
	I	−	−	−	−	−	−
γ-Metachromasia	R	−	−	−	−	+ +	−
	I	−	−	−	−	−	−

Notes:
[a] R, Rice Institute variety; I, Irish variety.
[b] −, negative reaction; +, weak positive reaction; + +, moderate positive reaction; + + +, strong positive reaction.

181

reticulum (Fig. 7.11) and abundant free ribosomes and polysomes. The chemical composition of the embryophore in *Hymenolepis* is still undetermined. In the taeniid cestodes (see below) the embryophore gives strong histochemical and biochemical reactions for keratin but these are much weaker in *Hymenolepis*, although histochemical reactions of -SS-bonds are fairly intense (*440*). Nevertheless, the fact that it can be weakened by organic solvents attacking electrovalent linkages, and is susceptible to digestion by proteolytic enzymes, indicates that the structure is not keratinised (*442*).

5. The oncospheral membrane

This structure (Fig. 7.14) has received much attention due to the role it plays in the hatching process (see below). It is not a typical unit membrane but resembles a membranous lamina and consists of a layer of regularly arranged granules bounded on both sides by a number of lamina (*442*). Its chemical composition has not been determined but, in taeniid cestodes, there is some histochemical evidence that it may be a lipoprotein (*442*). It is apparently formed by the delamination of the inner part of the inner envelope, detaching from it as a thin, separate layer.

6. The polar filament layer (*H. nana*)

This layer appears to be unique to *H. nana*. It forms a layer between the oncospheral membrane and the oncosphere (Fig. 7.11). It is apparently formed by the delamination of the epithelial covering of the oncosphere into two layers, separated from each other by membranes – the outer polar filament layer and the inner embryonic epithelium. It has been suggested that the polar filaments in *H. nana* are reminiscent of the tendrils of the egg cases of elasmobranchs and 'may serve to delay expulsion of the oncosphere from the mammalian intestine by becoming entangled amongst the intestinal villi or mucous lining of the gut. This may further serve to bring the oncosphere into close contact with the gut wall for successful penetration to take place' (*204*).

SUBGROUP II.ii: FORMATION OF THE *TAENIA*-TYPE EGG

This type of egg is formed essentially in the same way as that of the previous group, only one vitelline cell becoming associated with the fertilised ovum. The embryonic capsule is very thin and almost invisible when laid; it is normally lost in faecal eggs. As pointed out earlier, lack of protection by an egg shell is compensated for by the development of a thick embryophore made up of keratin blocks (Fig. 7.1(*c*)) held together by a cementing substance (*442*). This gives the egg its characteristically radially striated appearance.

Egg: formation and structure

Variation in egg structure

The above division of the eggs into different groups is an oversimplification of a more complex situation, for in many species the egg structure is further modified in relation to its life cycle. In *Mesocestoides* spp., all the eggs in a proglottis are enclosed in a thickened sac, the *paruterine organ* (*151*), which provides additional protection (Fig. 10.11, p. 279). In the linstowiid cestode *Oochoristica anolis*, the eggs shed in gravid proglottides are enclosed in 'uterine capsules', which are formed from the uterus and which break down shortly after their formation, although the individual cells do not apparently break down. The uterine capsule is therefore, essentially epithelial in nature (*150*). The uterus has also been implicated in the formation of the egg shell/capsule in *Dipylidium* (*451*) and *Cittotaenia* (*143*).

One of the structures which shows the greatest degree of variation is the embryophore, especially in cyclophyllideans, where it has been most widely studied. Even in species of the same genus, the form of the embryophore can vary greatly. In *H. nana*, after the embryophore is formed within the inner envelope (Fig. 7.11), it does not grow inwards but remains a thin, discontinuous, peripheral layer of the inner envelope. In *H. diminuta*, in contrast, it increases in thickness and moves towards the inner region of the inner envelope (Fig. 7.11) (*204*).

Egg production

As Kennedy (*389*) points out, for cestodes a terrestrial habitat is more demanding than an aquatic one, since in the former the mortality rates are higher, transmission losses are higher and probabilities of infection lower. This situation is reflected in the level of egg production. The pseudophyllidean *Triaenophorus nodulosus* (with a three-host cycle) is patent for only one to two months and a single adult produces about 1.75×10^6 eggs during its life span. In contrast, the cyclophyllidean *H. diminuta*, with only two hosts, is probably patent for at least 18 months and a single adult produces 2.5×10^5 eggs *per day* (*391*) – i.e. it produces nearly as many eggs in a week as a single *T. nodulosus* does throughout its entire life span.

Embryonation and viability

PSEUDOPHYLLIDEAN EGGS

Viability
These eggs normally require water and probably oxygen for embryonation as desiccation rapidly kills them. The eggs of many species can survive and

The biology of the egg

embryonate after storage (in water in a refrigerator) at 2–4°C, e.g. *Spirometra mansonoides, Diphyllobothrium* spp. and *Triaenophorus lucii* (*796*), but eggs of *Schistocephalus solidus* do not embryonate after even a 'short' period at this temperature (*796*). On the other hand, the eggs of *Bothriocephalus gowkongensis*, have been reported as being viable after freezing at −4°C for 72 h (*981*).

Embryonation

As can be expected, most physico-chemical factors which are known to affect biological reactions generally will influence embryonation. The effect of temperature has been most studied, but the responses vary greatly with species. For example, eggs of *D. latum* require 8–9 days to embryonate at 18–20°C, but at the same temperature, eggs of *T. lucii* require only 5 days (*284*). Some other embryonation times at about the same temperatures are: *D. dendriticum*, 6–8 days; *D. cordatum*, 9 days; *Bothriocephalus opsariich-thydis*, 3–4 days (*852*); *B. acheilognathi*, 2.5–18 days (*266*, *655*); *Polyonchobothrium ophiocephalina* (*852*), 3 days. The temperature of embryonation and at hatching can affect the survival time of coracidia. Thus, in *B. acheilognathi* (Fig. 7.12) at 20°C the peak coracidium motility of about 35% occurred at 4 days, but at 30°C coracidium motility peaked at

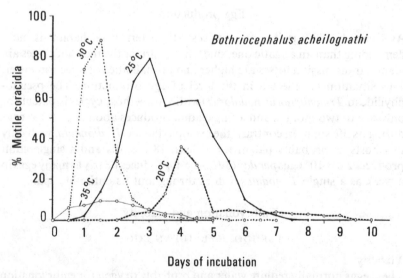

Fig. 7.12. Percentage of motile coracidia of the pseudophyllidean *Bothriocephalus acheilognathi*, from the freshwater grass carp, following incubation of the eggs at different temperatures. (Modified from Granath & Esch, 1983.)

Egg: formation and structure

Table 7.7. Taenia saginata: *survival of eggs in the field. (Data from Bürger, 1984)*

Country	Medium	Season of contamination	Survival (days)	Infectivity proven
Denmark	Grass, proglottides	Summer	58	Yes
Denmark	Grass, proglottides	Winter	159	Yes
Great Britain	Pasture	Autumn	30	Yes
Germany	Pasture and meadows	Spring	160	Not tested
Germany	Pasture	Spring	108	Yes
Australia	Pasture		100	Yes
Kenya	Open pasture	February	364	Yes
Soviet Union	Pasture	Spring	50–200	?
Soviet Union	Pasture	Spring	77–93	Not tested
Soviet Union (Moscow)	Pasture	Summer	35–45	Not tested
Soviet Union (Moscow)	Pasture	Autumn	80–97	Not tested
Soviet Union (Moscow)	Pasture	Winter	128–142	Not tested
Soviet Union (Kubinsky)	Soil, shade	Spring	49–90	Not tested
Soviet Union (Kubinsky)	Soil, shade	Summer	9–15	Not tested
Soviet Union (Kubinsky)	Soil, shade	Autumn	53–64	Not tested
Soviet Union (Kubinsky)	Soil, shade	Winter	100–229	Not tested
Soviet Union (Kubinsky)	Soil, sun	Spring	27–68	Not tested
Soviet Union (Kubinsky)	Soil, sun	Summer	4–8	Not tested
Soviet Union (Kubinsky)	Soil, sun	Autumn	20–38	Not tested
Soviet Union (Kubinsky)	Soil, sun	Winter	73–207	Not tested

90% after 1.5 days. At 35°C, although embryonation was the most rapid, coracidium motility did not exceed 10% (*266*). The effect of other factors, such as pO_2, pCO_2, pH, and Eh on hatching or mobility do not appear to have been examined and could provide valuable information.

CYCLOPHYLLIDEAN EGGS

These are normally embryonated when laid so that embryonation is not affected by external environmental factors. On account of their important role in the epidemiology of cestodiasis, the effect of environmental factors has been much studied. In particular, the effect of various physico-chemical factors on the viability of eggs of *Echinococcus* spp. and *Taenia* spp. in soil, on pastures or in sewage has been extensively reviewed (*82, 106, 252, 253, 638, 825, 963, 968, 969*). Results of some of these studies are summarised in Tables 7.7 and 7.8).

The biology of the egg

Table 7.8. Taenia saginata: *survival of eggs in sewage and sludge. (Data from Bürger, 1984)*

Medium	Conditions	Survival (days)	Infectivity remaining
Sewage	Laboratory, 18°C	16	Yes
Sewage	Exp. plant, trickling filter	42	No
Sewage	Septic tank	40	Not tested
Sewage	Plant, raw waste water	Yes	Yes
Sludge	Lab, anaerobic digestion, 25–29°C	200	Not tested
Sludge	Lab, anaerobic digestion, 35°C	< 5	No
Sludge	Plant, anaerobic digestion, 26–28°C	56	Not tested
Sludge	Plant, anaerobic digestion, cold	38	Not tested
Sludge	Exp. plant, activated sludge	42	Not tested
Effluent	Plant activated sludge	Yes	Yes
Effluent	Plant trickling filter + lagoon	No?	No?
Effluent	Plant lagoons	No	No

It is, however, very difficult to give meaningful figures for the survival of eggs at a given temperature unless other relevant factors, such as situation (in soil, in sludge, etc.,), exposure to sun, relativity humidity etc., are known. There is no doubt, however, that eggs of some species can withstand great extremes of temperature or chemical treatment. The eggs of *Echinococcus multilocularis* were found to be infective after 54 days at 26°C and after 24 h at 51°C (*796*). Eggs of *E. granulosus* buried at a depth of 90 mm have survived for over a year (*779*).

In Kenya (Table 7.7) eggs of *T. saginata* have been shown to survive in soil for a year (*106*). Modern sewage treatment appears to be relatively ineffective in destroying cestode eggs and those of *T. saginata* have been shown to survive at almost all stages of sewage treatment (Table 7.8).

As well as being distributed by the natural elements, eggs of many species, e.g. *T. saginata*, are dispersed by insects or earthworms both externally (by adherence) and internally by being ingested and passed in the faeces; the insect or worms, may, of course, also be ingested (*256, 257, 449, 450*). The egg size is an important factor limiting ingestion of eggs; eggs larger than 0.05 mm cannot be ingested.

Ovicides
The wide distribution of eggs of cestodes of all classes in soil, sewage or water makes the application of ovicides as a method of control generally impractical. Various aspects of the effects of ovicides and/or fertilisers on

eggs have been reviewed by Gurbanov & Simonov (*270*), Nakajima & Egusa (*586*), Owen & Stringer (*614*), and Prokopič & Jelenová (*659*). Ovicides for pseudophyllidean eggs have been little studied, an exception being the eggs of *Bothriocephalus gowkongensis*, a serious parasite of cyprinid fish, whose eggs are killed by the compound Pollena Jod K, at a concentration of 0.75 parts per 1000 (*981*).

Much more attention has been paid to the search for ovicides against cyclophyllidean eggs, but the presence of a thick, keratin embryophore (Fig. 7.1(*c*)) in taeniid eggs of zoonotic importance means that such eggs are extremely resistant to chemical attack. For, example, eggs of *E. granulosus* and *T. pisiformis* have been shown to be viable after two weeks in 40% (v/v) formaldehyde (*796*). Potential ovicides can affect eggs by either (*a*) penetrating the embryophore and acting directly on the oncosphere or (*b*) affecting the embryophore to prevent hatching (*614*). Viability of eggs after chemical treatment can be tested by disrupting the embryophore with sodium hypochlorite or enzymes (see Hatching, p. 193) and testing the viability of released oncospheres with tetrazolium salts (Table 7.9). Taeniid species whose eggs have been most commonly used for laboratory experiments include, *T. saginata*, *T. hydatigena*, *T. pisiformis*, and *Multiceps multiceps*. Compounds for which some ovicidal effects have been reported include Dichlorophen (*614*) and Carbathion (*270, 779*).

Hatching

The hatching of a cestode egg provides an interesting model for certain kinds of biological reactions – especially responses to physico-chemical factors such as light, temperature, pO_2, pCO_2 and enzymes. It is clearly of survival value for a species (*a*) if hatching is inhibited by conditions within the definitive host so that it does not occur prematurely (but see *H. nana*, p. 191) and (*b*) if hatching occurs externally, i.e. in water, it is stimulated to do so by ecological factors which are also favourable to the intermediate host.

Pseudophyllidean eggs

INFLUENCE OF LIGHT

In trematodes (*810*), the opening of the operculum is believed to occur as the result of a light-released enzyme attacking the opercular seal. It is not known if a similar mechanism operates in those pseudophyllidean cestodes whose eggs require light for hatching. Not all eggs require light, however, and other factors must operate.

Effect of light intensity
This has been examined experimentally in *Diphyllobothrium latum*.
Hatching occurred within 1 min after exposure to white light of an intensity
of 1000 lx and almost all eggs which were capable of hatching were hatched
within 5–6 min. The minimum light intensity which caused maximum
hatching lay between 50 and 100 lx, acting for 30–60 s. The minimum
amount of energy for causing maximal hatching has been calculated to be
10^{-9} J/s per egg.

Effect of wavelength
Experiments, to determine whether or not a particular wavelength stimulates maximum hatching of the eggs of *D. latum* (*263*) have shown that there
appear to be two peak values of hatching at about 300 nm (ultraviolet) and
600 nm (yellow) (Fig. 7.13). Infrared rays failed to initiate hatching, and the
red region of light, 693–704 nm, showed only a weak stimulating effect.
Peaks of hatching in the yellow and ultraviolet bands could possibly be
related to light penetration in shallow and deep water. Studies have shown

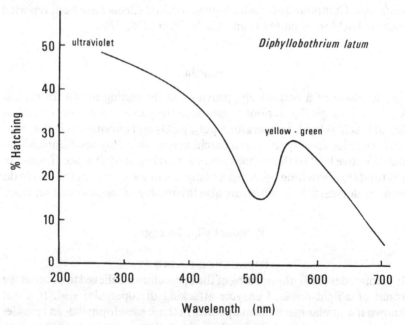

Fig. 7.13. Influence of wavelength on the hatching of the embryonated eggs of
Diphyllobothrium latum. (After Grabiec *et al.*, 1963.)

Hatching

that yellow and green rays (500–600 nm) penetrate farthest into deep water but ultraviolet light, although penetrating poorly into deep water, carries the greatest energy into shallow water.

Ecological implications of hatching mechanism

Since the eggs of *D. latum* tend to be deposited mostly in shallow water near the shores of closed water bodies, it can be assumed that they would normally be exposed to the peak wavelengths shown above. The hatching phenomenon may thus be regarded as an adaptation probably related to the ecology of the copepod intermediate hosts. It can be speculated that under the influence of sunlight, a large number of eggs may hatch at once, and the same light conditions could be associated with a similar release of copepod larvae, or a phototropic migration of larvae towards the egg-hatching area. In contrast, eggs of some species, e.g. *Bothriocephalus* spp. (*585*), *D. dalliae*, *D. ursi* and *Triaenophorus nodulosus* (*796*) are capable of hatching in the dark. This hatching behaviour also undoubtedly has survival value, for eggs of these species may occur in deep lakes or muddy pools.

Cyclophyllidean eggs

Hatching of the cyclophyllidean egg involves two processes: (*a*) the rupture of the egg shell (when present) and the passive digestion and/or disruption of the embryophore, (*b*) 'activation' of the hexacanth embryo to become motile and rupture its enclosing oncospheral membrane. These processes frequently require different stimuli for their initiation. The hatching process has been reviewed by Lethbridge (*442*), and Ubelaker (*888*).

HATCHING IN CYCLOPHYLLIDEA OTHER THAN TAENIIDAE

Hatching of the eggs of many species in this group have been reported to occur (sometimes at very low percentages) in very simple solutions, such as distilled water, saline or solutions containing various ions, especially Na^+ and Cl^-, e.g. *Dipylidium, Cittotaenia, Davainea, Raillietina,* and *Hymenolepis* (*796*). In invertebrate hosts, breaking of the egg shell is probably brought about largely by the mechanical action of the mouthparts, a process which exposes the remaining embryonic membranes to the host's digestive juices. *In vitro*, this process can be simulated by agitating the eggs with glass beads.

The hatching process has been most thoroughly studied in the Hymenolepididae and has been described for *H. diminuta* (Fig. 7.14) (*331, 439, 441, 906*) *H. microstoma, H. citelli,* and *H. nana* (*204, 796*). In the natural host, the factors which stimulate oncosphere activity are not clear.

189

Valuable methods for the *in vitro* hatching and activation of the eggs of various *Hymenolepis* species are available and serve as useful experimental models for laboratory studies for students (*74, 198, 204, 440, 474, 749*).

The hatching process as it occurs in *H. diminuta* (*331*) may be summarised as follows (Fig. 7.14). As indicated above, hatching essentially

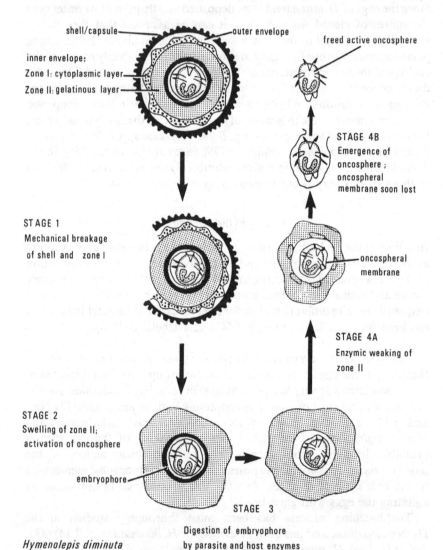

shell/capsule

outer envelope

freed active oncosphere

inner envelope:
Zone I: cytoplasmic layer
Zone II: gelatinous layer

STAGE 4B
Emergence of oncosphere ; oncospheral membrane soon lost

STAGE 1
Mechanical breakage
of shell and   zone I

oncospheral membrane

STAGE 4A
Enzymic weaking of zone II

STAGE 2
Swelling of zone II;
activation of oncosphere

embryophore

STAGE 3

Hymenolepis diminuta

Digestion of embryophore
by parasite and host enzymes

Fig. 7.14. *Hymenolepis diminuta*: stages involved in the hatching of the egg. (Modified from Holmes & Fairweather, 1982.)

Hatching

involves rupture of the egg shell by the insect's mouthparts, digestion of the remaining egg-layers in the mid-gut lumen, and subsequent escape of the activated oncosphere. Mandibular damage is usually confined to a small area and results in an irregular hole in the shell. Simultaneously, the exposed surface of the cytoplasmic layer is also ruptured. The latter layer eventually expels the enclosed embryo, now partly activated, as evidenced by the 'swimming' hook movement, but still enclosed in its remaining layers. The second stage in hatching involves the destruction of the gelatinous layer by the insect's enzymes, followed by a change in the refractivity and elasticity of the embryophore. The latter becomes distorted by the out-pushing of the hooks, eventually ruptures and the oncosphere escapes, still enclosed in the oncospheral membrane, which is soon lost.

It is important to note that hatching in *H. nana* is somewhat different from that of *H. diminuta* (*204*) and this may be related to the fact that hatching not only takes place in the insect intermediate host but also in the intestine of the definitive rodent host. It has been pointed out earlier (p. 183) that the embryophore in *H. nana* (in contrast to that of *H. diminuta*) is thin and discontinuous (Fig. 7.11) and this may make it more readily vulnerable to the host's enzyme action and hence facilitates intestinal hatching. The presence of polar filaments in this species may serve to delay expulsion of the hatched oncosphere from the rodent intestine by becoming entangled in the intestinal mucus and villi. This process would also serve to bring the oncosphere into close contact with the gut wall and hence facilitate successful penetration (*204*).

HATCHING IN THE TAENIIDAE

The main difference between the taeniid egg and that dealt with above is the presence of a greatly thickened embryophore in the former (Fig. 7.1).

The most detailed hatching studies have been carried out on *E. granulosus* (*251, 314*), *T. saginata* (*614, 796, 824*), *T. ovis* (*251, 314*), *T. pisiformis* (*314, 796*), *T. solium* (*727, 796*), *T. hydatigena* (*251, 614*), *T. multiceps* (*251*), *T. serialis* (*314*) and *T. taeniaeformis* (*796*).

The first stage in hatching is the disintegration of the keratin blocks (Fig. 7.1(c)) comprising the embryophore. It has been thought that disintegration of the embryophore is the result of the enzymic dissolution of a 'cementing' substance holding these blocks together. TEM studies (*796*) show a clear space between the embryophore blocks – a result difficult to explain if a cement substance is present. On the other hand, some workers have described the presence of a highly refractive yellow-brown cement substance. Nevertheless, the fact that hatching is brought about by digestive enzymes suggests that some cementing material is dissolved.

191

The biology of the egg

Table 7.9. *Hatching: eggs of* Taenia saginata. *Artifical hatching solutions and technique. (Based on Owen, 1985, Owen & Stringer, 1984; Smyth, 1969; Stevenson, 1983)*

Storage. Gravid proglottides should be stored in normal saline containing antibiotics, penicillin G (100 i.u./ml), streptomycin sulphate (100 μg/ml) and myostatin (100 μg/ml), as soon as possible after being passed.

Solution A. Artificial gastric fluid (AGF)

pepsin	1.5g
HCl (conc.)	1.0 ml
0.85% NaCl[a]	100 ml

Step 1: Transfer washed eggs into 5 ml of solution A in a capped 15 ml centrifuge tube in a water bath for 1 h at 38°C. Shake vigorously for 10 min. Sediment by centrifugation and remove all but 0.5 ml of the supernatant fluid.

Solution B. Artificial intestinal fluid (AIF)

trypsin (Sigma, Type II; 10 000–13 000 BAEE units/mg)

	0.05 g
$NaHCO_3$	1.18 g
0.85% NaCl[a]	100 ml

[1 ml of whole ox bile (fresh or deep frozen) should be added to 2 ml of AIF, just before use.]

Step 2: After step 1, add 2 ml of solution B to the tube of eggs; replace the cap immediately and shake vigorously. Replace in water bath and shake vigorously every 10 min. Wash hatched oncospheres three times in saline before use. Oncosphere viability can be tested for using 0.25% (w/v) MTT (tetrazolium salt) in glucose-saline (*613*).

Notes:

[a] Throughout these procedures, Hanks, or other isotonic saline, can be used in most cases instead of 0.85% (w/v) NaCl. Different species may, however, differ regarding their ionic or pCO_2 requirements.

The experimental conditions required to bring about hatching in different species have been shown to vary remarkably. In some, pancreatin alone can bring about hatching, e.g. *T. pisiformis* (*796*), but many species require treatment with both artificial gastric fluid (AGF) (Table 7.9) i.e. pepsin at pH 2.0, and artificial intestinal fluid (AIF) containing trypsin, bile and $NaHCO_3$. A variety of different protocols have been used by different workers, all varying slightly in the details of concentrations of reagents, type of vessel, gas phase, etc. (*247, 314, 430, 614, 727, 824*). Most successful techniques have involved the use of tightly capped containers, such as screw-topped tubes, combined with vigorous shaking, at intervals. The latter action may be related to maintaining an appropriate pCO_2 level, which is well known to play a role in the hatching of metacercarial cysts or nematode eggs (*427, 800, 810*). The eggs of some species hatch very readily

192

whereas others (e.g. *T. saginata*) require more precise treatment – a result which may explain the narrow host specificity of this species. A modified procedure for the hatching of the eggs of *T. saginata*, which may be applicable to other *Taenia* spp., is given in Table 7.9.

Sodium hypochlorite can also be used to disintegrate the embryophore and provides a simple, rapid technique for obtaining oncospheres (*430*, *614*). However, after this procedure, additional treatment with AIF (Table 7.9) is needed to free the oncospheres from their oncospheral membranes (*824*). Taeniid eggs have also been hatched by using sodium sulphide to disintegrate the embryophore (*75*).

The hypochlorite technique is as follows: place washed ova in a 15 ml centrifuge tube in 1 ml of 0.85% NaCl. Add 1 ml NaOCl (5%, w/v) and cap tube tightly. Shake gently for 3 min and stop the reaction by flooding the tube with saline. Rinse at least three times in saline and treat with AIF (Table 7.9) as above.

Before becoming infective, a released oncosphere must further become 'activated' – a phenomenon about which very little is known. When hatched artificially, an oncosphere is usually still wrapped in its oncospheral membrane (Figs. 7.1 and 7.14); the latter is often very closely applied to the oncosphere and difficult to observe except under phase microscopy. Sometimes, only a very small number of free oncospheres show any activity. It is only when bile is added to the hatching medium that an appreciable proportion of embryos show movement. The physiological basis of this onset of activity is not known, but it may be related to the surface-active properties of bile. Related to this activity is an increase in the permeability of the oncospheral membrane in the presence of bile and pancreatin. This effect may be demonstrated by the use of dyes, such as Nile blue sulphate, neutral red, methylene blue, alizarin, or Janus B green, which stain the hexacanth embryo after, but not before, activation (*796*). The fact that the oncospheral membrane stains in Nile blue sulphate, but not in other stains, points to its being a lipoidal membrane – a conclusion supported by electron microscopical studies. Diazo salts, such as MTT (0.25% in glucose saline) are especially useful for distinguishing living oncospheres from dead ones (*613*). In view of the marked variation in the composition of bile in different species of vertebrates (p. 49; Table 3.5), it is likely that bile may play an important part in determining intermediate host specificity by selectively activating some species of hatched oncospheres but not others.

Metabolism

No doubt it is the technical difficulties involved that have caused the metabolism of cestode eggs to remain almost unstudied. Since cyclophyllidean eggs develop within a proglottis while still within the host, direct examination of the metabolism is virtually impossible. Pseudophyllidean eggs, which develop in water, are more suitable for study. Limited work on the respiration and/or histochemistry of pseudophyllid (*D. latum*) and cyclophyllid (*H. diminuta*) eggs suggest an essentially aerobic metabolism (Fig. 8.7, p. 205) (*39*, *796*). Nevertheless, an anerobic component must be present, at least in some species, e.g. *Taenia saginata*, the eggs of which are able to survive prolonged periods in very anaerobic sewage treatment systems (Table 7.8). The metabolism of the hatched coracidia of pseudophyllideans, surprisingly shown to be anaerobic, is discussed in Chapter 8.

8

Developmental biology of larvae

General account

Problems of special physiological interest in larval development include those related to: (*a*) entry into intermediate hosts, (*b*) migration to the site(s) of development within these hosts, (*c*) the physiological and biochemical changes taking place during the passage from the external environment into the intermediate host, or – in the case of more than one intermediate host – those taking place on transference from one intermediate host to another. These problems are clearly interwoven with the ecological relationships between definitive and intermediate host(s). For example, if the first intermediate host of a pseudophyllid cestode is a freshwater copepod, it can be assumed that: (*a*) the free-swimming coracidium will be osmotically adapted to fresh water and low temperatures; (*b*) in the copepod host, it will still be adapted to freshwater temperatures but now osmotically adapted to the copepod haemocoele; (*c*) in the second intermediate host (normally a fish), the osmolarity adaptation will have changed again (but not the temperature adaptation); (*d*) if the defintive host is a bird or mammal, further adaptations in the the osmolarity and temperature tolerance will occur. Associated with these physiological changes, of course, will be the related switching on and off of the appropriate enzymes adapted to the internal environmental conditions in the host concerned and, probably, major respiratory changes.

It is not intended here to give an account of the various forms of cestode larvae or to discuss their morphology or ultrastructure in detail, except where it is especially relevant to the physiology of the stage concerned; various aspects of larval biology have been reviewed by Freeman (*234*), Kuperman & Davydov (*421*) (Table 8.1), Šlais (*778*), Ubelaker (*889*) and Voge (*903*).

Data on the development of larval forms relate largely to the Pseudo-phyllidea and Cyclophyllidea, on which most work has been carried out, with scattered observations on other groups.

Developmental biology of larvae

Table 8.1. *Fine structure of glands in larval pseudophyllidean cestodes: species investigated by Kuperman & Davydov (1982a). For glands in adult cestodes, see Table 2.2*

Cestode spp.	Stage of development	Host spp.
Eubothrium rugosum	Procercoid	*Cyclops vicinus*
	Plerocercoid	*Acerina cernua*
E. salvelini	Plerocercoid	*Gasterosteus aculeatus*
Bothriocephalus gowkongensis[a]	Coracidium	
	Procercoid	*C. vicinus*
Diphyllobothrium latum	Coracidium	
	Procercoid	*C. strenuus*
	Plerocercoid	*Esox lucius*
D. dendriticum	Plerocercoid	*Osmerus eperlanus*
Diphyllobothrium spp.	Plerocercoid	*Oncorhynchus nerka*
Spirometra erinacei	Plerocercoid	*Rana ridibunda*
Schistocephalus solidus	Plerocercoid	*G. aculeatus*
Ligula intestinalis	Plerocercoid	*Abramis brama*
Triaenophorus nodulosus	Procercoid	*C. vicinus*
	Plerocercoid	*Perca fluviatilis*
T. crassus	Plerocercoid	*Coregonus albula*

Notes:
[a] Synonym of *B. acheilognathi* (Pool & Chubb, 1985).

Pseudophyllidea

First larval stage: coracidium

CORACIDIUM: GENERAL BIOLOGY

Although the biology of many pseudophyllid species has been studied in detail, most studies have dealt with the structure, ultrastructure (Figs. 8.1 and 8.5), histochemistry and population dynamics of the various stages in the intermediate and definitive hosts. However, presumably on account of its small size and brief life span, the physiology of the coracidium, either as a free-swimming larva or during the process of copepod, has been little investigated.

Some relevant studies on the coracidium are those on: *Bothriocephalus acheilognathi* (*654, 655*), *B. claviceps*, *B. gowkongensis* (*= acheilognathi**) (*189, 421*), *B. opsariichthydis* (*= acheilognathi**) (*852*), *B. rarus* (*360*); *Diphyllobothrium latum* (*268–264, 285, 421, 796*); *D. dendriticum*, *D. ditremum* (*235, 265, 971*); *Ligula intestinalis* (*285*); *Proteocephalus exiguus*, *P. percae* (*971*), *P. filicollis* (*850*); *Schistocephalus pungitii* (*850, 971*); *Spirometra mansonoides* (*461*); *Triaenophorus crassus* (*711*); *T. nodulosus* (*262, 285, 849*).

*Synonyms according to Pool & Chubb (*656*).

196

Pseudophyllidea

Most of the work on the invasion of the first intermediate host deals with the susceptibility of different species of copepods to infection by coracidia. The evidence points to the conclusion that procercoid specificity is generally euryxenous (i.e. with a broad host spectrum). A striking example is seen with *D. latum* in which no less than 17 species of copepods are susceptible (*89*). Some species, however, show a tendency to be stenoxenous (i.e. with a narrow host spectrum), e.g. *B. claviceps*, with only two copepod species being highly susceptible (*189*).

Useful practical techniques for embryonation and hatching of pseudophyllidean eggs and infections of copepods by coracida have been developed by a number of workers (*189, 573*). Coracidia appear to swim in random directions, like gas molecules, spiralling slightly around the central direction of motion, although some species may be phototactic, which results in coracidia rising to the surface. Copepods do not appear to pursue coracidia actively, even when close to them; nor does visual recognition appear to be involved. Apparently, a coracidium must 'collide' with a

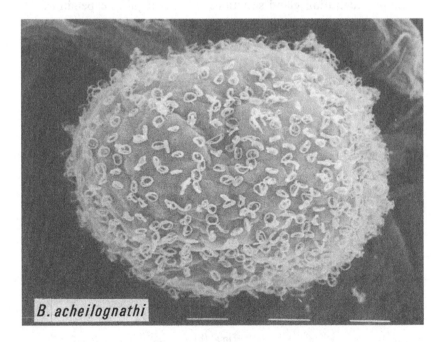

Fig. 8.1. Scanning electron micrograph of a coracidium of *Bothriocephalus acheilognathi* 5 h after hatching. The coracidium has swollen as a result of water absorption and the cilia are coiled and non-functional. The surface is covered by numerous protruberances of unknown function. Scale bar = 4.9 μm. (After Pool, 1984.)

197

copepod in order to be detected as food and eaten. Energy for cilia activity appears to be provided by phospholipid and polysaccharide granules in the embryophore and, surprisingly, the metabolism appears to be anaerobic (p. 205).

CORACIDIUM: SURVIVAL AND INFECTIVITY

The abiotic and biotic factors affecting the survival (= 'longevity') and infectivity of coracidia have been examined in only a few species. Activity and survival is temperature dependent (Figs. 7.12 (p. 184), 8.2 and 8.3); data for some species are given in Table 8.2. This result is predictable as – assuming that coracidia are incapable of absorbing external nutrients at this stage – the energy reserves in the embryophore would be more rapidly utilised at higher temperatures. In *B. acheilognathi*, the time (after hatching) when coracidia remain infective is also temperature dependent (Fig. 8.3). This result is likely to be related to the availability of energy reserves which will be more rapidly depleted at higher temperatures. Infectivity is also likely to be related to the availability of these reserves, as, presumably, release of penetration gland secretions (p. 202) requires expenditure of energy.

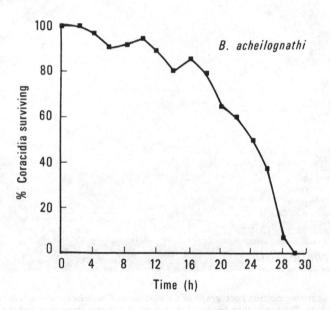

Fig. 8.2. *Bothriocephalus acheilognathi*: survival of coracidia at 18 °C. (After Pool, 1985.)

Pseudophyllidea

Table 8.2. *Survival of coracidia of pseudophyllidean cestodes at different temperatures*

Species	Temperature (°C)	Survival time (h)	References
Schistocephalus solidus	5–8	96–120	184
	16–18	48	
	22–25	<24	
Ligula intestinalis	24–28	24–36	184
	35	<4	
Triaenophorus nodulosus	5–7	240	418
	18–20	24–72	
	29	1	
Bothriocephalus acheilognathi[a]	10	38	655
	14	31	
	18	30	
	25	15	
	30	11	

Note: [a] See Fig. 8.3.

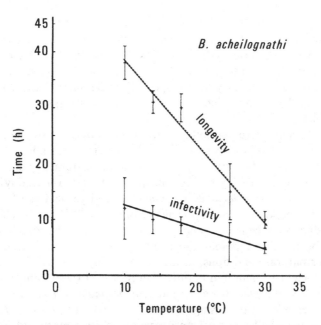

Fig. 8.3. *Bothriocephalus acheilognathi*: effect of temperature on infectivity (i.e. the period a coracidium remains infective) and longevity. (After Pool, 1985.)

Developmental biology of larvae

CORACIDIUM: INVASION OF COPEPOD

It is well established by numerous workers that, when a coracidium is ingested by a copepod, three types of events may occur: type A – the coracidium may penetrate the gut wall almost immediately and undergo further development; type B – it may penetrate the gut wall but not develop fully or degenerate or be encapsulated; type C – it may remain for some time in the copepod gut and either become lost or eventually penetrate the wall and develop.

Examples of these various procercoid–copepod systems have been studied in *B. claviceps* (a parasite of the eel *Anguilla anguilla*) (*189*). Thus, coracidia were found to infect two species of copepods (*Acanthocyclops robustus*, *Macrocyclops albidus*) and procercoid development proceeded readily; this clearly represents a type A system. Type C susceptibility was represented by the copepod *Macrocyclops fuscus*, in which coracidia had difficulty in penetrating the alimentary canal but some penetrated later, although the development was variable. Type B susceptibility was represented by *Eucyclops serrulatus*, in which infection took place readily, but procercoids degenerated or failed to develop fully.

CORACIDIUM: HOST SPECIFICITY

Very little appears to be known regarding the factors which determine the host specificity of coracidia in copepods. Clearly, the chief factors which may affect development are likely to be the nature of the digestive juices, the structure of the alimentary canal (especially its cellular thickness?) and the nature of the haemocoele, including the reactions of phagocytic or encapsulating cells.

Measurement of susceptibility is further complicated by the fact that species of copepods which may readily be infected experimentally in the laboratory may only serve as poor hosts under natural conditions. For example, in the laboratory, *Cyclops scutifer* was found to be readily invaded by the coracidia of *Triaenophorus nodulosus* and yet in Lake Lergi, this species of copepod showed a prevalence of only 0.28% (*849*). This result may be due to the behaviour and/or the physiology of the coracidia or the host or both, being different under different environmental conditions.

Under natural conditions, in fact, remarkably low invasion rates are often sufficient to maintain an infection in a host (fish) population. For example, *Schistocephalus pungitii* and *Proteocephalus filicollis* are the commonest cestodes parasitising the nine-spined stickleback, *Pungitius pungitius*, in the USSR and yet, until recently, all attempts to find, in nature, copepods infected with these species were unsuccessful. In a marathon

200

study, which involved examining 195 000 copepods in a small lake, it was found that, although six copepod species were present, only one, *Mesocyclops oithonoides*, was infected, the invasion rate being only 0.078% for *S. pungitii* and 0.283% for *P. filicollis*. (*850*). Presumably as a result of water temperature, maximum invasion of both species was recorded in August (Fig. 8.4). From these data, Yakushev *et al*. (*971*) has shown that the prevalence dynamics of infection in copepods could be best approximated by a parabolic polynomial of the fifth power as follows:

$$y = 0.04 - 0.009x + 0.001x^2 - 0.00004x^3 + 0.0000005x^4 - 0.000000003x^5$$

The correlation coefficient ($r = 0.419$), however, indicated poor coincidence between empirical and theoretical curves.

CORACIDIUM: PENETRATION GLANDS

It has been assumed that, unlike cyclophyllidean oncospheres, the oncosphere contained in the coracidium of pseudophyllideans does not possess 'penetration glands'. Their apparent absence has always been difficult to explain because some species of both groups penetrate the alimentary canal of the same copepod host (*721*). Ultrastructure studies have, however,

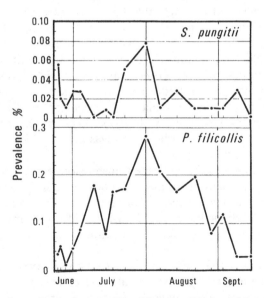

Fig. 8.4. Seasonal prevalence of procercoids of *Schistocephalus pungitii* and *Proteocephalus filicollis* in copepods in Lake Koljushkovoe, USSR. (After Sysoev, 1985.)

shown that 'penetration glands' do occur in pseudophyllids and have been described in *Bothriocephalus clavibothrium* (*845*), *Diphyllobothrium latum* and *B. gowkongensis* (= *acheilognathi*) (*421*). The position and fine structure of these glands closely resemble the corresponding glands in cyclophyllids (Fig. 8.16). The cytoplasm is filled with round, electron-dense, secretory granules which are larger in *Diphyllobothrium* than in *Bothriocephalus*. As with the cyclophyllid oncosphere, it is assumed that the gland secretions are histolytic and aid the oncosphere in penetrating the copepod tissues, although there is, as yet, no direct evidence for this supposition.

CORACIDIUM: INTENSITY OF INFECTION AND CYCLOPID MORTALITY

Attention has been drawn above to the very low prevalence of procercoids in copepods under natural conditions. This result is probably directly related to the fact that heavy infections of a copepod by numerous coracidia (say > 10) may result in the death of the former, with the result that only those with relatively low infections may survive. Experiments with *Triaenophorus crassus* in *Cyclops bicuspidatus thomasi* (*711*) showed that the first peak of mortality occurred when coracidia penetrated the gut into the body cavity. A second and larger peak of mortality coincided with the completion of most of the procercoid growth (Table 8.5). In this case death was probably related to nutritional or physiological stress or mechanical pressure on internal organs. The net effect of the above mortality resulted in the surviving copepod population at 28 days post infection (p.i.) being made up of uninfected or lightly infected (one to five procercoids) hosts. The resulting infection gives a parasite frequency which can be described by a negative binomial. This distribution is similar to that reported in Cyclophyllidea, in which Keymer (*391*), working with *Hymenolepis diminuta* in *Tribolium confusum* found that the overdispersion was attributable to mortality of hosts with high parasite burdens.

It is well recognised that parasite transmission is dependent on many ecological factors and it may be that copepods with a single procercoid may survive better and have larger and more fully differentiated procercoids than hosts with larger infections. Low-level infections may thus be an important adaptive strategy for transmission to the next host (*711*).

THE CORACIDIUM/PROCERCOID TRANSFORMATION

The ultrastructural changes which take place during the coracidium/procercoid transformation have been studied largely in the Pseudophyllidea (*265, 461, 485*), but some limited data are available for the coracidia in other orders (*420*). The following account is based largely on *D. dendriticum* (*265*) and *Spirometra mansonoides* (*461*).

The embryophore (Fig. 8.5), which is essentially a ciliated syncytium enclosing the oncosphere, is covered by a thin fibrous layer resembling a glycocalyx (but appears to be periodic acid–Schiff (PAS) negative), the remainder being divided into four zones. Characteristic of zone 1, is the presence of electron-dense granules which may represent the phospholipid and polysaccharide granules reported in the embryophore by other workers

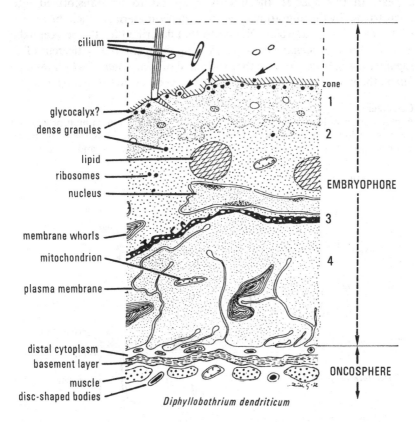

Diphyllobothrium dendriticum

Fig. 8.5. *Diphyllobothrum dendriticum*: fine structure of the coracidium showing the embryophore with its zones, 1 to 4, and part of the oncosphere. The embryophore is covered with a glycocalyx-like fibrous layer. The very dense zone 1 contains dense granules, some of which appear to pass out into the fibrous zone (arrow). The oncosphere plasma membrane is extensively folded. (Reprinted with permission from *International Journal for Parasitology*, **3**: Grammeltvedt, A.-F., Differentiation of the tegument and associated structures in *Diphyllobothrium dendriticum* Nitsch (1824) (Cestoda: Pseudophyllidea). An electron microscopical study, © 1973, Pergamon Journals Ltd.)

(*262*) and thought to provide energy reserves for the swimming coracidium (see p. 205). Details of the other zones are shown in Fig. 8.5.

The embryophore is shed immediately the coracidium is taken into the copepod intestine, a process which initiates procercoid differentiation. Microvilli develop as early as day 3 p.i. probably from granules in zone 1 (*461*), the first step in the formation of a villus being a bulge on the distal cytoplasm (*265*). Well-developed microvilli can be recognised as early as day 6 p.i. in *Haplobothrium* (*485*) and by 9 days developing microtriches appear. In this species, the microvilli appear to be transformed into microtriches by the migration of electron-dense material to their distal tips. This appears to be a process similar to that described for the polycephalic larva of *Paricterotaenia paradoxa* (*483*; Fig. 8.6), and the cysticercus of *T. saginata* (*740*). In contrast, in the procercoid of *S. mansonoides* (*461*) and *D. latum*, the majority of microtriches have been described as arising *de novo*.

Cercomer
A major morphological change which occurs during the coracidium/procercoid transformation is the elongation of the body and the separation of the hooks into a distinct appendage, the *cercomer*. The nature of this structure and especially its

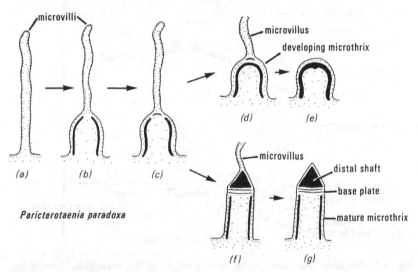

Fig. 8.6. *Paricterotaenia paradoxa*: development of microtriches from microvilli in the polycephalic larva (not to scale). (*a*) The original microvillus. (*b*) and (*c*) The microvillus base swells and contains peripheral electron-dense material. (*d*) A developing microthrix on the posterior neck region of larva; the original microvillus is still attached. (*e*) A mature, modified microthrix on the posterior neck region. (*f*) A developing microthrix on the anterior neck region, the original microvillus is still attached. (*g*) A mature microthrix on the anterior neck region. (After MacKinnon & Burt, 1984.)

possible homologies in other groups have given rise to much controversy. This is crystallised in the form of the Janicki Cercomer theory, first postulated by Janicki in 1920 (*480*), which proposed that the tail of a cercaria (Digenea), the opisthaptor of oncomiracidia and adult Monogenea and the cercomer of a cestode larva were all homologous structures.

This theory may have implications for the physiology of the oncosphere and procercoid, for, since the hooks in the oncosphere are considered to be 'anterior' and those in the procercoid (i.e. in the cercomer) 'posterior', it has been suggested that a reversal of polarity may occur, with a subsequent change in the metabolic gradient. It is questionable, however, if the terms anterior and posterior (as adjudged by direction of movement) have any significance in larval forms and it is likely that the cercomer is not a vestigal organ, but simply the remains of the oncosphere being shed (*480*).

CORACIDIUM: METABOLISM

In spite of being in an aerobic environment, the metabolism of coracidia, as studied in *D. latum, Ligula intestinalis* and *Triaenophorus nodulosus* appears to be wholly anaerobic (*262, 285*), although the eggs are aerobic (Fig. 8.7). This is demonstrated by the absence of oxygen consumption, the slight activity of oxidation–reduction enzymes, and the reduction of the amounts of phospholipids and polysaccharides in the embryophore. Especially significant is the drop in the level of ATP in older coracidia correlated with the decrease in the amounts of phospholipids and polysaccharides, indicating that the most probable source of energy for the coracidia is the anaerobic decomposition of reserve substances in the embryophore. Cytochemical studies on the oxidoreductases in the above species (*285*) show the presence

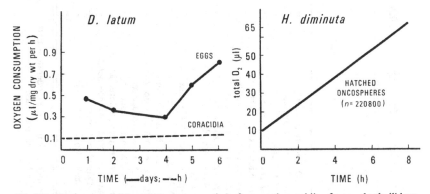

Fig. 8.7. Comparison of the oxygen consumption of eggs and coracidia of a pseudophyllidean (*Diphyllobothrium latum*) with the hatched oncospheres of a cyclophyllid (*Hymenolepis diminuta*). Note that although the pseudophyllidean eggs consume a little oxygen, the hatched coracidia are virtually anaerobic; in contrast, hatched cyclophyllidean oncospheres are highly aerobic. (Adapted from Grabiec *et al.*, 1964; Moczoń, 1978*b*.)

Table 8.3. *Oxidoreductases in coracidia of pseudophyllidean cestodes. (Data from Guttowa & Moczoń, 1974)*

| | Intensities of histochemical reactions | | | | | |
| | D. latum | | L. intestinalis | | T. nodulosus | |
Enzyme	Embryophore	Oncosphere	Embryophore	Oncosphere	Embryophore	Oncosphere
Isocitrate : NAD oxidoreductase	+	++	++	++	++	++
Isocitrate : NADP oxidoreductase	±	±	±	±	±	±
Succinate dehydrogenase	+++	+++	+++	+++	+++	+++
L-Malate : NAD oxidoreductase	+	++	+++	+++	+++	++
NADH : tetrazolium oxidoreductase	+++	+++	+++	+++	+++	+++
NADPH : tetrazolium oxidoreductase	+++	+++	+	+	±	+++
NADH: tetrazolium oxidoreductase[a]	+++	+++	+++	+++	+++	+++
NADPH: tetrazolium oxidoreductase[a]	++	++	++	++	+	+
NADPH : NAD transhydrogenase	-	-	-	-	-	-
Cytochrome oxidase	-	-	-	-	-	+
L-Glycerol-3-phosphate : NAD oxidoreductase	+	+	+	+	+	+
L-Lactate : NAD oxidoreductase	-	+	-	+	-	+
Glucose-6-phosphate : NADP oxidoreductase	-	+	+	+	+	++
6-Phosphogluconate : NADP oxidoreductase	-	±	-	±	-	±
L-Glutamate : NAD oxidoreductase	±	±	+	+	±	±
L-Glutamate : NADP oxidoreductase	-	-	-	-	-	-
L-Methionine : tetrazolium oxidoreductase	-	-	-	-	-	-
3-Hydroxybutyrate : NAD oxidoreductase	-	-	-	-	-	-
Alcohol : NAD oxidoreductase	-	-	-	-	-	-
Choline : tetrazolium oxidoreductase	-	-	-	-	-	-
Peroxidase	-	-	-	-	-	-

Notes: + + + high activity, + + moderate activity, + low activity, ± very low activity, − no activity.
[a] As soluble 'menadione reductase'.

of some 11–13 enzymes in the embryophore and oncosphere (Table 8.3). Two oxidoreductases operating in the pentose-phosphate cycle, glucose-6-phosphate:NADP oxidoreductase and 6-phosphogluconate:NADP oxidoreductase are active in the oncosphere only but not in the embryophore. The inability to utilise oxygen is reflected in the absence of cytochrome oxidase or other terminal oxidase. Coracidia cannot oxidise 3-hydroxybutyrate, choline, L-methionine and L-glutamine. Mitochondria were found not to be positive for cytochromes of the *a* and *c* types or peroxidase. Energy is probably obtained largely by glycolysis, as suggested by the activity of the enzymes lactate dehydrogenase and L-glycerol-3-phosphate:NAD oxidoreductase (*285*).

Second larval stage: procercoid

PROCERCOID: GROWTH AND DEVELOPMENT

General account
Early studies (*796*) have examined the growth rate of procercoids of *D. latum* and *S. solidus* in copepods and those of *Archigetes limnodrili* in freshwater oligochaetes. More recent studies have dealt with *B. rarus* (*360*), *B. acheilognathi* (*654, 655*), *Triaenophorus crassus* (*711*), and *B. claviceps* (*189*). Although there is some considerable variation between different species, as far as procercoid–copepod systems are concerned, some generalisations can be made, as follows.

(*a*) Growth of a procercoid is density dependent, the maximum growth occurring in single procercoid infections.

(*b*) Development (i.e. procercoid differentiation) appears to be variable, being density independent in some species, e.g. *B. claviceps* (*189*) but strongly density dependent in others, e.g. *D. latum* (*796*), *T. crassus* (*711*).

(*c*) In *T. crassus*, procercoids grow larger in female copepods than in males (Table 8.4), a result which probably reflects the greater supply of nutrients in the female haemocoele required for ovum development. Mortality is frequently heavy; for example, more than 80% of *Cyclops b. thomasi* infected with *T. crassus* died within 28 days (Table 8.5).

(*d*) Development appears to be largely independent of the stage of copepod development, although some stages are more readily infected (at least experimentally) than others.

Table 8.4. Triaenophorus crassus: *growth of procercoids in adult males and females of the copepod* Cyclops bicuspidatus thomasi. *Comparison of mean length and width of 21- to 28-day old procercoids. (Data from* Rosen & Dick, 1983)

Intensity of infection	Length (μm)			Width (μm)		
	Mean ± s.d.	Range	N	Mean ± s.d.	Range	N
1 (♂)	233.14 ± 40.55	142–303	29(29)	95.17 ± 11.64	64–112	29(29)
(♀)	295.00 ± 37.91[a]	236–374	25(25)	109.52 ± 13.02[a]	79–131	25(25)
2 (♂)	188.46 ± 34.60	112–232	22(11)	86.32 ± 9.43	64–101	22(11)
(♀)	229.64 ± 30.60[a]	157–273	14 (7)	95.36 ± 11.24[a]	82–120	14 (7)
3 (♂)	175.67 ± 32.00	112–228	24 (8)	72.71 ± 9.40	49–90	24 (8)
(♀)	198.38 ± 39.38[a]	112–302	24 (8)	83.75 ± 9.76[a]	60–101	24 (8)

Notes: Intensity of infection is the number of procercoids per host. *N*, number of procercoids, with number of hosts in parentheses.
[a] Significant at *P* < 0.05. Student's *t*-test on log transformed data.

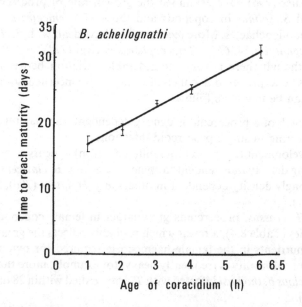

Fig. 8.8. *Bothriocephalus acheilognathi*: the relationship between age of coracidia ingested by *Cyclops viridis* and time taken for procercoids to reach maturity. (After Pool, 1985.)

208

Pseudophyllidea

Table 8.5. Triaenophorus crassus: experimental infection of copepods. Mortality of infected and uninfected Cyclops bicuspidatus thomasi and prevalence in cyclopids by host sex. (Data from Rosen & Dick, 1983)

Comparison of treatments	Cyclopid mortality, day 14 p.i.	Cyclopid mortality, day 28 p.i.	Prevalence of infection, day 1 p.e.	Prevalence of infection, day 28 p.e.
Control group	4/200	8/200		
Low level of exposure (L)	67/200[a]	173/200[a]	—	—
Control group	4/200	8/200		
High level of exposure (H)	105/232[a]	204/232[a]	—	—
Male hosts, L	24/101	84/101	101/138	17/49
Male hosts, H	38/91[a]	78/91	91/110	13/2
Female hosts, L	43/99	89/99	99/155	10/63
Female hosts, H	67/141	126/141	141/181[a]	15/51

Note: L, low level exposure: 58 coracidia per copepod. H, high level of exposure: 116 coracidia per copepod. Mortality, no. control dead/total controls and no. of infected dead/total infected. Prevalence, no. infected/no. exposed; p.i., post infection; p.e., post exposure.
[a] Significant at $P < 0.05$ (chi-square test).

(e) The effect of age of coracidia on their development has not been much studied. In *B. acheilognathi*, the time taken for a procercoid to develop fully is inversely proportional to its age when the copepod is infected, being 17 days for a 1 h coracidium but 30 days for a 6 h coracidium (Fig. 8.8), a result presumably related to the greater availability of energy reserves for growth in the younger coracidium (655).

PROCERCOID: GLANDS

Procercoids are typically provided with well-developed glands, which open by anterior ducts (Fig. 8.9). Light and electron microscopical studies have been carried out on the cytology and ultrastructure of these glands in *Eubothrium rugosum*, *T. nodulosus*, *B. acheilognathi* and *D. latum* (Table 8.1; Figs. 8.9 and 8.10) and *Haplobothrium globuliforme* (485).The glands consist of 10 to 12 secretory cells, measuring about 4 μm, lying in the central parenchyma, with cellular extensions usually opening in the frontal evagination but also at the lateral surfaces. Electron microscopical studies show that the cytoplasm of the glands is filled with small, rounded, electron-dense secretory granules and membrane complexes. The ducts are well developed in some species, especially in *D. latum*, where the ducts are swollen with secretion (Fig. 8.10). Presumably this means that adequate

209

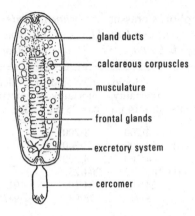

gland ducts

calcareous corpuscles

musculature

frontal glands

excretory system

cercomer

Fig. 8.9. A typical pseudophyllidean procercoid. (After Smyth, 1969.)

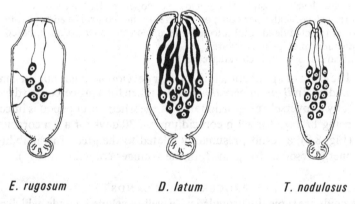

E. rugosum   *D. latum*   *T. nodulosus*

Fig 8.10. Gland cells in the procercoids of *Eubothrium rugosum, Diphyllobothrium latum,* and *Triaenophorus nodulosus*. (Source: see Fig. 8.11.)

secretions are immediately available on ingestion by the fish host and serve to aid passage through the tissues. In *Haplobothrium*, in contrast, ducts appear to be lacking and, presumably, only develop and become functional in the fish host; the reasons for this are not clear. As with the coracidia, nothing appears to be known regarding the nature of the gland secretions, which can be assumed to be cytolytic. The functions of the secretions of the penetration glands in cyclophyllidean oncospheres are similarly unknown (*442*).

210

Pseudophyllidea

Third larval stage: plerocercoid

PLEROCERCOID: GROWTH, DIFFERENTIATION AND DEVELOPMENT

The development of plerocercoids shows a tendency towards progenesis (advanced development of genitalia in a larva without maturation) and neoteny (maturation of gonads in a larva). The full range of postoncospheral development includes procercoid, plerocercoid and adult worm. However, in the procercoid of the most 'advanced' neotenic genus, *Archigetes*, the procercoid may reach maturity in the body cavity of its first intermediate host (a tubificid oligochaete), although maturation may also occur in a fish (*796*). In the genus *Bothriocephalus*, the procercoid develops in a copepod (p. 200) and when ingested by a fish or other host, passes through a transitional plerocercoid stage in the gut (i.e. without a second intermediate host being required) before developing into an adult. In a fully extended life cycle, as exemplified by *Diphyllobothrium* spp., the plerocercoid develops in fish muscles or viscera in a second intermediate host and the adult usually in the gut of a warm-blooded vertebrate.

Nothing appears to be known regarding the factors controlling the differentiation of somatic and germinal tissues in these larvae. The common occurrence of neoteny in the group suggests that in this group – and possibly in cestodes in general – these growth processes may operate independently of each other, although normally they are co-ordinated. This is borne out by the fact that 'monozoic' forms (i.e. unsegmented strobila with genitalia) of *Echinococcus* spp. may sometimes appear during culture *in vitro*, illustrating the apparent independence of these processes (Fig. 10.8, p. 275).

Although there are few data available on the establishment or growth rate of plerocercoids in intermediate hosts, the morphology, ultrastructure, cytology and host–parasite relationships of the commonly available species have been studied extensively. The biochemistry is discussed in Chapter 4.

Early work has been reviewed in the first edition; relevant recent references include: *Diphyllobothrium dendriticum, D. latum* and *Diphyllobothrium* spp. (*172, 421*); *Ligula intestinalis* (*269, 421*); *Schistocephalus solidus* (*255, 269*), *Triaenophorus nodulosus* (*269*); *T. crassus* (*269, 712, 713*); *Spirometra mansonoides* (*641*); *S. erinacei* (*269, 330, 424–426, 972*); *Eubothrium* spp. (*269*); *Haplobothrium globuliforme* (*484*); *Penetrocephalus* sp. (*119*).

THE PROCERCOID/PLEROCERCOID TRANSFORMATION

In spite of the major differences in the environments encountered, the observed changes in the structure of plerocercoids are relatively small.

211

Some data are available for *D. dendriticum*, *D. ditremum* and *D. latum* (*13, 798*); *Schistocephalus solidus* (*120*); *L. intestinalis* (*120*); *Spirometra mansonoides* (*461*); *S. erinacei* (*972*); and *Proteocephalus ambloplitis* (*369*).

The main morphological changes involve the loss of the cercomer, development of bothria, growth in size and an increase in the number of microtriches. In *Spirometra*, the microtriches are reported to be formed *de novo* via condensation of a granular matrix, producing the dense cap (*461*). The plerocercoid brush border includes microtriches of the (conoid) procercoid type and, in the older spargana, increasing numbers of an elongated (digitiform) type. The latter especially would greatly increase the surface area in relation to the increase in volume and the nutritional requirements of the plerocercoid. In some forms, e.g. *Diphyllobothrium* and *Haplobothrium*, there is some regional differentiation of microtriches, presumably in relation to function (*13, 484*).

PLEROCERCOID: GLANDS

The occurrence of glands in plerocercoids has been reviewed by Kuperman & Davydov (*421*). Glands have been found in all species investigated (Table 8.1) to date: *Eubothrium rugosum*, *E. salvelini*, *Triaenophorus nodulosus*, *D. dendriticum*, *D. latum*, *Spirometra erinacei*, *Schistocephalus solidus* and *L. intestinalis*. The distribution of some of these is shown in Fig. 8.11. Although their distribution, cytology and ultrastructure have been investigated, their function is still largely unknown. Since several species accumulate in a paratenic host when the latter preys on the original intermediate host, it might be expected that these glands would release histolytic secretions which could facilate passage through host tissues. This has been demonstrated only in the sparganum of *Spirometra erinacei* by Kwa (*425*), using both biochemical and histochemical methods (Fig. 8.12). Since spargana are well known to penetrate the tissues of many hosts, frogs, mice etc., this function would be compatable with their known life cycles.

The histochemical method used (*425*) involved cutting frozen sections of spargana and mounting them on (blackened) photographic plates previously exposed to light and developed. Controls consisted of (*a*) drops of Ringer and (*b*) sections of scolex heated in Ringer for 10 min before sectioning and mounting on the same photographic plate. Plates were incubated at 37 °C in a humid atmosphere and removed at half-hour intervals over a period of 2 h. Plates were then immediately dehydrated, cleared and mounted in Canada balsam. The site of the protease(s) showed up as clear areas where the gelatine had been digested; both controls were negative. Attempts to digest collagen films by the same techniques proved negative, suggesting (but not unequivocally proving) the absence of a collagenase. The biochemical assay was based on the digestion of casein and ninhydrin colorimetric analysis (Fig. 8.12).

Pseudophyllidea

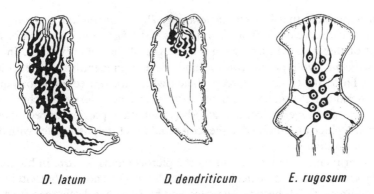

D. latum   **D. dendriticum**   **E. rugosum**

Fig. 8.11. Location of gland cells in plerocercoids of *Diphyllobothrium latum, D. dendriticum and Eubothrium rugosum.* (Reprinted with permission from *International Journal for Parasitology,* **12**, Kuperman, B. I. & Davydov, V. G., The fine structure of glands in oncospheres, procercoids and plerocercoids of Pseudophyllidea (Cestoda), © 1982, Pergamon Journals Ltd.)

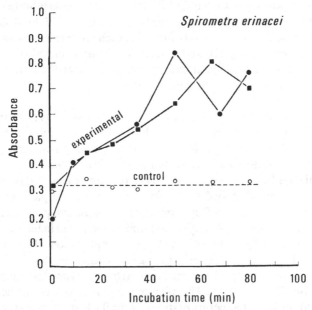

Fig. 8.12. Biochemical demonstration of release of proteolytic enzyme(s) by the plerocercoids ('spargana') of *Spirometra erinacei.* Graph represents reading of colorimetric analysis, showing relationship between incubation time and colour yield. Solid lines represent increasing amounts of amino acids liberated, indicating protein digestion (two replicates). Dashed line represents control (heated preparation), showing no digestion. (Reprinted with permission from *International Journal for Parasitology,* **2**, Kwa, B.H., Studies on the sparganum of *Spirometra erinacei.* II. Proteolytic enzyme(s) in the scolex, © 1972, Pergamon Journals Ltd.)

213

Developmental biology of larvae

It is possible that in some genera, e.g. *Diphyllobothrium*, some of the glands may serve other functions such as assisting in the initial establishment of the ingested plerocercoid in the gut of the definitive host. In *D. dendriticum*, two types of frontal glands have been identified in the scolex (*279*). These are: type I, the 'green' glands, which are located posteriorly and peripherally in the scolex and in the cortical and medullary parenchyma of the neck; and type II, 'golden' glands (which have previously been overlooked), which are confined to the scolex and are closely associated with the nervous system.

Experiments involving culturing the plerocercoids *in vitro*, in Medium 199 + serum, at 37 °C and *in vivo* (in golden hamsters) showed that within 3–4 h the green glands become activated and by 12 h have discharged most of their contents (*279*). In contrast, the golden glands remained unchanged. The nature of the green gland secretion was not determined, although it may contain sulphur compounds. It has been speculated that it may have adhesive properties which may be important during the migrating phase of the worm. It is known (see Chapter 9) that the worm first establishes itself in the caecum and then migrates forwards to the duodenum (*290*). It should be noted that the gland complex in *D. latum* is much more extensive than in *D. dendriticum*, where the green and golden glands are confined to the scolex and neck (Fig. 8.11). The functional differences between the glands in these two species needs to be investigated further.

PLEROCERCOID: SOMATIC/GENITAL DIFFERENTIATION

It has been stressed above that pseudophyllid larvae show a tendency towards progenesis and neoteny. In the genus *Diphyllobothrium*, the plerocercoid is almost entirely sexually undifferentiated, only traces of genital anlagen being present; such larvae require at least 7 days to reach maturity in a warm-blooded host (*796*). In contrast, the plerocercoids of *Ligula*, *Schistocephalus* and *Digramma* are progenetic and possess sufficient food reserves, so that a relatively simple stimulus (e.g. rise in temperature) is all that is required to induce rapid sexual maturation. This has been unequivocally demonstrated in *Ligula* and *Schistocephalus*, by inducing plerocercoids of these species to mature *in vitro* by raising the temperature to 40 °C (*786, 787, 790, 802*). This phenomenon is discussed further below and in Chapter 10. A temperature stimulus similarly induces maturation in *Digramma interrupta*; and, by raising the aquarium temperature, maturation of the plerocercoid, whilst still within its fish host, has been achieved (*184*).

The effect of temperature on the cell population kinetics of *Diphyllobothrium latum* and *D. osmeri* has also been examined. In a mara-

214

Pseudophyllidea

Table 8.6. *The composition of the cell population of plerocercoids of* Diphyllobothrium osmeri. *(Data from Wikgren, 1966)*

	Tegument	Outer paren- chyma	Muscle layer	Inner paren- chyma
Total no. of cells	2586	2088	972	1271
Percentage of flame cells	0	6.5	0	0
Percentage of calcareous corpuscle cells	0	6.6	0	5.2
Total no. of mitotically inactive cells	2586	274	0	66
No. of parenchymal cells	0	1814	972	1205
No. of mitoses	0	52	129	92
Percentage of mitoses	0	2.9	13.3	7.6

Note: The number of cells in a quarter of a section of each of 10 plerocercoids was counted. The percentages of flame cells associated with the calcareous corpuscles were determined from separate sections. The plerocercoids were treated for 5h with colchicine at 38°C.

thon study, Wikgren (*955*) counted the number of cells in a single plerocercoid of *D. osmeri* and by colchicine treatment examined the percentage of mitoses (Table 8.6). The rate of entry of cells into mitosis was very sensitive to temperature, varying from 0.1 cell/100 cells per h at 10°C to 1.9 cell/100 cells per h at 38°C. This pattern closely follows the pattern observed in other biological material. The doubling time of the whole cell population was estimated to be 54 h at 38°C but 960 h at 10°C.

PLEROCERCOID: HOST–PARASITE INTERACTIONS

Temperature-related adaptations
Schistocephalus solidus has its plerocercoid stage in an ectotherm (the fish *Gasterosteus aculeatus*) and its adult stage in an endoderm (a fish-eating bird) and it serves as an excellent model for the study of temperature adaptation in parasites. Walker & Barrett (*922, 923*) have studied the effect of temperature on (*a*) the activities of the mitochondrial enzyme adenosine triphosphatase (ATPase) and (*b*) the physical state of mitochondrial membranes in adult and larval *S. solidus*.

Theoretically, if the plerocercoid behaves similarly to a typical ectotherm, it would be expected to possess biological mechanisms which would enable the plerocercoid (whilst within the fish host) to cope with variations

215

Developmental biology of larvae

in the ambient environmental temperature (seasonal compensation) and also to carry out immediate compensation to negate the metabolic disruption by the drastic rise in temperature (to 40–41 °C) when the fish is eaten by a bird host. Once in the final host, however, as sexual maturation develops, slower changes (long-term compensation), which are analogous to a typical endotherm, could be expected. Where a biological process is characterised by having a Q_{10} value of 1.0 or substantially less than expected, there is no change in the reaction velocity as the temperature changes and the process can be said to exhibit immediate temperature compensation.

The Q_{10} of mitochondrial ATPase (a membrane-bound enzyme) from *S. solidus* plerocercoids was found to show increasing values with increasing temperature and this could result in a degree of rate stability when the temperature is abruptly increased (*922*); the Q_{10} of mitochondrial ATPase from adult *S. solidus* showed a similar increase. When the apparent K_m for the ATPase of the plerocercoid was examined, it was found to increase with temperature and this could be a significant factor in immediate temperature compensation during infection. The fact that adult *S. solidus* ATPase shows an activation energy lower than that of the corresponding plerocercoid enzyme suggests that a distinct difference in the functioning of the adult and plerocercoid enzymes may operate. Evidence has also been produced to indicate that a physical change occurs in the mitochondrial membrane at bird body temperature (*923*). This would be in agreement with the view that by altering the composition of the membrane lipids, in response to temperature changes, organisms are able to modulate the response of their membrane-bound enzymes.

Physiological/endocrinal interactions

It is well known that plerocercoids can produce disease in fish, especially as a result of migratory damage in their tissues. In trout, for example, the plerocercoids of *Triaenophorus crassus* cause extensive lesions in muscles, with the formation of granulomas, some of which are fatal to the fish and may also affect its swimming behaviour (*712, 713*). It is beyond the scope of this text to discuss this and similar pathological effects, but it is worth while pointing out that very little is known regarding the physiology of the pathological processes related to plerocercoid infections in fish or other hosts.

In the family Diphyllobothriidae, however, certain species produce substances which can exert hormone-like effects – a phenomenon of fundamental interest to biologists other than parasitologists. The species most studied are reviewed below.

Pseudophyllidea

Table 8.7. *Effect of plerocercoids ('spargana') of* Spirometra erinacei *in mature mice. Body weight of mice in weeks (w) after infection. Mean ± S.D. (After Shiwaku et al., 1982)*

Group	Weight gain (g)					
	1 w	2 w	3 w	4 w	5 w	6 w
Control	1.8 ± 1.1	2.6 ± 1.7	3.3 ± 1.6	3.6 ± 1.6	3.8 ± 1.5	4.4 ± 1.5
1 plerocercoid	1.7 ± 0.9	3.1 ± 1.4	4.4 ± 1.3	5.1 ± 1.6	5.3 ± 1.8	6.0 ± 1.8[a]
5 plerocercoids	2.6 ± 0.8	5.0 ± 1.1[b]	6.4 ± 1.5[c]	7.2 ± 1.7[c]	8.1 ± 2.2[c]	8.8 ± 1.8[c]
10 plerocercoids	1.8 ± 1.1	4.0 ± 1.7	6.9 ± 1.8[c]	7.6 ± 2.2[c]	8.8 ± 1.8[c]	9.3 ± 2.4[c]
20 plerocercoids	2.4 ± 0.5	5.9 ± 1.0[c]	8.6 ± 1.0[c]	9.6 ± 1.6[c]	10.6 ± 1.5[c]	11.4 ± 1.5[c]
30 plerocercoids	3.3 ± 1.7[a]	7.0 ± 2.1[c]	9.5 ± 2.1[c]	11.0 ± 2.1[c]	12.1 ± 2.0[c]	13.0 ± 1.9[c]

Note: The significance of the differences between the means for experimental groups and control group was tested using Student's *t*-test.
[a] $P < 0.05$ [b] $P < 0.01$ [c] $P < 0.001$.

Spirometra spp.

Perhaps the most remarkable physiological effects associated with infection of plerocercoids was first described for *Spirometra mansonoides* by Mueller (*574, 575*). He made the unexpected observation that when plerocercoids (spargana) of this species were injected subcutaneously into young mice, the host gained weight at a rate which could not be accounted for by the weight of the parasites or the associated tissue reaction. A similar (but not identical) stimulating effect was later described for the oriental species. *S. erinacei* (Fig. 8.13; Table 8.7) (*329*) and other species (*603*).

The early literature on this unusual phenomenon has been reviewed by Mueller (*576*) and Smyth (*796*). Most workers concluded that the phenomenon was due to the release by the parasite of a 'growth factor' consisting of one or more substance(s) with growth-hormone-like effects on certain rodents. It was originally thought that the effect was due to an insulin-like hormone (*796*), but more recent studies point to it being a substance resembling a mammalian growth hormone (*642, 762*), which, although stimulating growth, differs from it in not possessing anti-insulin/ diabetogenic activity. The close similarities between the activities and molecular characteristics of the plerocercoid growth factor and those of the mammalian growth hormone are clearly evident from Tables 8.8 and 8.9.

Physiological differences between the effects produced by *S. mansonoides* and *S. erinacei* (*762*) have been observed. The major effect of *S. mansonoides* appears to be the stimulation of lipogenesis resulting in

217

Table 8.8. Spirometra mansonoides: *comparison of the activities of the plerocercoid growth factor (PGF) to those associated with mammalian growth hormone (GH). (Data from Phares, 1987)*

	GH	PGF
Growth-related activities		
Weight gain	Increase	Increase
Skeletal growth	Increase	Increase
Somatomedin activity	Increase	Increase
Endogenous GH	Decrease	Decrease
Metabolic activities		
Insulin-like in:		
Hypophysectomised rats	Yes	Yes
Normal rats	No	Yes
Lipogenic rats	No	Yes
Anti-insulin-like in:		
Lipolytic rats	Yes	No
Diabetogenic rats	Yes	No
Other endocrine processes		
Regulation of receptors for PRL and oestrogen	Increase	Decrease
Puberty onset	Normal	Delayed
Growth rate of hormone-sensitive mammary cancer	Increase	Decrease

Note: PRL, prolactin.

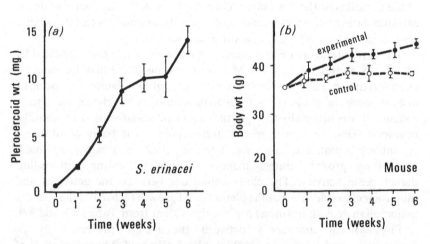

Fig. 8.13. Growth of plerocercoids ('spargana') of *Spirometra erinacei* in mature mice and their effect on the body weight of their hosts. (After Shiwaku *et al.*, 1982.)

Pseudophyllidea

Table 8.9. Spirometra mansonoides: *comparison of some molecular activities of human growth hormone (hGH) with those of plerocercoid growth factor (PGF). (Data from Phares, 1987)*

	hGH		PGF
Molecular weight	22 000		22 000[a]
Isoelectric point (pH)	4.9		4.7
Reacts with hGH specific monoclonal antibody (%)	100		61[b]
Reacts with monoclonal antibodies directed against epitopes shared equally by hGH and placental lactogen (hPL) (%)	100	35	49[c]

Notes:

[a] Obtained after immunoblot analysis using a hGH-specific mAb (QA68) with PGF purified by receptor affinity chromatography. Other forms identified by QA68 are a 27.5×10^3 M_r glycoprotein and a 17×10^3 M_r form.

[b] Immunopotency of PGF relative to hGH in radioimmunoassay (RIA) using the hGH-specific mAb QA68.

[c] Immunopotency of PGF relative to hGH in RIA using *two* separate anti-hPL mAbs which react with nonoverlapping epitopes shared equally by hGH and hPL.

obesity, although stimulation of protein synthesis in muscle and skeletal structures has also been reported (*820*). In contrast, the plerocercoids of *S. erinacei* were found not to cause obesity, although they caused a definite growth-producing effect (*330*). The growth of *S. erinacei* in mice is shown in Fig. 8.13(*a*). and the effect of an infection of 10 plerocercoids on the body weight of mice is shown in Fig. 8.13(*b*). That the growth-promoting effect increases with the number of plerocercoids present is clear from Table 8.7, which shows the effect of burdens of 1–30 plerocercoids in mice over a period of six weeks. It has also been shown (*760*) that the plerocercoid stimulates cell division in muscle, liver and epiphyseal cartilage and that weight gains commence from day 10 p.i. That a growth-promoting substance is present in the serum of infected mice (but not rats) has also been demonstrated by injecting serum from mice infected with the plerocercoids of *S. erinacei* into Snell normal and dwarf mice; significant increases in the weights of liver and spleen resulted (*762*).

Since the biological, immunological and molecular characteristics of the plerocercoid and mammalian growth factors are so similar (Tables 8.8 and 8.9), Phares (*642*) has put forward the intriguing hypothesis that the plerocercoid factor is the product of a human growth factor gene which has been sequestered by the parasite, a situation which he refers to as 'genetic theft'. He suggests that the most plausible mechanism to explain this acquisition of genetic material is that of viral transduction, although alternative mechanisms are also possible. A further interesting question is from what cells the growth factor is produced within the parasite and how it is transported to the outside? These and many related problems await further investigation.

Ligula intestinalis

The adult of *L. intestinalis* is a parasite of fish-eating birds. Its plerocercoid is better-known than the adult worm, as the former occurs in some 70 species of fish of the family Cyprinidae; its large size – up to 1 m in extreme cases – makes it especially noticeable (Fig. 8.14). The plerocercoid frequently causes parasitic castration in fish, as evidenced by a severe reduction in the development of the gonads. The ovary is reduced to the level found in a 'spent' fish with only oogonia and early primary oocytes present, whereas the testis contains only germ cells and some spermatogonia. These effects are associated with the reduction in size and granulation of the basophil cells in the middle glandular region (transitional lobe) of the pituitary (*22, 390*). This effect is accompanied by a reduction in the level of gonadotrophic hormones secreted (*796*). The mechanism of this phenomenon is not understood and it was thought that *Ligula* plerocercoids might secrete a steroid with androgenic and/or oestrogenic properties which could interact with the hypothalmus or pituitary gland or suppress gonadotrophin production. However, experiments using bioassay, thin-layer chromatography and nuclear magnetic resonance failed to identify the secretion of a sex steroid (*26*) and the mechanism of gonad supression is still not understood.

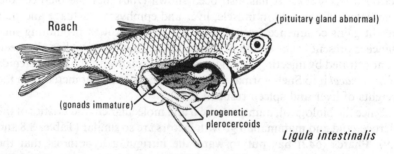

Fig. 8.14. Roach infected with plerocercoids of *Ligula intestinalis*. The presence of the larvae causes a reduction in size of the basophil cells in the transitional lobe of the pituitary and suppresses gonad maturation causing 'parasitic castration'. (After Smyth, 1976.)

220

Pseudophyllidea

Schistocephalus solidus

The plerocercoid of this species is a parasite of the three-spined stickleback, *Gasterosteus aculeatus*, a fish which occurs in a wide variety of freshwater, estuarine and marine habitats. Unlike *Ligula*, the presence of *Schistocephalus* does not affect the pituitary of the fish (*390*), nor does it cause parasitic castration. Nevertheless, some suppression of growth and maturation of the gonads does occur and the relative weights of the gonads (expressed as the percentage of fish consisting of gonad) are substantially less in parasitised fish than in normal fish (Fig. 8.15) (*518*). Swimming behaviour is also affected (*255*). Ecological evidence from Canada suggests that *Schistocephalus* plerocercoids have been selected to delay adverse effects on their fish (intermediate) host until after the latter has reproduced (*503*). This delay would shift the 'temporal appearance' of heavily infected sticklebacks to the time of peak abundance of the bird predators during the fall migration. (*503*). This would increase the probability that infected sticklebacks would be eaten by birds, and thus that the parasite would complete its life cycle.

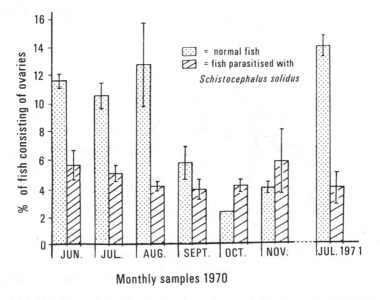

Monthly samples 1970

Fig. 8.15. Monthly variation in the ovarian content of the three-spined stickleback, *Gasterosteus aculeatus*, parasitised with the progenetic plerocercoids of *Schistocephalus solidus*. (After Meakins, 1974.)

221

Developmental biology of larvae

Cyclophyllidea

Oncosphere form and development

MORPHOLOGY

General account

The process of hatching in the Cyclophyllidea, which results in the release of the oncosphere, has been discussed in Chapter 7. The basic structure of a mature oncosphere (Fig. 8.16), which will not be discussed in detail here, does not vary substantially between species. It is composed essentially of (*a*) a thin covering epithelium with cytoplasmic extensions, (*b*) an additional complex system of muscles operating the three pairs of hooks, (*c*) a pair of large 'penetration' glands, (*d*) a small core of germinative cells from which the next larval stage develops and (*e*) a primitive nervous system.

Early ultrastructural and developmental studies on the oncosphere have been reviewed in detail (*442, 888, 889*). More recent studies are those on:

(*a*) Species with invertebrate intermediate hosts – *Anomotaenia constricta* (*243, 245*), *Aploparaksis filum* (*176*), *Catenotaenia pusilla* (*834*), *Dipylidium caninum* (*661, 662*), *Hymenolepis diminuta* (*439, 541, 560, 888, 889*), *H. nana* (*205*), *Oochoristica* sp. (*847*), *Polycercus paradoxa* (*175*).

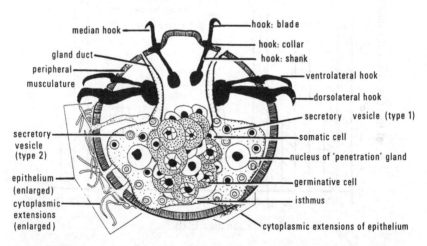

Fig. 8.16. Diagrammatic representation of the main features of a cyclophyllidean oncosphere, based on various species, especially *Hymenolepis diminuta*. (Modified from Davis & Roberts, 1983*a*, copyright © 1983. Reprinted by permission of John Wiley & Sons, Ltd.)

Cyclophyllidea

(b) Species with vertebrate hosts – *Echinococcus granulosus* (*307, 311, 314, 837, 839*), *Taenia crassiceps* (*131*), *T. pisiformis* (*38, 307, 314*), *T. ovis* (*299, 314*), *T. hydatigena* (*314*), *T. serialis* (*307, 314*), *T. saginata* (*471, 741, 822*).

It is not intended here to describe the ultrastructure of the oncosphere in detail as this is well covered in the above accounts but the following organ systems involved in the process of penetration require further consideration.

'Penetration' glands

Although secretions from the (so-called) penetration glands undoubtedly play a role in penetration, their nature and function are largely unknown. The glands have been particularly studied in *H. diminuta* and *H. nana* (*205, 888, 889*). In most species, glands have been described as syncytial (Fig. 8.16) or multicellular, but in *H. nana* they appear to be unicellular (*205*). One or two pairs of ducts may be present and these are supported by microtubules. Secretion has been described as being holocrine in some species (as in trematode miracidia and cercariae) but apocrine and merocrine secretory mechanisms have also been proposed for others (*205, 242, 443*).

Although the penetration gland secretions are released at the time of penetration and are widely assumed to cause lysis of the host tissue at the site of penetration, in fact, there is no direct evidence for this. The indirect evidence is based on (*a*) (What appears to be) visible lysis of tissue at the penetration site and (*b*) the reduction of the gland contents after penetration (Fig. 8.17). Tests for proteolytic activity in hatched oncospheres of *Taenia saginata* have proved to be equivocal (*287, 288*) and other workers have failed to detect lytic enzymes in other species (*442, 548*). The penetration gland contents stain vitally with neutral red, are strongly PAS positive (Fig. 8.20), Alcian blue positive, and β-metachromatic with toludine blue. This makes the oncosphere relatively easy to follow during and after penetration (Figs. 8.17 and 8.18).

Using neutral red as a marker and measuring the area (= volume) of the penetration glands by photographic techniques, Lethbridge (*439*) found that in intact eggs the (stained) penetration glands accounted for 33.9% of the total hexacanth area and in hatched oncospheres (i.e. before penetration) about the same (35.4%). After penetration, however, the area fell to 16% (Fig. 8.17). Rather surprisingly, Lethbridge & Gijsbers (*443*) found that identical secretory rates were recorded for *H. diminuta* hexacanths incubated in either Tyrode's saline or Tyrode + adult *Tenebrio molitor* gut extract. In both media, the reduction in the stained gland contents (Fig.

223

Developmental biology of larvae

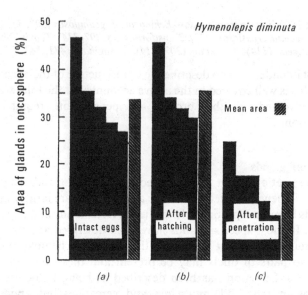

Fig. 8.17. *Hymenolepis diminuta*: penetration gland secretions in (*a*) intact (= unhatched) eggs; (*b*) after hatching of hexacanths and (*c*) after penetration through the gut wall of adult *Tenebrio molitor*. Volume of the glands estimated by area of gland stained in neutral red. Before hatching, the glands accounted for 33.9% of the total hexacanth area, recorded photographically (mean of five determinations); after hatching, the position was virtually unchanged at 35.4% (four determinations). After penetration, however, the area corresponds to only 16% (five determinations) demonstrating that substantial secretion had taken place during gut penetration. (After Lethbridge, 1971*a*.)

8.18) was equivalent to that observed in hexacanths penetrating *in vivo*. This result was interpreted as indicating that a chemical stimulus from the *Tenebrio* gut is not required to initiate gland secretion.

As several types of secretory cells are present in some species (Fig. 8.16) it is possible that two or more types of secretions may be produced (*443*) and the functions of the glands may be more complex than is at present recognised. Considering that oncosphere secretions have been shown to be strongly antigenic and have been used as vaccines (Chapter 11), it is surprising that so little is known regarding their composition and this is an area that clearly calls for further work.

Other possible functions suggested for the gland secretions have been (*a*) lubricating the passage of the oncosphere through the gut wall (*38*), (*b*) protecting the oncosphere against the digestive enzyme of the host and (*c*) assisting with adhesion before penetration (*205*).

Hooks, muscles and nervous tissue

The initial penetration of the oncosphere in *H. diminuta* appears to be largely mechanical as a result of the destruction of the columnar epithelium of the insect gut by hook movements (*888*), but as indicated above, lysis may also be involved. The same pattern is likely to be followed by other species in other invertebrate and vertebrate hosts.

The cyclophyllidean oncosphere is well supplied with musculature and the general pattern of muscles appears to be similar in most species. So in both *Hymenolepis citelli* (which penetrates an invertebrate gut) and *E. granulosus* (which penetrates a vertebrate gut) there are 16 somatic muscle cells (*146, 839*). However, *E. granulosus* has 16 hook muscle cells but *H. citelli* has only 13. In *E. granulosus*, the hook muscles have been shown to insert at the collar and base of the hooks and at the basal lamina of the embryonic epithelium. Each pair of hooks has three muscle systems associated with it: (*a*) a *protractor* system, for hook extensions; (*b*) an *abductor* system, which draws the hooks together and (*c*) a *retractor* system which pulls the hooks into the body (*839*).

The mechanism involved in the co-ordination of the muscular system is not understood. The presence of a nervous system in oncospheres has long been a matter of debate. Early workers demonstrated what appeared to be

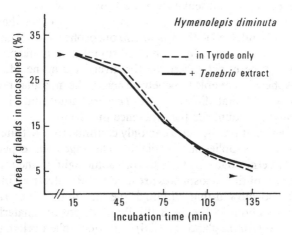

Fig. 8.18. *Hymenolepis diminuta*: rate of secretion of neutral red-positive material from the penetration glands of newly hatched hexacanths in Tyrode's solution alone (---) and in Tyrode's + adult *Tenebrio molitor* midgut extract (—). (Modified from Lethbridge & Gijsbers, 1974.)

acetylcholinesterase (AChE) activity (755) but later work suggested the esterase activity was not due to AChE (205). Further critical electron microscopical studies (205) have revealed the presence of nerve cells in the oncosphere of *H. diminuta*. These cells were all paraldehyde-fuchsin positive and show much secretory activity, as evidenced by the presence of large numbers of dense-cored vesicles produced by the Golgi complex. The cells are regarded as *possible* neurosecretory cells whose most likely main function is a motor one, stimulating and controlling the rhythmic movements of the active oncosphere. The neurosecretion may exert a more general influence on oncosphere development and control the differentiation of the somatic cell types from the germinal cells.

PENETRATION AND DEVELOPMENT: INVERTEBRATE INTERMEDIATE HOSTS

Most of the studies on penetration and subsequent growth and development have been carried out on *H. diminuta* but other species have also been studied (for references, see p. 222). Oncospheres appear to be able to penetrate the gut wall of insect hosts and reach the haemocoele in about 30–120 min (442). In *H. diminuta*, Lethbridge (439) found that, in adult *Tenebrio*, oncospheres penetrated the gut within 45–75 min, the majority of oncospheres penetrating across the epithelial diverticula (which project into the haemocoele), thus apparently avoiding the dense muscle layer; larvae which arrived at the musculature were only occasionally successful in penetrating. This could account for the well-known fact that *Tenebrio* larvae cannot be infected with *H. diminuta* oncospheres, whereas adults are readily infected. However, Moczoń (549) found that, in his preparations, very few oncospheres penetrated via the diverticula in the adult and that most oncospheres were able to penetrate across the muscular region of the gut. He suggested that differences in *Tenebrio* adult and larval muscle structure might account for the difference in infectivity.

It is evident that oncospheres can only continue to penetrate as long as sufficient energy supplies are available. The ingenious experiments of Anderson & Lethbridge (16) have thrown some light on the rate at which energy reserves of the oncospheres are used up. Results from these experiments (16) also showed that, although the maximum life of *H. diminuta* oncospheres is about 11 h (Fig. 8.19), the PAS-positive material (largely glycogen, except for the gland material) in the oncosphere cells had substantially declined by 7 h (Fig. 8.20). Moreover, it was speculated that the functionally active life of an oncosphere was probably only about 2 h, for by that time most of the hook movements had ceased. It is possible, of course, that these *in vitro* experiments do not exactly mirror the situation *in vivo*

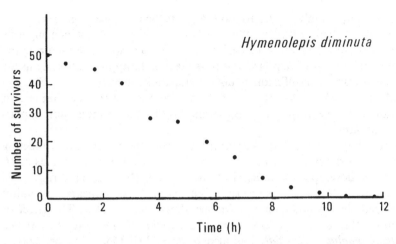

Fig. 8.19. *Hymenolepis diminuta*: survival of oncospheres in Tyrode's saline alone and in Tyrode + different levels of glucose. As no significant differences were detected between the various media, the survival results were pooled and averaged for each time interval (six replicates). (After Anderson & Lethbridge, 1975.)

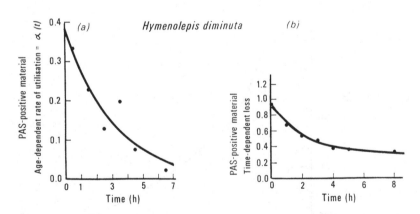

Fig. 8.20. (a) Age-dependent rate of utilisation of periodic acid–Schiff (PAS) positive material $\alpha(t)$ in oncospheres of *Hymenolepis diminuta* incubated in Tyrode + glucose at 25°C. Solid line represents the best fit for exponential model $\alpha(t)\beta \exp \{\gamma t\}$, where $\beta = 0.3822$ and $\gamma = -0.3259$. Dots, observed points. (b) Time-dependent loss of PAS positive material. Comparison of observed and predicted results. Solid line, predicted results; dots, observed results. (After Anderson & Lethbridge, 1975.)

Rather surprisingly, it was found (*16*), that the presence of glucose in the maintenance saline had no effect on survival (Figs. 8.19), which suggests that (like some trematode cercariae) the larvae are non-feeding during the hexacanth penetration phase. It is possible that the appropriate systems for active transport or diffusion do not develop until after penetration. Since so little is known regarding the metabolism of the oncosphere (see Chapter 7) it is not profitable to speculate further on this question without more experimental data.

As might be predicted, temperature is one of the major factors affecting the rate of development in the intermediate host. The cysticercoids of *H. diminuta* develop in *Tribolium* in 5 days at 37°C (Fig. 8.21) but require 65 days at 15°C; a temperature higher than 37°C appears to inhibit cysticercoid development and induce abnormalities (*796*). The effect of temperature on the development of *Dipylidium caninum* in the cat flea *Ctenocephalides felis felis* has also been studied (*662*). No perceptible growth occurred at 20°C but development was accelerated for every 5 deg. C increment from 20°C to 35°C; it was also affected to some extent by the relative humidity. The most satisfactory temperature for development appears to be 32°C, at which temperature the flea hosts developed normally

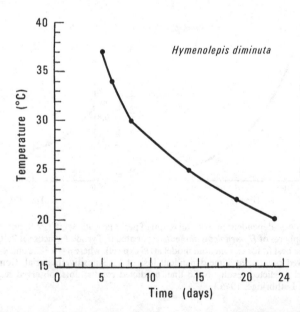

Fig. 8.21. Time required for the development of mature cysticercoids of *Hymenolepis diminuta* in adult *Tribolium confusum* at different temperatures. (After Voge & Turner, 1956.)

228

Cyclophyllidea

and the cestode larvae developed fully in 12 days. This is in contrast to the situation at 35°C, where, although the larvae developed in 9 days, all the flea hosts died at pupation. The ability of the host to mount an effective host reaction resulting in encapsulation of the larvae, also appeared to depend, to some extent, on the temperature (662).

PENETRATION AND DEVELOPMENT: VERTEBRATE INTERMEDIATE HOSTS

Penetration

In vertebrate intermediate hosts, oncospheres of *E. granulosus*, *H. nana*, *T. pisiformis*, *T. ovis*, *T. hydatigena*, *T. taeniaeformis* and *T. saginata* appear to be able to penetrate the epthelial border of the intestinal villi within 30–120 min (Fig. 8.22) (38, 307, 442). According to Heath (307) the route followed after penetration depends to some extent on the size of the oncosphere and suggested that the larger size of the oncosphere of *E. granulosus* enabled it to

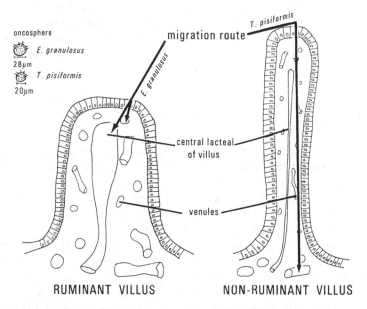

Fig. 8.22. Jejunal villi of ruminant and non-ruminant showing paths followed by oncospheres of *Echinococcus granulosus* and *Taenia pisiformis* (and some other *Taenia* spp.). The larger diameter of the *E. granulosus* oncosphere probably enables it to reach the lymphatic before being translocated in a venule. (Reprinted with permission from *International Journal of Parasitology*, 1, Heath, D. D., The migration of oncospheres of *Taenia pisiformis*, *T. serialis* and *Echinococcus granulosus* within the intermediate host, © 1971, Pergamon Journals Ltd.)

penetrate the lacteal vessels. In contrast, the species with smaller oncospheres (such as *T. pisiformis*), after penetrating the epithelium, progressed down the villus until a venule of sufficient diameter was encountered and penetrated; this allowed active transport to the liver. Some species, e.g. *T. serialis*, appear to be able to pass through both liver and lungs before reaching the muscles (*307*).

Migration to site of development
It is not known if the final site in which the larvae develop is reached by a random process, determined by characteristics such as the size of the oncosphere (as suggested above) or whether in hosts such as vertebrates some sort of tropism is involved. In vertebrate hosts, most species appear to show a predilection for a particular site. For example, hydatid cysts are found most frequently in man and domestic animals in the liver and lungs, but, in fact, almost any site can be invaded. Similarly, *T. saginata* cysticerci (*C. bovis*) are usually looked for in the heart and masseter muscles of cattle and these have long been regarded as the most important predeliction sites, but, again, they can occur in almost any organ (*969*). This probably explains how the infection may be easily overlooked during abattoir inspection. This is confirmed by experimental cysticercosis in cattle in which larvae were found in lungs (78.5%), liver (42.8%), parotid gland (28.5%), kidney (14.2%) and intestinal wall (7%) (*84*).

Although in many cases, oncospheres are carried to their final site by the blood or lymphatic systems (Fig. 8.22), in some cases the pattern of development involves migration through a dense organ such as the liver. This can be seen, for example, in *T. pisiformis* where the postoncospheral stage involves migration through the liver, from which it emerges about 15 days p.i. to complete its development in the body cavity. During *in vitro* culture of this species Heath & Smyth (*314*) noted that not only did the oncosphere exhibit strong muscular movements after about 4 days of culture, but within 5 days the larvae began secreting droplets at their anterior ends. Subsequent cytological examination of the anterior end (*757*) revealed the presence of large secretory cells (Fig. 8.23) which were PAS positive (amylase fast), stained with paraldehyde fuchsin and were toluidine blue β-metachromatic. It is tempting to assume that the secretion of these cells is cytolytic and assists the passage through the dense liver tissue, which, with its extensive collagen fibril network, must serve as a formidable barrier to a migrating larvae. The possible cytolytic nature of this secretion has not been examined in this species. Proteolytic activity, however, has been shown in larval *Taenia saginata* (*Cysticercus bovis*) using ^{131}I-labelled casein digestion and activity was found to be significantly higher in the developing (migrating?) larvae than in those with a developed scolex (*287*). This result

Cyclophyllidea

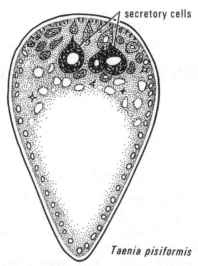

Fig. 8.23. Eight-day (postoncosphere) larva of *Taenia pisiformis* migrating through the liver of a rabbit, showing the presence of anterior secreting cells. (Reprinted with permission from *International Journal for Parasitology*, **3**, Shield, J. M., Heath, D. D. & Smyth, J. D., Light microscope studies on the early development of *Taenia pisiformis* cysticerci, © 1973, Pergamon Journals Ltd.)

suggests that, like *Taenia pisiformis*, the larva of this species may develop a transitory secretory gland during its migratory phase. Since a single infection of a rabbit with *T. pisiformis* produces an absolute resistance to further infection, it is likely that the secretions produced by migrating taeniid larvae may be strongly antigenic. It should be noted that, in the Pseudophyllidea, the secretory cells in the sparganum of *Spirometra erinacei* (p. 212), which also migrates through tissues, has been shown to release proteolytic enzymes (Chapter 6) (Fig. 8.12) (*425*).

NUTRITION AND METABOLISM

The study of the nutrition and/or the metabolism of a larva *in situ*, within a vertebrate intermediate host, clearly raises difficult technical problems and apparently no direct studies have been made. Most information is based on *in vitro* studies, special attention being paid to the changes which take place when larval/adult transformation takes place. Since the larval metabolism is so intimately related to that of the adult, it is discussed, with the latter, in Chapters 4–6.

Although relatively little is known regarding the mechanisms of membrane transport in cyclophyllidean larvae, it should be pointed out that pinocytosis has been shown, albeit indirectly, to take place in the bladder of *T. crassiceps* (*188, 880, 881*).

9

Development within definitive host

Mechanism of invasion

In spite of the vertebrate intestine being a biologically hostile environment for parasites, cestodes have been very successful in developing a wide range of morphological, physiological and biochemical adaptations which enable them to become established and to reproduce there.

With rare exceptions (e.g. *Archigetes*, *H. nana*), a host becomes infected by ingesting an intermediate host containing a cestode larva which may be (i) free, (ii) encysted, (iii) encapsulated, or (iv) both encysted and encapsulated. The term *encysted* is used when a cyst is produced by the parasite, *encapsulated* when it is produced by the host. In the latter case, the cyst may sometimes (e.g. *Echinococcus*) consist of two layers, the outer of which is derived from the host and the inner from the larva. The larval scolex is normally invaginated or withdrawn – a measure which presumably protects the scolex until it is stimulated to evaginate in the appropriate region of the host gut. To become established in the gut, therefore, a larva must (*a*) free itself from its surrounding membranes (i.e. excyst), (*b*) evaginate its scolex, (*c*) become 'activated' and (*d*) become attached to the intestinal mucosa.

Excystment and evagination

General account

ROLE OF ENZYMES

Although excystment and evagination are theoretically separate processes, in many cases the second follows so closely after the first that they cannot readily be separated. In some species, a particular enzyme may be required to digest the cystic membranes completely before the larva escapes; in others, the same enzyme may so activate the larva that it breaks out of its enveloping membranes before they are fully digested. Requirements for

232

Excystment and evagination

Table 9.1. *Requirements for excystment and/or rapid[a] evagination of some cyclophyllidean larvae; for these results to be of value the duration of treatment and the physico-chemical conditions of the medium must be taken into account*

Species	Pepsin	Trypsin	Lipase	Pancreatin	Bile salt	Reference
Echinococcus granulosus	+	−	·	−	+	*801*
E. multilocularis	+	−	·	−	+	*801*
Hymenolepis diminuta	E	+	·	·	+	*796*
H. citelli	E	+	·	·	+	*796*
H. nana	E	+	·	·	+	*796*
Mesocestoides lineatus	−	+	−	−	−	*382*
M. corti	−	−	−	−	−	*607*
Oochoristica symmetrica	−	−	·	·	+	*796*
Paricterotaenia paradoxa	+	·	·	+	−	*745*
Raillietina cesticillus	−	−	·	·	+	*672*
R. kashiwarensis	−	+	−	+	−	*732*
Spirometra mansonoides	−	−	−	−	+	*72*
Taenia pisiformis	−	−	·	·	+	*796*
T. saginata	−	−	·	·	+	*93*
T. solium	+	−?	·	·	+	*796*
T. taeniaeformis	+	−	·	−	−	*796*

Notes:
+, Essential; −, non-essential; ·, no information; E, enhance effect.
[a] Some species (e.g. *E. granulosus*) will evaginate in saline alone; but *rapid* evagination requires the presence of bile (see Fig. 9.1).

excystment and evagination in some species are summarised in Table 9.1. For these results to be meaningful, the precise experimental conditions under which they were obtained must be known. In some species, e.g. *Echinococcus granulosus*, pepsin is necessary to digest the thick wall of the (hydatid) cyst, before the larvae are released. In other species, however, pepsin merely has an enhancing effect but is not necessary for evagination or excystment. In many species, trypsin and/or pancreatin in the presence of bile is required for these processes.

Some species, e.g. *E. granulosus*, will evaginate their scolex in saline alone, but this process may be slow or only temporary, and such a response would be inadequate to bring about the rapid evagination and subsequent attachment to the mucosa which is required if the larva is to avoid being swept away by the continuous peristaltic movements of the gut. The addition of bile, however, produces almost instant evagination in most

233

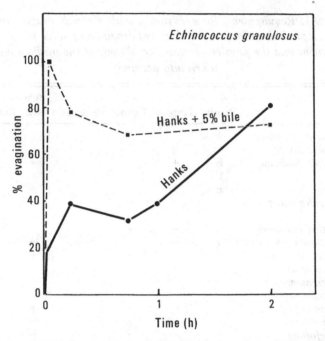

Fig. 9.1. Evagination of protoscoleces of *Echinococcus granulosus* with and without bile treatment. Although bile stimulates immediate evagination, the percentage of worms finally evaginating is almost the same, with or without bile. (After Smyth, 1967.)

species (Fig. 9.1). The pseudophyllidean scolex appears to evaginate more readily than the cyclophyllidean scolex; one small species of *Diphyllobothrium* will evaginate in bile-less media, such as saline or broth, although larger species, e.g. *Spirometra mansonoides*, may require bile and enzyme treatment to bring about evagination (72).

Establishment

Establishment and site selection

ROLE OF SCOLEX

The dominating feature in the process of establishment is the morphology of the scolex. In the Pseudophyllidea, this can vary from a structure that is little more than a shallow groove – as in *Schistocephalus* or *Ligula* – to a well-developed bothrium, as in *Diphyllobothrium* spp., which, when

Establishment

evaginated, is such a powerful adhesive organ that *in vitro* it readily attaches to a hard, inanimate surface, such as the wall of a culture tube. In the case of *Schistocephalus* and *Ligula*, the bothia are so poorly developed that they are (probably) incapable of attaching to the mucosa. This appears to be compensated for by the fact that the longitudinal musculature is exceptionally well developed, which would enable the worms to flex themselves against the intestinal wall and remain there long enough to become sexually mature; the success with which this is achieved is reflected in the different positions in which *Schistocephalus* establishes itself in different hosts (Fig. 9.2).

Establishment is greatly facilitated in these two species by the fact that the plerocercoids are progenetic (p. 211) and rapidly reach maturity after only 48 h (*Schistocephalus*) or 72 h (*Ligula*) in their bird hosts – a process which has been successfully completed *in vitro* (Chapter 10).

In species with strong, muscular scoleces, however, two main patterns of establishment appear to have evolved. In the first group (e.g. *Echinococcus* and *Taenia*) the scolex attaches firmly to the mucosa, and (as far as is known) remains permanently attached in the same site. In the second group, the scolex only attaches lightly to the mucosa, and then undergoes a migration. Thus, in the case of *D. dendriticum*, in the hamster, most worms become established initially in the second or third third-part of the small intestine (Fig. 9.3) but undergo a forward migration so that by 66 h all worms are found in the anterior third (*290*). In contrast, in the fowl

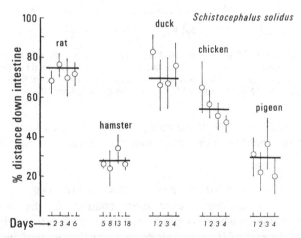

Fig. 9.2. Position of the pseudophyllidean cestode, *Schistocephalus solidus* when established in different laboratory hosts. (After McCaig & Hopkins, 1963.)

235

Development within definitive host

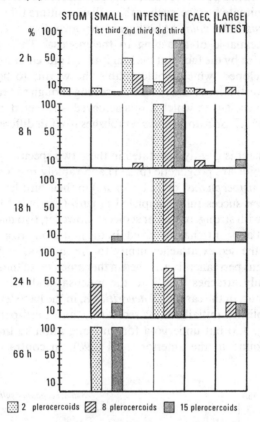

Fig. 9.3. Establishment of *Diphyllobothrium dendriticum* in the gut of the golden hamster. Most worms initially become established in the second or third 'third' of the small intestine, but then migrate forwards, so that by 66 h all worms are found in the anterior third. STOM, stomach; CAEC, caecum; INTEST, intestine. (After Halverson & Anderson, 1974.)

intestine, *Raillietina cesticellus* becomes established initially in the anterior small intestine and then undergoes a posterior migration during the next 2–17 days (*796*).

MIGRATION IN THE HYMENOLEPIDIDAE

The migratory behaviour is even more complex in the case of the Hymenolepididae, in which the phenomenon has been extensively studied, especially in the case of *Hymenolepis diminuta*, by the fact that two interacting patterns of migration take place (*20*). One is an age-dependent forward migration and the other a diurnal (circadian, diel) migration. Since it is

236

likely that at least the diurnal migration is related to the feeding pattern of
the host, the effect of altered feeding regimes has been much investigated.

The results of different workers do not always agree, which probably
reflects the different protocols adopted in different laboratories. Apart
from differences in the 'strain' of *H. diminuta* used – which is not always an
easily identifiable problem – a major factor appears to be the worm burdens
used. In single-worm infections, *H. diminuta* was found to attach initially
30–40% down the length of the small intestine (*95, 800*), but from days 7–14
migrated forwards to the 10–20% region. In 10-worm infections, on the
other hand, 5-day worms were found to be attached in the 39% region, 7-
day worms in the 15% region, but, after 7 days, there was a spreading effect
and worms could be found anywhere in the anterior 79% region (*115, 800*).

As indicated above, this long-term migratory pattern is overlaid by
a diurnal migration. That this latter pattern is related to feeding is

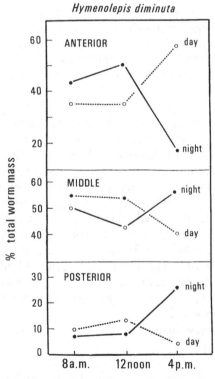

Fig. 9.4. *Hymenolepis diminuta*: the effects of day versus night feeding on the distribution of
worms in the small intestine. Each point is the mean of four rats. (After Read & Kilejian, 1969.)

demonstrated clearly by the fact that, when the feeding schedule of the rats was changed from day to night, the migratory pattern was reversed (Fig. 9.4). This suggests that the anterior daily migration enables a worm to move to the region of the intestine most favourable for absorption (*160*). The explanation may not be as simple as this, however, as in glucose-fed rats the worms did not appear to follow the highest levels of glucose concentrations (*525*).

An alternative approach to studying the migration phenomenon has been adopted by Hopkins & Allen (*333*). These workers severed the scolex of worms (one-worm infections) on day 14, without disturbing the attached scolex from the mucosa. They found that, within 48–72 h, the scolex had moved back to a position where a young worm of that size would normally be found. They suggested that the position of the scolex is, in part, determined by the position of the strobila, which monitors information about its position from all over its surface, 'balancing the input of adverse information from its tail and head ends' (*333*). They concluded that the preferred site for *H. diminuta* was 30–50% down the small intestine.

There has been much speculation as to the nature of the factors inducing migration in cestodes. It is well established that the physico-chemical characteristics of the vertebrate intestine vary substantially before, during and after feeding (Chapter 3), and that these characteristics are further modified by the presence of the worms (*527*). Characteristics particularly involved are the levels of Na^+, HCO^-_3 and bile, and the pO_2, pH, Eh, hormones, enzymes and the microfloral composition (Fig. 3.1, p. 37). Any or all of these could be involved in acting as biochemical or physiological cues or triggers for the migratory behaviour of cestodes. In the case of *H. diminuta*, there is some evidence that the levels of 5-hydroxytryptamine (5-HT, serotonin) in the mucosa and the intestinal lumen may play a part in the migratory pattern (*529*). By monitoring the levels in normal infected rats and those treated with 5-HT inhibitors, it was shown that the migration pattern was substantially altered. Thus, methysergide, which is a specific 5-HT antagonist, resulted in 90% inhibition of the worm migratory response and $MgSO_4$ a 67% inhibition. From these results it was concluded that there appeared to be a direct relationship between 5-HT intestinal levels and the distribution and migratory behaviour of this species in the rat intestinal lumen. Whether 5-HT is involved in the migratory patterns of other species has yet to be investigated.

Migratory behaviour in other species of *Hymenolepis* has been reviewed by Arai (*20*).

The host–parasite interface

The host–parasite interface

Fixation: difficulties *in vivo*

WARM-BLOODED HOSTS

It is self-evident that in a species in which a diurnal and/or a long-term migratory pattern occurs, the scolex must be able to free itself readily from its attachment to the mucosa, and it also follows that this attachment will be more superficial than in those species which remain permanently attached and do not migrate (e.g. *E. granulosus*). The form of attachment to the intestine and the *host–parasite interface* resulting has been studied in only a very few species. This is understandable, because as soon as an autopsy has been performed on the host and the tissue fixed, the resultant shock and associated biochemical, enzymic and cytological changes – as well as the action of the fixative – causes the scolex to contract away from the host tissue almost instantaneously. Thus, it is extremely difficult to obtain a realistic cytological picture of the host–parasite interface which is likely to approximate to the situation *in vivo*.

The above is especially true of warm-blooded hosts, as the temperature homeostasis is instantly upset on death. In *Echinococcus*, a situation approaching 'instantaneous' fixation of worms in the dog gut has been obtained (*794*) by (*a*) killing the dog host with a humane killer, (*b*) immediately opening the abdomen in a matter of seconds, (*c*) tying the appropriate region of the gut, and (*d*) fixing the worms *in situ* by injecting a hot (60 °C) fixative such as Zenker. This gives cytological pictures showing the scolex deeply embedded in the crypts of Lieberkühn (Fig. 9.5) with the suckers attached to the epithelium with very little contraction from the gut wall. A similar technique was used for *E. multilocularis*, in this case the gut being fixed in an anaesthetised dog, which was later killed (*866*). In *E. granulosus*, freshly evaginated protoscoleces establish themselves, at least initially, within the crypts of Lieberkühn, but later move out towards the lumen as the scolex grows in size. The majority finally become established in the mucosa with the rostellum pushed into a crypt with the suckers grasping the base of the villi (Fig. 9.5). A small proportion of worms breaks through into the lamnia propria. In spite of this, *Echinococcus* generally evokes no marked cellular reaction, although antigen is clearly reaching the bloodstream, as anti-*Echinococcus* antibody can be detected in canine serum after 14 days (*368*; see Chapter 11).

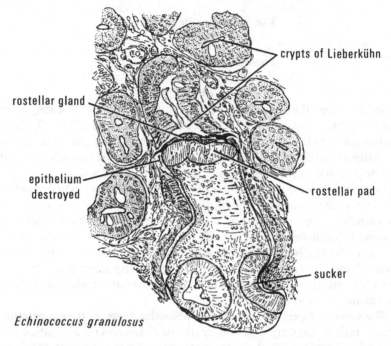

Echinococcus granulosus

Fig. 9.5. Scolex of *Echinococcus granulosus* in the duodenum of the dog; tissue fixed rapidly with hot Zenker's fluid. The rostellum is extended into a crypt of Lieberkühn with the hooks lightly penetrating the epithelium, some cells of which have become flattened and destroyed. The adult worm must therefore be considered to be both a *tissue* and an *intestinal* parasite.

COLD-BLOODED HOSTS

Rapid fixation of the host–parasite interface is somewhat easier to achieve in cold-blooded hosts, especially fish, where a temperature drop is not a problem. Descriptions are available for *Bothriocephalus scorpii* (*481*), *B. acheilognathi* (*750, 751*) and three species of Tetraphyllidea (*506*). In *B. scorpii* in the Red Cod (in New Zealand) (*481*) some 76% of the worms were found in the first third of the intestine. The scolex of large worms penetrated the gut as far as the muscle underlying the mucosa. In the area of attachment, the intestinal villi disappeared and there was a marked fibrosis, forming a swelling. The host reaction consisted chiefly of a deposition of collagenous connective tissue around the scolex. Vesicles present in the scolex tip suggested that secretory activity might be taking place, presumably from rostellar gland cells. In *B. acheilognathi* in the carp (in USSR), the zone of contact between the scolex and the host tissue has been described as being so close that a junction line could not be distinguished in some areas

(*750*). In many places, the plasmalemma between the microtriches and that of the epithelial cells of the host intestine were lacking, so that the matrix of the tegument was in direct contact with the cytoplasm of the host cells. Lysosomes were observed in the substance surrounding the microtriches embedded in the host cytoplasm.

In heavily infected fish, the cestodes were so closely packed that individuals were (remarkably!) reported as being partially or totally fused together! These zones of fusion gave strong histochemical reactions for -SH groups and appeared to be areas of intense metabolic activity. This suggested (*750*) that, in heavy infections, parasites which fail to find room for attachment on the host mucosa may derive their nutriment from another individual worm. This unusual observation clearly requires confirmation.

PATHOGENESIS

Although, in many species, the adult scolex does little damage to the mucosa, some species (e.g. *D. latum*, *H. microstoma*, *Multiceps* sp.) can produce marked pathogenetic effects. Consideration of these is beyond the scope of this book. The literature in this field has been reviewed by Arme *et al.* (*25*).

Growth rate

Most investigations of growth have been concerned with length/ weight/time or area/time studies and, with a few exceptions (e.g. *H. diminuta*), little information is available on the physiological problems related to growth, such as the change in chemical composition during development. The time required for maturation (the pre-patent period) in the definitive host varies substantially between species and often with the strain of the same species or host. Data for some common cestodes of warm-blooded hosts are given in Table 9.2. Before considering the growth rate, it is important to clarify what is implied by *growth* as distinct from *development* and *maturation*; more detailed discussions are given by Roberts (*696*) and Clegg & Smyth (*138*). A simple definition of growth, which avoids the special problems of animals with few cell divisions (such as nematodes) is 'an increase in the total mass of protoplasm'. Development is a combination of growth and differentiation, and is characterised by the production of progressively different groups of cells or tissues, which begin to synthesise different structural proteins or enzymes. It is widely recognised that differentiation is a central problem in biology and cestodes provide unusually interesting material in this field. For example, an adult cestode is essentially made up of a string of embryos of increasing degrees of maturity, so that all

Development within definitive host

Table 9.2. *Pre-patent periods of some cestode species in warm-blooded hosts*

Species	Host	Pre-patent period (days)	Reference
Pseudophyllidea			
Diphyllobothrium dendriticum	Hamster/gull	6–12	290
D. latum	Hamster	17	178
D. latum	Man	16	379
Ligula intestinalis	Bird	67 (hours)	800
Schistocephalus solidus	Bird	48 (hours)	800
Spirometra mansonoides	Cat	12–13	573
Cyclophyllidea			
Echinococcus granulosus	Dog (Australia)	40	793
E. granulosus	Dog (Switzerland)	37	873
E. granulosus	Dog (Russia)	35–49 (summer)	982
E. granulosus	Dog (Russia)	39–54 (winter)	982
E. multilocularis	Dog	30	378
Hymenolepis diminuta	Rat	12	19
H. microstoma	Mouse	14	199
H. nana	Rat	11–16	800
H. nana	Mouse	14–25	800
Mesocestoides lineatus	Hamster	??	151
M. corti	Cat	10	383
M. corti	Dog	14	606
Moniezia expansa	Lamb	50	111
Pseudanoplocephala crawfordi	Pig	30	412
Taenia crassiceps	Fox	31–32	692
T. hydatigena	Dog	48	211, 306
T. multiceps	Dog	38–43	963
T. ovis	Dog	35–56	312
T. pisiformis	Dog/cat	42	572
T. saginata	Man	87	821, 969
T. solium	Man	62–75	969
T. taeniaeformis	Cat	20	589
T. taeniaeformis	Cat	30–35	572

development stages can be studied in one organism. Again several species, e.g. *E. granulosus* and *Mesocestoides corti* are capable of differentiation into asexual or sexual stages depending on the conditions provided by the external environment (*806*). Moreover, under certain conditions *in vitro* somatic growth can be suppressed experimentally but genital development is unaffected. In *E. granulosus* and *E. multilocularis* this results in the production of remarkable unsegmented, but sexually mature 'monozoic' forms (*797, 807*; Fig. 10.8, p. 275). Differentiation in cestodes is further discussed below and in Chapter 10.

Some relevant studies on growth and/or maturation in the two major orders, Pseudophyllidea and Cyclophyllidea are as follows.

Pseudophyllidea
D. dendriticum (*290, 887*), *D. latum* (*259, 379*), *Spirometra* spp. (*73, 604, 605*), *Ligula intestinalis* (*723*).

Cyclophyllidea
E. granulosus (*350, 415, 733, 795, 863, 872, 873, 982*), *E. multilocularis* (*378, 867, 868*); *H. diminuta* (*88, 525, 526, 697, 699*), *H. nana* (*352, 417*), *H. microstoma* (*199*), *Diploposthe laevis* and *D. bifaria* (Hymenolepididae) (*829*), *Echinatrium filosomum* (Hymenolepididae) (*902*), *Moniezia benedeni* (*111*), *Mesocestoides lineatus* (*151, 382*), *M. corti* (*47, 383, 607, 739, 871*), *Taenia crassiceps* (*692*), *T. hydatigena* (*211, 306*), *T. multiceps* (*724*), *T. ovis* (*312*), *T. pisiformis* (*572*), *T. saginata* (*821*), *T. taeniaeformis* (*589, 771*).

It is well recognised that the growth rate can be affected by numerous intrinsic and extrinsic factors related to the host, such as host species or strain (*733*), host nutrition, endocrine status or environmental conditions (e.g. temperature), as well as the worm burden present.

The growth rate has been studied most thoroughly in the Cyclophyllidea, especially in the Hymenolepididae. Much of the basic work in this field has been reviewed by Roberts (*697*), but some data are also available for the Pseudophyllidea (*796*).

Influence of worm load – the crowding effect

Cyclophyllidea

Most of the basic studies on the influence of worm loads have been carried out on the Cyclophyllidea.

HYMENOLEPIS DIMINUTA

The growth rate is very markedly influenced by the population density (= worm load) within the gut and there is an extensive literature on the effect of 'crowding' on cestode development. Examples from the Cyclophyllidea and the Pseudophyllidea are given below. In general, it can be said that the size of the worms at maturity are approximately inversely proportional to the worm load. Early work on *H. diminuta* suggested that if there are fewer than four worms per host, the size attained is independent of the number present, since the maximum size per worm is reached in

243

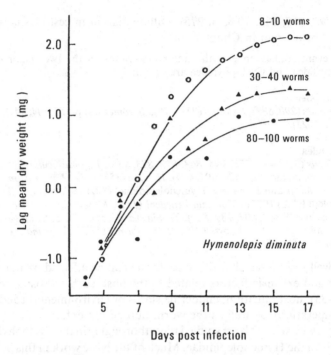

Fig. 9.6. The 'crowding' effect in *Hymenolepis diminuta*. The growth is measured as dry weight at different worm loads. (After Roberts, 1961.)

populations of that size (*696, 697*). In heavier loads, worm size is independent of load during the first 8–10 days of infection, but thereafter the effect of crowding becomes evident (Fig. 9.6).

Other workers (*325*), however, found that when the crowding effect was assessed by measuring the egg output, an effect became evident even at the two-worm level. Thus, in a one-worm infection, the egg production per worm was found to be about three times that of a two-worm infection, and in 5–20-worm infections, each worm produced only about one-sixth to one-ninth of a two-worm infection, independent of the number of worms harboured (Fig. 9.7; Table 9.3).

As might be expected, crowding has a marked effect on the chemical composition of worms, especially their carbohydrate content; this topic is considered further in Chapter 5.

The number of worms recovered is also greatly influenced by the population density (*325*). In rats given 1–20 cysticercoids, the mean recovery was 100–65% (Table 9.3), while in rats given 40–200 cysticercoids the recovery

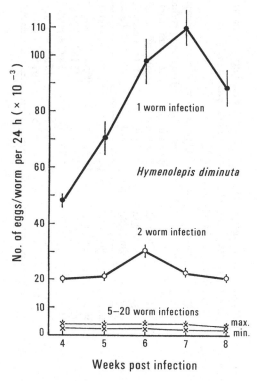

Fig. 9.7. Egg output in *Hymenolepis diminuta* in the rat at different population levels. (After Hesselberg & Andreassen, 1975.)

range was 13% at the 40 cysticercoid level but only 2% at the 400 cysticercoid level. On the other hand, Keymer *et al.* (*392*) found that, in the range 1–160 cysticercoids, the number of worms recovered and the prepatent period appeared to be independent of the cysticercoid dose, although density-dependent decreases in both worm weight (dry wt) and egg production were recorded. Early work (*796*) suggested that competition for available carbohydrate appeared to be the major factor concerned in the crowding effect. So, where *single* strobilas of both *H. citelli* and *H. diminuta* were present in the same host fed on an unlimited starch diet, both species were affected. *H. citelli* was affected proportionally more than *H. diminuta*, being reduced to 30% of its size when present alone.

There is, however, increasing evidence that the crowding effect (at least in *H. diminuta*) cannot be explained simply in terms of interworm competition for carbohydrate. In rats given exclusively a diet containing 3% (w/w)

245

Table 9.3. *The influence of population density on recovery, length, wet weight, dry weight, position and egg production of Hymenolepis diminuta in Wistar male rats eight weeks after the infection, calculated from the actual number of worms found at autopsy. (Data from Hesselberg & Andreassen, 1975)*

Number of rats	Number of cysticercoids given	Mean recovery (%)	Mean length (mm ± s.d.)	Mean wet weight (mg ± s.d.)	Mean dry weight (mg ± s.d.)	Mean position % from pylorus (± s.d.)	Total egg production per worm from the beginning of the 3rd week to the end of the 8th week (± s.d.)
5	1	100	1094 ± 134	1448 ± 249	302 ± 51	14.4 ± 5.8	2 859 900 ± 200 193
5	2	100	887 ± 28	838 ± 68	173 ± 14	18.9 ± 8.9	935 050 ± 68 259
5	5	96	523 ± 78	292 ± 99	62 ± 20	19.6 ± 4.1	145 167 ± 10 111
4	10	100	305 ± 12	131 ± 31	27 ± 7	34.9 ± 13.8	116 887 ± 6078
4	12	79	319 ± 22	139 ± 22	29 ± 3	35.1 ± 14.1	115 026 ± 8167
4	20	65	225 ± 60	85 ± 43	14 ± 7	38.1 ± 16.5	101 596 ± 6482

mannose for four weeks, worms resulting from a 10-cysticercoid infection were, on average, much heavier (46 mg) than those from rats on diets containing an equivalent of either galactose, glucose or fructose which had an average weight of 18 mg (*392*). One explanation of this result might be that the intestinal wall absorbed the other monosaccharides, but not mannose, faster than the worm. On the other hand it was found (*392*) that worms recovered from rats (infected with 10 cysticercoids) on a diet containing 4% mannose (w/w) for three weeks were, on the average, more than twice as heavy (37 mg) as those from rats on diets containing 1, 2 or 8% mannose, which weighed 16 mg. As indicated, this result clearly suggests that the crowding effect cannot be explained simply in terms of interworm competition for carbohydrate. One explanation of this result could be that there is a saturation level for uptake by the worm at about 4% and beyond this (e.g. at 8%) no more can be absorbed.

It is unlikely, however, that the explanation is as simple as this and there is now considerable experimental evidence that worms themselves may secrete or 'crowding factors' which inhibit DNA synthesis and hence growth (*348, 699, 979, 980*). When a 'worm-conditioned saline' (WCS) was prepared by incubating 10-day-old *H. diminuta* from infections of different populations (10, 50, 100 worms per host) it was found that WCS from more crowded populations inhibited [^{3}H]thymidine incorporation into DNA to a greater degree than did WCS prepared from less crowded populations. WCS prepared from adult (20–30 days old) from crowded (50-worm) infections, also inhibited DNA synthesis in 10-day-old worms from 10-worm infections (*699*). The evidence suggests that succinate, acetate, D-glucosaminic acid and the cyclic nucleotide cGMP may be the chief putative inhibitory (crowding) factors in *H. diminuta* infections (*979, 980*).

Hymenolepis microstoma, H. nana
A crowding effect is also well documented for *H. microstoma* (*373*) and *H. nana* (*254*). In the latter, the worm burden which developed from cysticercoid-derived infections was greater than that from egg-derived infections (Fig. 9.8), although the crowding effect was similar in both.

EGG PRODUCTION IN HYMENOLEPIDIDAE
Evidence from the studies reported above indicate that the mean egg production decreases with increasing worm burden in *H. diminuta, H. microstoma* and *H. nana*. An exponential model (*254*) describing the relation between mean egg output and worm numbers is shown in Fig. 9.9. This is based on the simple exponential equation:

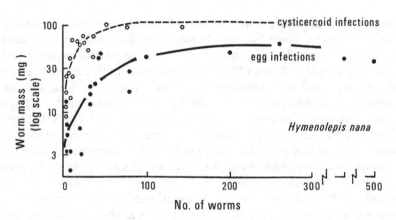

Fig. 9.8. *Hymenolepis nana*: worm biomass and worm numbers in egg-derived and cysticercoid-derived infections. (After Ghazal & Avery, 1974.)

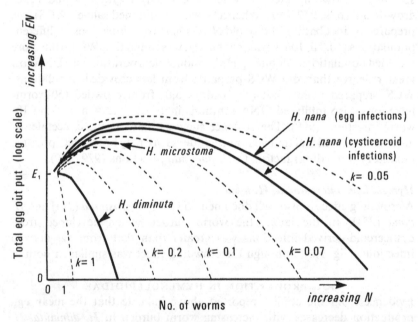

Fig. 9.9. Total egg output ($\bar{E}N$) in relation to worm numbers (N) for different values of k (see text). The approximate positions of the empirical curves for *H. nana*, *H. microstoma* and *H. diminuta* are shown, all drawn from the same arbitrary origin (E_1) (After Ghazal & Avery, 1974.)

248

Strobilisation and maturation

$$\bar{E} = E_1 \exp[-k(N-1)],$$

where $\bar{E}$ is the mean egg output, E_1 is the egg output from a single worm infection and N is the number of worms. From this model, it can be seen that when k is small (i.e. in *H. nana*, in egg-derived infections, $k \approx 0.055$; in cysticercoid-derived infections, $k \approx 0.06$), the total egg output initially rises with increasing worm burdens, reaches a peak and subsequently falls. On the other hand, when k is large, the total decreases and there is no peak.

Pseudophyllidea

DIPHYLLOBOTHRIIDAE

In the Pseudophyllidea, the effects of population density have been studied largely in the Diphyllobothriidae. When hamsters were given infections of 2–15 plerocercoids of *D. dentriticum*, there was a higher rate of recovery from hamsters given many (8 or 15) plerocercoids than from those given few (1–4) (*290*). The same order of result was obtained when gulls (the natural hosts) were used instead of rodents and it seems clear that there is an increased rate of establishment – and thus less larval death – when larger groups of plerocercoids enter the host in aggregation. This result could be a protective effect of a decreased surface/mass relationship connected with the aggregation of plerocercoids when passing through the host stomach (*290*). This probably reflects the ecological situation, for those fish which are most heavily infected are most likely to be caught by a gull. This is in keeping with the well-known model (*15, 517*) in which the infection process in the fish produces an overdispersed distribution of parasites within a host population. The fact that a crowding effect is also operating is reflected in the fact that the smallest individuals were found in the dense worm populations.

Strobilisation, sexual differentiation and maturation

Induction of strobilisation and sexual differentiation

With the exception of a few species with progenetic plerocercoids (e.g. *Schistocephalus*) it is only within the definitive host that cestode tissue develops in a strobilar direction, and it is self-evident that strobilar differentiation must be induced by factors present in the intestinal environment. Clearly, any number of known parameters of the intestine (Fig. 3.1, p. 37)

249

acting singly or together could act in this capacity. It is well recognised, for example, that in nematodes the high intestinal pCO_2 serves as a trigger which switches the organism from the larval to the adult metabolism (*796*). In cestodes, however, the factors inducing strobilisation are largely unknown, although, in recent years, the development of successful techniques for *in vitro* cultivation (Chapter 10) have enabled these to be investigated experimentally in a few species.

In general, a whole range of factors appears to be involved and there is little in common between the triggers utilised by different species. Thus, in *Echinococcus granulosus*, *in vitro* contact of the rostellum with a suitable protein substrate induces sexual differentiation and strobilar growth (*795–806*). This was first demonstrated with the sheep isolate of this species (*795*) and since then with isolates fom other commonly infected hosts, buffalo, camel, cattle, goat and man (*505*). Unexpectedly, however, it was found that this stimulus did not operate for *E. granulosus* of horse origin, a result which, in due course, led to the recognition of this isolate as a metabolically distinct strain (*501, 808*; see Chapters 4–6). The factors stimulating sexual differentiation in this isolate *in vitro* remain unknown. Since this form develops readily in dogs (*874*), an unidentified factor – perhaps an intestinal hormone or an enzyme (such as trypsin, see *Mesocestoides*, below) – may be involved. In contrast to *E. granulosus*, the closely related species *E. multilocularis* does not require the presence of a solid substrate in its culture system and readily strobilates in a liquid medium (p. 274).

Contrasting results have also been obtained with different species of *Mesocestoides*. In *M. corti*, the chief factors inducing strobilar differentiation *in vitro* appear to be anaerobic conditions (Fig. 10.12, p. 280) and a pH greater than 7.4 (*606, 607, 806*). Yet in *M. lineatus*, the presence of trypsin in the culture media is required for sexual differentiation (*382*). These phenomena are discussed in further detail in Chapter 10. Whatever the nature of the stimulus which initiates sexual differentiation, there is now strong experimental evidence that it operates via the neurosecretory system. It was speculated, many years ago, that in *E. granulosus*, the contact stimulus operated via a neurosecretion which in turn induced the release of a strobilisation organiser (*796*).

An elegant demonstration that the larval/adult transformation, resulting in sexual differentiation, is, in fact, associated with a neurosecretory mechanism has been made by Gustafsson and co-workers (*207, 278, 280, 281, 283*). They showed that a clear activation of the peptidergic neurones took place when plerocercoid larvae of *Diphyllobothrium dendriticum* were transferred from the poikilothermic intermediate fish host to the final homeothermic bird host. The effect could also be reproduced by cultivating plerocercoids *in vitro* at bird body temperature (40 °C). Release of the peptidergic neurosecretory material took place beneath the basal lamina of the tegument after this change of hosts. Although the actual nature of the material released has not yet been determined, results suggest that a number of mammalian peptides may be involved (*207*; see also Chapter 2). It has already been pointed out that in the case of the progenetic plerocercoids of *Schistocephalus solidus* and *Ligula intestinalis* (p. 262) temperature is again the stimulus which initiates sexual differentiation and it is important to note that several neurohormonal peptides and serotonin have been reported in *Schistocephalus* (Table 2.3, p. 31).

Stages in maturation

Organogeny and maturation involve the production and differentiation of large numbers of cells, the subsequent aggregation of these into organs, and the maturation of the gonads and related structures. It is convenient to divide the process of maturation into a series of stages which can serve as recognisable criteria for assessing development. Identification of these stages has been of particular value in assessing the success (or otherwise!) of *in vitro* culture. The stages in cyclophyllid development as exemplified by *Echinococcus* are shown in Fig. 9.10. Developmental stages of some other species (e.g. *Mesocestoides*) are considered in Chapter 10.

Briefly, the stages in *Echinococcus* spp. are summarised below.

Stage 0: *Undeveloped protoscolex*. This stage is represented by the freshly evaginated, but undeveloped, protoscolex.

Stage 1: *Segmentation*. The appearance of the first proglottis of the developing strobila represents an easily recognisable stage in development. This stage is foreshadowed by the appearance of a clear protoplasmic 'band' where the division separating off the first proglottis will appear. The excretory canals also become very prominent early on and the calcareous corpuscles disappear.

Stage 2: *Second proglottis*. This stage too is clearly defined by the appearance of an interproglottid 'partition' (but see p. 5).

Stage 3: *Early gametogony*. Characterised by the appearance of the testes as clumps of undifferentiated cells.

Stage 4: *Genital pore formation*. Characterised by the appearance of the lateral genital pore. Although this pore appears small and contracted in whole mount preparations, in the living worm (as viewed *in vitro*) it appears larger and is in a constant state of activity, opening and closing rhythmically – an activity probably related to copulatory activity.

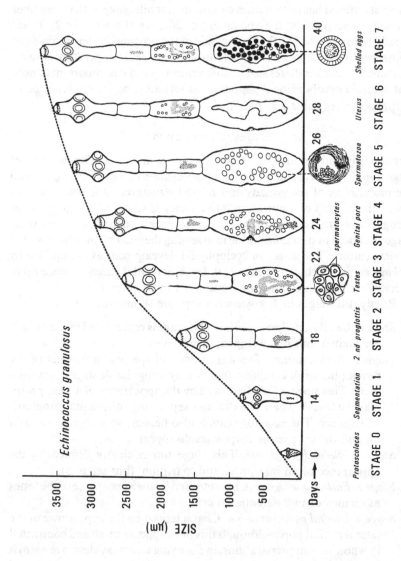

Fig. 9.10. Criteria for the various stages of maturation of *Echinococcus granulosus*. (After Smyth *et al.*, 1967.)

Strobilisation and maturation

Stage 5: *Late gametogony*. Characterised by the maturation of the male and female genitalia. This stage is recognised by the presence of spermatozoa which can be readily identified by preparing routine aceto-orcein squashes (*796*).

Stage 6: *Uterus dilation*. Characterised by the appearance of the uterus as a thin-walled sac in the centre of the maturing proglottis.

Stage 7: *Oncosphere formation*. Characterised by the appearance, in the uterus, of embryonated eggs containing fully formed hexacanth embryos.

Factors affecting growth and maturation

The growth rate and the onset of maturation are likely to be influenced by a number of abiotic and biotic factors related to the physiology of the host intestine, many of which have been discussed earlier (Chapter 3) or are discussed later (Chapters 10 and 11).

These factors can be summarised briefly as follows.

Physico-chemical:	temperature, pH, pCO_2, pO_2, Eh.
Biochemical:	composition of bile, enzymes.
Morphological:	micromorphology of mucosa.
Nutritional:	host diet.
Immunological:	previous immunological history.
Hormonal:	sex and hormonal status of host.
Behavioural:	whether host is stressed or normal.

INFLUENCE OF HOST OR PARASITE SPECIES OR 'STRAIN'

A major factor – and one which often introduces uncertainty into experimental work – is the influence of the species or strain of host or the cestode isolate used. This question clearly overlaps the whole question of host-specificity which may have a morphological, physiological or immunological basis.

There are many examples of cestode species which, although capable of infecting different hosts, may either fail to remain established or develop at a different rate. The best-known example is probably *H. diminuta*, which becomes mature in rats in 12–19 days and normally survives for the life span of the host, at least in infections of 10 or fewer worms per rat (*353*). In contrast, in mice, *H. diminuta* is expelled within 7–14 days without reaching maturity. In most mice strains, growth ceases abruptly about day 10 and worms are expelled. This phenomenon has been much investigated (*19*).

There is also substantial evidence that workers in different laboratories may be dealing with different 'strains' of *H. diminuta* (see Chapter 5). This is not surprising, for it is such a widely used experimental model that many years of passage through different strains of rats and probably different intermediate hosts is likely to lead to different selection pressures in different institutions and intraspecific variations would result. The related species, *H. nana*, has long been recognised to exist as a number of 'strains', although in contrast to other species (e.g. *Echinococcus*; see Chapters 4–6) little work appears to have been carried out on the possible physiological and/or biochemical differences between these strains. The strains recognised are: a 'human' strain adapted to man; a 'mouse' strain adapted to mice and a 'rat' strain adapted to rats. It has been shown experimentally (by volunteers!) that the mouse strain is infective to man and that the human strain is infective to mice. On the other hand, the mouse strain is not infective to rats. It does not appear to be known if the rat strain is infective to man or mice.

As mentioned earlier, the developmental biology in *E. granulosus* and *E. multilocularis* is complicated by the fact that it is now well established (*501*) that several (and perhaps many) strains of *E. granulosus* and possibly *E. multilocularis* exist, not all of which may prove to be infective to man. Some of these strains may show striking metabolic and biochemical differences (see Chapters 4–6). Although *E. granulosus* has been reported from a wide range of carnivores, the host range of *E. multilocularis* is much narrower (*814*). Thus, *E. multilocularis* establishes itself in both dogs and cats, but the latter is a 'poor' host and few worms develop (Fig. 9.11). It is not known, however, whether this inhibited growth in the cat is the result of physiological or immunological differences – or both – between the cat and dog. In spite of this low establishment, cats have been shown to play a role in the epidemiology of alveolar hydatid disease.

Aberrant asexual/sexual differentiation in *Mesocestoides*

In the cyclophyllidean *Mesocestoides corti*, uniquely, an asexual development pattern occurs in the definitive host intestine. Both the recently ingested larva (a tetrathyridium) and the adult worm are capable of undergoing asexual multiplication. This is illustrated dramatically by an experiment in which 2000 tetrathyridia were fed to a dog and 45 days later some 15 000 worms were recovered on autopsy (*739*). This unusual development was first shown by Eckert *et al.* (*192*) and has since been studied by a number of workers (*383, 607, 739, 871, 907*).

Although there is not agreement in all details of the life cycle, the overall pattern of development (Fig. 9.12) appears to be as follows. When a

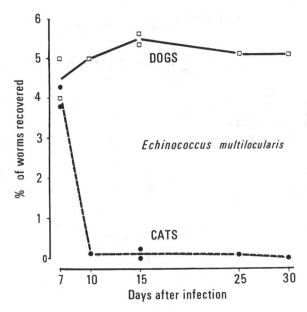

Fig. 9.11. Comparative development of *Echinococcus multilocularis* in the cat and dog. (After Kamiya *et al.*, 1985.)

tetrathyridium enters the gut, after shedding its posterior tissue, its scolex and body divide longitudinally, somewhat unevenly, resulting in two forms, one small and one large, each with two suckers. The smaller form regenerates two suckers and may divide again; the larger form regenerates two suckers, strobilates and develops into an adult worm. The adult worm appears to be capable of further asexual division by either longitudinal splitting of the scolex to produce a two-sucker form which can regenerate, as above, or a bud can arise from the strobila. It is not clear whether a freshly ingested tetrathyridium can develop *directly* into an adult worm (Fig. 9.12), i.e. without first dividing asexually. Both asexual and sexual differentiation of *M. corti* can now be carried out *in vitro* (Chapter 10), which makes this species a valuable model for experimental studies on asexual/sexual differentiation (*607, 806*).

It should be noted that the closely related species, *M. lineatus*, does *not* undergo asexual reproduction in either the tetrathyridium or adult stage; it too, has been successfully cultured to maturity *in vitro* (*382*). The possible factors inducing asexual/sexual differentiation in both these species, and in other cestodes, is further considered in Chapter 10, where *in vitro* systems are reviewed.

255

Development within definitive host

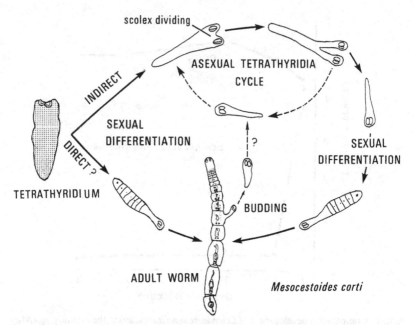

scolex dividing

ASEXUAL TETRATHYRIDIA CYCLE

SEXUAL DIFFERENTIATION

INDIRECT

DIRECT?

TETRATHYRIDIUM

SEXUAL DIFFERENTIATION

?

BUDDING

ADULT WORM

Mesocestoides corti

Fig. 9.12. *Mesocestoides corti*: simplified diagram of the asexual and sexual processes occurring in the intestine of the definitive host (cat/dog). (After Smyth, 1987*b*.)

10

Cultivation of cestodes *in vitro*

General considerations

A major development in parasitology within recent years has been the increased interest in attempts to culture parasites outside their hosts. A stimulus to this has been the introduction of efficient antibiotics and the general availability of commercially prepared media – both problems which greatly discouraged most early workers from working in this area. Substantial progress in this field has been made and a number of species can now be cultured to maturity or near maturity and several species (e.g. *Hymenolepis diminuta*) have been cultured through their entire life cycle *in vitro*. Early work in this field has been reviewed elsewhere (*796, 800, 905*). More recent work has been reviewed by Arme (*24*), Evans (*198*) and Howell (*339*).

It is well recognised that the value of *in vitro* techniques is that not only do they allow experiments to be carried out without the use of laboratory animals (which is desirable in itself) but also they allow the nutrition, physiology and biochemistry of a parasite to be studied in isolation from the interacting physiology of its host. Such techniques are also of especial value in studying immunity of cestodes, for they allow the collection of excretory/secretory (E/S) antigens to be undertaken without the problem of contamination by host molecules and this has greatly facilitated research in the development of possible vaccines. It is intended here to discuss only the general problems underlying *in vitro* culture and to deal with those species which have been cultured most successfully.

Terminology

A number of terms are in use to designate particular culture techniques. Cultivation in the absence of other organisms (i.e. a 'pure' culture) is termed axenic (Greek *a* = free form: *xenos* = stranger). A culture with one other species of organism present is said to be monozenic; when more than one species is present, the term polyzenic is used. The term *in vitro* (meaning literally 'in glass') has long been used to describe a culture utilising a liquid

257

or solid medium in a glass tube or similar container and, for convenience, this more general term is used here.

Basic problems of cultivation

General comments

Cultivation of cestodes presents a number of problems, many of which are shared with other parasitic helminths, such as trematodes and nematodes (*796, 800, 810*). Some problems, however, are unique to cestodes, particularly those related to the tapeworm's lack of an alimentary canal. Since cestodes normally inhabit the intestine of vertebrates, a major stumbling block to culture attempts in the past has been the initial establishment of sterility. Availability of antibiotics such as gentamycin, penicillin and streptomycin has largely eliminated this difficulty, and there now seems no reason why any species from any host habitat cannot be obtained in a sterile condition.

Once sterility is established, however, further problems arise, the chief of which can be summarised as follows.

PHYSICO-CHEMICAL PROBLEMS

The physico-chemical characteristics of the habitats of adult and larval cestodes are rather poorly quantified. Although the broad characteristics of the vertebrate gut are known (Chapter 3) little information is available on the precise conditions prevailing in specific sites such as, for example, the crypts of Lieberkühn, in which a cestode scolex might be embedded. Some data on tissue sites inhabited by larvae, such as the liver, muscles, body cavity, bloodstream, and brain in vertebrates, are now more generally available (*388, 796, 800*). Information necessary to establish culture conditions, ideally, should contain data on the pH, pO_2, pCO_2, Eh, amino acid and sugar levels, temperature, osmotic pressure and concentrations of the common physiological ions, enzymes and proteins (Fig. 3.1, p. 37) of the natural habitat. Much of this information is very difficult to obtain with accuracy, as any technique used is likely to interfere with the very characteristic (e.g. pO_2) it is trying to measure.

NUTRITIONAL REQUIREMENTS

Results from *in vitro* work, reviewed below, have shown that individual species may have very specific nutritional requirements, and what is satisfactory for one species may not be so for another, even though it utilises apparently the same site in the same host. Theoretically, such nutritional

258

differences would have survival advantage in that they would separate the nutritional or metabolic niches of competing species (*803, 805*). The nutritional materials available to cestodes in intestinal or tissue sites are complex in composition and difficult to replace by defined media. Nevertheless, as mentioned, the wide range of commercially available artificial media now available has considerably alleviated this problem.

HOST–PARASITE SPATIAL RELATIONSHIP
The surface topography of intestinal and tissue sites is usually elaborate and difficult to reproduce *in vitro*, as are the spatial relationships between the adult worms and the mucosa. These relationships have been shown to be especially important in some species (e.g. *Schistocephalus*, *Ligula*, *Echinococcus*), for, if they are not reproduced within reasonable limits *in vitro*, insemination may not take place (See Fig. 10.2).

REMOVAL OF METABOLIC WASTE
In their natural habitat, the metabolic waste products of adult or larval cestodes are readily removed from the vicinity of the organism as a result of the natural circulation of body fluids. A successful *in vitro* method must similarly provide conditions which promote rapid removal of toxic waste products.

PROVISION OF 'TRIGGER STIMULI'
The complex nature of cestode life cycles often involves up to three hosts, and specific stimuli may be required in each host to trigger the organism to develop to the next stage. The nature of these triggers has been resolved for very few species, those concerned in inducing strobilisation and sexual differentation probably causing most problems.

Criteria for development and maturation

Before considering the 'cultivation' techniques it is important to distinguish between *in vitro* culture conditions which provide mere *survival* and those which allow growth, development and maturation to occur. The term survival is here interpreted as implying merely the maintenance of an organism *in vitro* at a metabolic level sufficient to keep its cells and tissues alive, but not sufficient to allow growth and differentiation to occur at the appropriate rate and in a normal manner.

In order to obtain an accurate assessment of these processes *in vitro*, it is essential to have detailed information on the normal pattern of development in the natural host. Without such information, it is impossible to

Cultivation of cestodes in vitro

assess accurately the degree of success of culture attempts. Unfortunately, detailed developmental patterns *in vivo* are known for very few cestodes, and are confined, for the most part, to species of *Hymenolepis, Schistocephalus, Ligula, Echinococcus, Diphyllobothrium, Taenia* and *Mesocestoides*. Although there is some variation between the various orders, the main criteria of development relate to segmentation, organogeny, gametogenesis and egg shell formation. The broad pattern of development for a cyclophyllidean cestode is shown in Fig. 9.10 (p. 252).

Species used

Since adult cestodes, freshly removed from a host gut, are invariably covered with a mucus film containing micro-organisms such as yeasts, bacteria and fungi, most workers have used larval stages, which invariably occur in sterile habitats, as initial culture material. Nevertheless, by application of antiobiotics certain species of adult cestodes (e.g. *H. diminuta, E. granulosus*) may readily be cultured *in vitro* after removal from the definitive host before the end of the pre-patent period and then maintained *in vitro*, when egg release is safely contained within the culture vessel (*736, 811, 867*).

By the use of larval cestodes as initial material, the following species have been cultured *in vitro* to maturity or near maturity.

Pseudophyllidea: *Schistocephalus solidus, Ligula intestinalis, Spirometra mansonoides.*

Cyclophyllidea: *Hymenolepis nana, H. microstoma, H. diminuta, H. citelli, Echinococcus granulosus, E. multilocularis, Taenia* spp., *Mesocestoides corti, M. lineatus.*

Pseudophyllidea

General comments

The Pseudophyllidea contain many well-known species which infect birds and mammals and whose plerocercoids occur in cold-blooded vertebrates, especially fish. Several have progenetic (p. 214) plerocercoids, e.g. *Schistocephalus* and *Ligula*, and furthermore these larvae contain sufficient food reserves (such as glycogen) to satisfy the energy and synthetic needs of maturation without additional external nutrients being provided. Such larvae provided useful experimental models for early work in this field, as maturation could be achieved *in vitro* once sterility and the appropriate environmental conditions were provided (*786, 787, 788*). Although these

Schistocephalus solidus

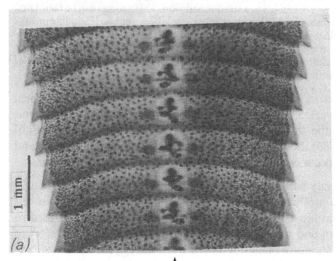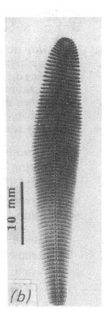

↑
genital anlagen

Fig. 10.1. Progenetic plerocercoid of the pseudophyllidean *Schistocephalus solidus*. (*a*) Enlarged view of plerocercoid showing genital anlagen. (*b*) Whole plerocercoid removed from fish.

early experiments did not initially provide much information on their nutritional requirements, they provided much basic data on the physico-chemical conditions under which maturation, insemination and fertilisation could take place. Perhaps, more importantly, they overcame a long-standing barrier in parasitology, demonstrating that vertebrate hosts could be replaced by *in vitro* systems.

Culture of progenetic plerocercoids to adults

SCHISTOCEPHALUS SOLIDUS
This was the first cestode species cultured to maturity *in vitro* (*786*) and it has proved to be a valuable experimental organism. The progenetic plerocercoid (Fig. 10.1) already contains well-developed anlagen of the

genitalia while still within the fish host and this enables maturation in the bird gut to be completed in 36–40 h (*786, 788*).

The *in vitro* technique for this species involves dissecting an infected fish aseptically and obtaining the plerocercoids in a sterile condition. This process can be facilitated by painting the surface of the fish with an aseptic solution (such as 1 % iodine in 90 % ethanol), before dissection and rinsing of the removed larvae in saline and antibiotics and then culturing. Early experiments (*786*) established that larvae became sexually mature adult worms, when incubated in an appropriate medium at 40 °C (bird body temperature).

Eggs produced in these early experiments were infertile, and histological examination revealed that (*a*) spermatozoa had not reached the receptaculum and (*b*) the testes showed some cytological abnormalities. Subsequent work (*788*) showed that, for normal maturation, insemination and fertilisation to occur *in vitro*, the following conditions were necessary.

(i) Use of highly buffered media to counteract the toxic effect of acidic metabolic products: 100% horse serum was used originally in this system, but most well-buffered tissue culture media can be used.

(ii) Cultivation under anaerobic conditions or a pO_2 sufficiently low to prevent premature oxidation of the phenolic egg shell precursors in the vitellaria (p. 172).

(iii) Gentle agitation of the culture media to assist diffusion of waste metabolites with renewal of media when appropriate.

(iv) Compression of cultured worms within narrow-bore dialysis tubing to assure that insemination and fertilisation takes place.

To establish these conditions and to simplify procedures, a culture tube with a ground glass assembly was eventually developed (Fig. 10.2). Compression of worms within the dialysis tubing permitted insemination to take place, the receptaculum becoming filled with spermatozoa and worms producing eggs with 80% fertility. In contrast, in worms cultured free in the medium, i.e. without compression, the receptaculum remained empty and the eggs infertile. Both these conditions can be provided within the same culture tube (Fig. 10.2) and clearly demonstrate the role of compression in the insemination process (*796, 788, 802*).

LIGULA INTESTINALIS

The plerocercoids of *L. intestinalis* (Figs. 8.14 (p. 220) and 10.3) are also progenetic and larvae are relatively large, usually about 100–200 cm in roach (*787*). This large size makes them especially suitable for experimental work and a number of valuable biochemical and metabolic studies have been carried out (*26, 502, 515*).

Pseudophyllidea

INSEMINATION TUBE

Schistocephalus solidus

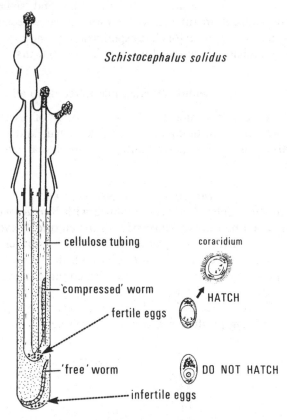

Fig. 10.2. Culture tube which enables the pseudophyllidean *Schistocephalus solidus* to undergo insemination during maturation *in vitro* at 40°C. Fertile eggs are produced only by worms compressed during maturation, a process which enables the cirrus to enter the vagina in each proglottis. Eggs from worms maturing 'free' in the medium produce only infertile eggs. (After Smyth, 1982.)

The same general culture conditions which were suitable for *Schistocephalus* (above) are also suitable for *Ligula*, except that the large size of the plerocercoids requires larger culture tubes and more frequent renewal of media. The removal of metabolic waste can be facilitated by perforating the dialysis tubing with tiny pricks from a sharp needle, thereby greatly increasing diffusion (*227, 787, 790*). Although the plerocercoid is progenetic, maturation takes slightly longer for *Ligula* than for *Schistocephalus*, about 63–72 h *in vitro*.

263

Cultivation of cestodes in vitro

Ligula is unusual in that fragments cut from the median regions of the plerocercoid will also mature *in vitro* (Fig. 10.3) and, since this can be achieved in small culture tubes, it is ideal for multiple (especially biochemical) experiments and useful for class experiments, as one large plerocercoid can provide individual cultures for a number of students.

Culture of undifferentiated plerocercoids to adults

Unlike *Schistocephalus* and *Ligula*, the majority of pseudophyllidean plerocercoids are morphologically undifferentiated, and must undergo considerable growth and differentiation before sexual maturity can be achieved.

SPIROMETRA MANSONOIDES

Very successful results have been obtained with *S. mansonoides* using a technique which had proved successful in the culture of cyclophyllidean cestodes (*72*). Starting with plerocercoids, which themselves had been cultured from procercoids *in vitro* (see below), the plerocercoids of this species in preliminary experiments were cultured through to early segmentation. Evagination of the scolex (with bile) appeared to be a necessary trigger for differentiation to be initiated. Later experiments (*73*) were more successful and egg-producing adults were eventually obtained. The gas

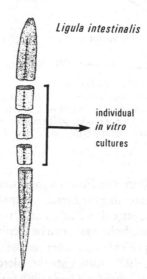

Ligula intestinalis

individual
in vitro
cultures

Fig. 10.3. Plerocercoid of *Ligula intestinalis* showing how small fragments may be cut from the larva and cultured individually *in vitro*.

Cyclophyllidea

phase (10% CO_2 in N_2) appeared to be particularly important; oxygen was reported as being detrimental to the culture system.

Culture of procercoids and plerocercoids

SPIROMETRA MANSONOIDES

Mueller (573–576, 796) in a series of detailed studies has developed a number of elegant techniques for obtaining procercoids of this species from copepods and subsequently cultivating them *in vitro* to infective plerocercoids. For the detailed technique, the original papers should be consulted. This method allowed large numbers of procercoids to be cultured to plerocercoids – organisms up to 30 mm being obtained in nine weeks. The plerocercoids were infective to cats and grew to mature *S. mansonoides*.

SCHISTOCEPHALUS SOLIDUS

The plerocercoids of this species, which ranged from 2 to 200 mg in the body cavity of the stickleback *Gasterosteus aculeatus*, have been grown to the infective stage in a relatively simple medium (467). In a liquid phase of 25% (v/v) horse serum, 0.5% (w/v) yeast extract, 0.65% (w/v) glucose in Hanks' saline (pH 7.1) and a gas phase of 5% CO_2 in air, dry weight increases of up to 500% were recorded in 8 days. These worms matured to adults when cultured *in vitro* at 40°C.

Cyclophyllidea

General comments

Many cyclophyllidean larvae will 'survive' in quite simple media, such as balanced saline plus glucose – provided sterility is maintained – for long periods, without undergoing further development. Early work (reviewed: *198, 796, 854, 855, 905*), which was rather uncritical and empirical, will not be discussed further here. The major advances have been made using various species of the genera *Hymenolepis, Echinococcus, Mesocestoides* and *Taenia* as experimental models.

Hymenolepididae: culture of larvae to adults

HYMENOLEPIS DIMINUTA AND H. NANA

Successful culture of both *H. diminuta* and *H. nana* was first achieved by Berntzen (70, 71), a result which represented a remarkable technical

Cultivation of cestodes in vitro

achievement. His method differed from those of earlier workers in that (a) it involved a continuous flow apparatus and (b) it utilised an extremely complex medium; the gas phase was not stated. There is some doubt about the exact composition of the medium used, as other workers have not succeeded in preparing it according to the instructions published. With this method, cysticercoids were grown to sexually mature adults, showing normal segmentation and well-developed gonads, which eventually developed gravid proglottides with oncospheres.

At first sight, the success of this method appeared to be due to the fact that (a) the medium used contained a wide range of metabolites known to occur in cellular metabolism, (b) the medium was continually flowing, so that worms were being continually exposed to new supplies of metabolites and at the same time the waste metabolic products were being continually swept away, (c) the gas phase was (apparently?) controlled, and (d) the oxidation–reduction potential was controlled, at least initially.

Although the initial cultivation of *H. diminuta* and *H. nana* represented a major step forward, some of the conclusions reached (*70, 71*) are open to question as being equivocal. Other workers, including Berntzen and his colleagues (*875*) failed to obtain consistent results with this method and it has also proved unsuccessful for other cyclophyllidean species (*198*).

Later work indicated that the requirements of both these species may be much less demanding than the above results suggest, for cysticercoids have since been cultured in much simpler systems without continuous flowing media (*736, 775*), see below.

H. diminuta grows well in a diphasic medium consisting of a blood-agar base (NNN) overlaid with Hank's saline and a gas phase of 3% CO_2 in N_2 (*736*). Worms were initially cultured in flasks in a shaking incubator, and after 6 days transferred to petri dishes with the same medium; additional glucose was added from time to time and further transfers made. Gravid worms developed and the contained oncospheres were infective to beetles. Worms developed *in vitro* differed from *in vivo* worms in that: (a) the pre-patent period was 24 days compared with 13 days in the rat; (b) the worms were 'miniature' in size compared with rat worms; (c) many proglottides were sterile, probably due to the failure of insemination to take place. Modification of this method, using 6-day-old worms from rats (rather than cysticercoids) was later developed (*700*) and this proved to be a most effective technique, producing gravid worms in 12 days. This method, which avoids the difficulties of growing the early postoncospheral stages has been widely used for experimental studies on *H. diminuta* nutrition (*198*). *H. nana* has also been grown in a relatively simple medium using a roller tube technique (Table 10.1) (*775*).

HYMENOLEPIS MICROSTOMA

Adults of this species have been grown successfully *in vitro* from cysticercoids by a number of workers. In early experiments, nearly mature adults

266

Cyclophyllidea

Table 10.1. *Liver-extract medium which supports growth of* Hymenolepis nana *cysticercoids to egg-producing adults in 14 days (medium changed every 3 days). (Data from Sinha & Hopkins, 1967)*

30% Horse serum
0.3% Glucose
0.5% Yeast extract
40% Hank's saline
10% Rat liver extract
Antibiotics (100 i.u. sodium penicillin G + 100 μg streptomycin sulphate) (Crystamycin-Glaxo)/ml of medium
NaHCO$_3$ (1.4%, w/v) + NaOH (0.2 M)
pH 7.2 Gas 95% N$_2$ + 5% CO$_2$ Roller tubes at 37 °C

with gravid proglottides were obtained, but no eggs were produced (*174*). Egg-producing adults were later grown (*197*) using a modified Eagle's medium, plus liver extract, bile and horse serum in a gas phase of 5% CO$_2$ in N$_2$ in a roller tube system. Since this species lives in the bile duct of rodents, it is not surprising to find that haemin is an essential nutritive requirement for strobilar development (*393, 747*). Seidel (*748*) developed a successful culture technique for this species, using a diphasic medium consisting of blood–agar with an overlay of Eagle's medium and horse serum.

Hymenolepididae: culture of oncospheres to cysticercoids

Four species of *Hymenolepis* have been grown from oncospheres to infective cysticercoids: *H. citelli* (*908*), *H. diminuta* (*904*), *H. nana* (*749*) and *H. microstoma* (*748*). The basic technique involves sterilising eggs in antibiotics and hatching according to standard procedures (*74*; see Chapter 7). The successful culture medium is the same for all species and is a modification of a medium used for growing cockroach cells (*198*). Air is a suitable gas phase, but *H. nana* grows better in 5% CO$_2$ in N$_2$. For technical details of culture, the original papers should be consulted.

Echinococcus spp.: culture of asexual (cystic) and sexual (strobilar) stages

GENERAL COMMENTS

The hydatid organisms, *E. granulosus* and *E. multilocularis*, present particularly interesting problems of morphogenesis, since the larval stages

267

Cultivation of cestodes in vitro

(protoscoleces) are exceptional in having the potential to differentiate in either of two directions depending on their location within the host (Fig. 10.4). Larvae ingested by a dog will develop in a sexual direction to form adult tapeworms in the gut. On the other hand, if a hydatid cyst ruptures while still within the intermediate host (which may happen during surgical removal), each released protoscolex is capable of differentiating asexually into a new hydatid cyst, i.e. 'secondary' hydatidosis results. This represents a degree of heterogeneous morphogenesis exceptional in a metazoan organism (see also *Mesocestoides*, p. 278), and makes *E. granulosus* an unusual model for differentiation studies (*796*). Clearly, factors present in the intestinal and tissue environments are responsible for initiating differentiation in these two directions. Elucidation of these factors and the reproduction of this phenomenon *in vitro* have been a challenging problem in parasitology. A comprehensive review of the history of research in this area and its present status has been given by Howell (*339*).

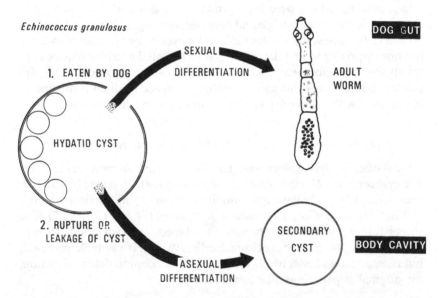

Fig. 10.4. Potential development of a protoscolex of *Echinococcus granulosus* or *E. multilocularis* in different habitats. 1, In the dog gut, the scolex evaginates, the organism attaches to the mucosa, differentiates in a strobilar direction and develops into an adult tapeworm. 2, *In vivo*, if a hydatid cyst bursts or leaks (as during a surgical operation) each protoscolex can differentiate in a cystic direction and form a (secondary) hydatid cyst. (After Smyth, 1987*b*.)

Cyclophyllidea

GROWTH PATTERNS OF THE PROTOSCOLEX *IN VITRO*
When cultured *in vitro*, a protoscolex can develop into a variety of different forms (*339, 791, 795*) and it is important to appreciate their interrelationships before considering the more general problems of asexual/sexual differentation (Fig. 10.5). Slightly different forms develop in the horse strain of *E. granulosus* (*706*). The following forms have been identified.

Unevaginated protoscolex (*Fig. 10.5* (*a*)). The protoscolex remains undifferentiated as in a hydatid cyst.

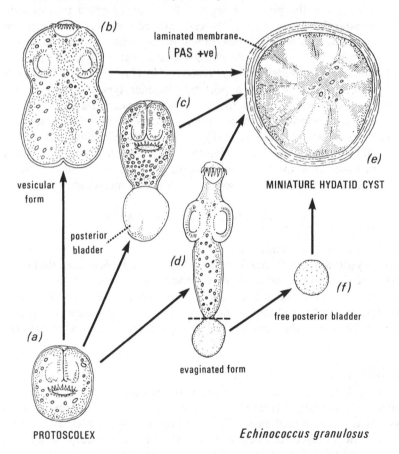

laminated membrane
(PAS +ve)

(b)

(c)

(e)

vesicular
form

MINIATURE HYDATID CYST

posterior
bladder

(d)

(f)

free posterior bladder

(a)

evaginated form

PROTOSCOLEX

Echinococcus granulosus

Fig. 10.5. The various forms developed from a protoscolex of *Echinococcus granulosus* when cultured in a monophasic (liquid) medium *in vitro*. ((*a*)–(*e*): after Smyth (1962, 1967); (*f*) after Rogan & Richards (1986).)

269

Cultivation of cestodes in vitro

Vesicular type (Fig. 10.5 (b)). A protoscolex swells and becomes 'vesicular', taking on a 'cottage-loaf' profile. This type becomes rounded, eventually secreting a laminated membrane (*313, 791*) and developing into a miniature hydatid cyst (Fig. 10.5 (e)). The laminated membrane is strongly periodic acid–Schiff (PAS) positive and is composed of a mucopolysaccharide (Chapter 4). The origin of this membrane was previously in doubt, but the fact that it is entirely of parasite origin is demonstrated by the fact that it develops *in vitro*, even in protoscoleces cultured in a (serum-free) defined medium (*313*).

Posterior bladder type (Fig. 10.5 (c)). A small 'bladder' or 'vesicle' develops in the posterior region of an unevaginated protoscolex, apparently arising from a few cells carried over from the germinal membrane to which it was previously attached within the brood capsule. This type also develops into a miniature hydatid cyst.

Evaginated protoscolex with posterior bladder (Fig. 10.5 (d)). This form appears to arise from a posterior bladder type, which becomes evaginated when the posterior bladder is at an early stage of development. In monophasic medium this will develop into the cystic type.

Free posterior bladder (Fig. 10.5(f)). In the horse strain (but not the sheep strain), some posterior bladders may become separated from protoscoleces and form independent bladders. These also develop into miniature cysts and secrete a laminated membrane (*706*).

CULTIVATION OF CYSTIC STAGES

Cystic *Echinococcus multilocularis*

As the cystic stage of *E. multilocularis* grows more rapidly than that of *E. granulosus*, it has proved to be more amenable to culture. Fragments of germinal membrane have been grown in complex media in which tissue proliferated and produced vesicles within 29 days; these contained protoscoleces at 55 days and proved to be infective to voles when injected (*670*).

Cystic *Echinococcus granulosus*

Early workers attempted the cultivation of protoscoleces of *E. granulosus* (after enzyme treatment) using a variety of media (*791*). Miniature hydatid cysts with a laminated membrane developed (Fig. 10.5(e)) but brood capsules or protoscoleces were not formed. This was not surprising for in the natural hosts brood capsules with protoscoleces require some 9–12 months to develop. In later experiments, the anlagen of brood capsules appeared in some eight-month cystic cultures (*804*).

Cyclophyllidea

IN VITRO CULTURE OF *ECHINOCOCCUS GRANULOSUS*

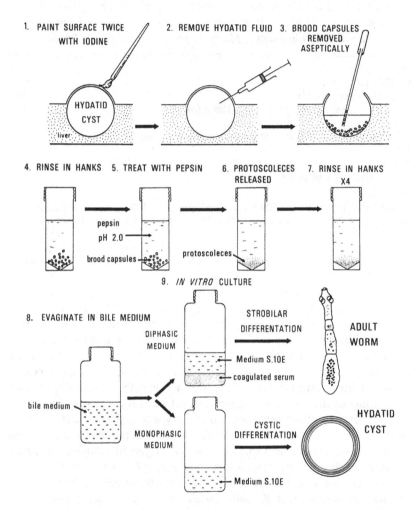

Fig. 10.6. Technique for the *in vitro* culture of *Echinococcus granulosus*. The composition of Medium S.10E is given in Table 10.2. For further practical details, the original papers should be consulted. (After Smyth, 1979, 1985*b*.)

Cultivation of cestodes in vitro

Table 10.2. *Medium S.10E. Liquid phase for the in vitro culture of* Echinococcus granulosus *or E. multilocularis. E. granulosus also requires a solid phase of coagulated serum (e.g. newborn calf) for strobilar development. (Modified from Smyth, 1979, 1985b)*

Basic medium	
CMRL 1066 (ml)	260
Foetal calf serum (ml)	100
5% (w/v) Yeast extract (in CMRL 1066) (ml)	36
30% (w/v) glucose (in dist. H_2O) (ml)	5.6
5% (v/v) dog bile or 0.2% (w/v) sodium taurocholate (in Hank's) (ml)	1.4
Plus:	
Buffer	
20 mM HEPES + 10 mM Na_2HCO_3	
Antibiotics	
100µg Gentamycin/ml [or 100 i.u. penicillin/ml and 100 µg streptomycin/ml]	

IN VITRO DIFFERENTIATION OF SEXUAL (= STROBILAR) STAGES

Echinococcus granulosus

Determination of the factors which induce the differentiation of a protoscolex in a strobilar direction has been one of the most fascinating problems of *in vitro* studies. The general failure of the more usual culture procedures to induce strobilar growth (*791*) led to the conclusion that some unsuspected and unusual requirement was missing from the culture conditions provided. Detailed study of the adult worm in the dog gut showed that the contact of the scolex was much closer than previously suspected, with the extended rostellum penetrating into a crypt of Lieberkühn (Fig.'9.5, p. 240). This led to speculation that the missing requirement was a solid substrate and contact with this in some way triggered strobilar differentiation, possibly via a neurosecretory mechanism (see Chapter 2).

Further experiments showed that this hypothesis generally held for *E. granulosus* (but contrast *E. multilocularis*, p. 274), as when protoscoleces were cultured in a diphasic medium with a suitable substrate (coagulated bovine serum) strobilar differentiation resulted and sexually mature adult worms developed (*795*). By varying the composition of the liquid phase (Table 10.2) and making other adjustments to the culture conditions, results gradually improved until a reasonably reliable technique was developed (Fig. 10.6). In its simplest form, this method utilises a roller tube system, but a more sophisticated system involving a circulating 'lift' system (Fig. 10.7) has also proved to be successful; this has the advantage that a medium

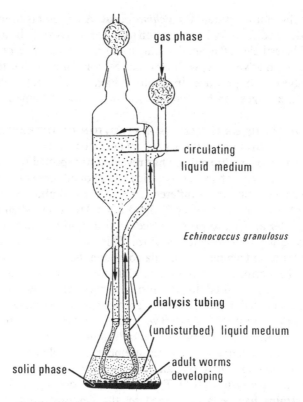

gas phase

circulating
liquid medium

Echinococcus granulosus

dialysis tubing

(undisturbed) liquid medium

solid phase

adult worms
developing

Fig. 10.7. A 'stationary' lift for the *in vitro* cultivation of adult *Echinococcus granulosus*, from protoscoleces. By use of cellulose tubing, the circulating medium is separated from the medium in the lower flask with the result that the interface between the worm and the solid substrate remains undisturbed; nutrients and waste materials can be exchanged via the cellulose membrane. Up to six such systems can be connected to a single gas pipeline. (After Smyth, 1969.)

change only once a week is required and up to six similar systems can be connected in series utilising one gas pipeline.

Some basic technical details are given in Table 10.2 and Fig. 10.6, but for complete protocols the original papers should be consulted (*809, 801, 804*).

Although this technique is successful for growing adult worms, fertile eggs have not been produced due to the failure of insemination to take place *in vitro* (*802, 812*). This result is reminiscent of early experiments with the culture of *Schistocephalus* (p. 262) in which failure of worms to become inseminated *in vitro* was overcome by compressing them during culture within semipermeable dialysis tubing (Fig. 10.2). This result suggests that some physical requirement (such as compression) for insemination was

273

lacking in the culture system for *Echinococcus*. All attempts to induce the appropriate conditions for insemination have so far proved to be unsuccessful (*802*). Nevertheless, the fact that only non-gravid worms are produced is somewhat of an advantage, with this species, for it means that the highly infective eggs are not produced in culture, which means that the physiology of the mature stages can be studied without danger of human infection.

ECHINOCOCCUS GRANULOSUS: DETECTION OF INTRASPECIFIC DIFFERENCES BY *IN VITRO* CULTURE

The application of *in vitro* culture has revealed unexpected biological and biochemical differences between hydatid isolates of different origins. All the early experiments on sexual differentiation described above were carried out on protoscoleces from sheep hydatid cysts. However, when this technique was later applied to protoscoleces of horse origin, surprisingly, the organisms failed to strobilate *in vitro*. This was at first attributed to suspected faults in technique or media components. When, however, after two years of experiments involving some 200 cultures, horse material failed to strobilate, it was realised that this result represented a new phenomenon, and it was concluded that isolates of *E. granulosus* from horse represented a different 'strain' from that of sheep with some unique (nutritional?) factor or culture condition for sexual differentiation at present unknown (*808*). This appears to be some unusual requirement, for isolates from buffalo, camel, cattle, goat and man have all been shown to differentiate sexually in the (sheep) *in vitro* system (*505*). That sheep and horse hydatids represent different strains has been confirmed by the demonstration of marked biochemical and physiological differences between them (Chapters 4–6; Table 4.1, p. 55; Table 5.10, p. 100).

ECHINOCOCCUS MULTILOCULARIS: *IN VITRO* DIFFERENTIATION OF SEXUAL STAGES

In striking contrast to *E. granulosus*, protoscoleces of *E. multilocularis* differentiate readily to sexually mature adults in the liquid phase of the culture system used for *E. granulosus*, i.e. the presence of a solid substrate is not essential for inducing sexual differentiation (*801, 804*). In early experiments, organisms showed a marked tendency to form 'monozoic' worms (see below; Fig. 10.8(*c*)), but as culture conditions improved, this tendency lessened and more normal worms developed (Fig. 10.8(*b*)). However, in some cultured worms with two to five proglottides (Fig. 10.8(*d*)) the cirrus region in each proglottis showed a characteristic 'bulge' (as in 'monozoic' forms; Fig. 10.8(*c*)) indicating that some inhibition of somatic development was still taking place. As with *E. granulosus*, insemination did not occur.

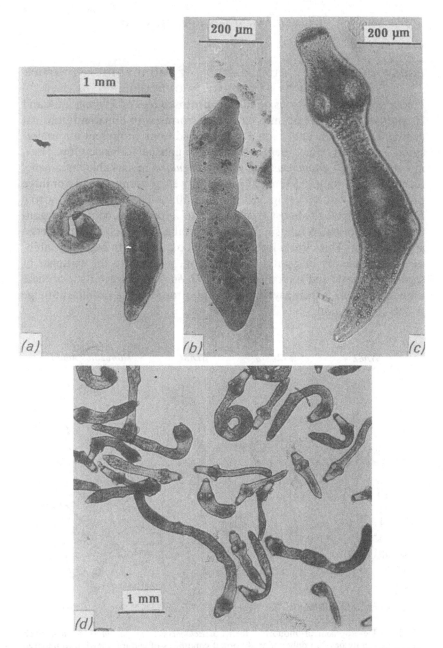

Fig. 10.8. *In vitro* cultivation of *Echinococcus granulosus* and *E. multilocularis*. (*a*) *E. granulosus*: sexually mature adult with three proglottides; 78-day culture. (*b*) *E. multilocularis*: normal adult with two proglottides, approaching sexual maturity; 40 day-culture. (*c*) *E. multilocularis*: 'monozoic' form present in 32-day culture. Note swelling ('cirrus bulge') due to development of genitalia without corresponding somatic growth. For morphology of this form, see Fig. 10.9. (*d*) *E. multilocularis*: 70-day culture showing a variety of types developed, including some monozoic forms together with adult worms with two to three proglottides. (After Smyth, 1979.)

Cultivation of cestodes in vitro

ECHINOCOCCUS SPP.: DEVELOPMENT OF 'MONOZOIC' AND OTHER ABNORMAL FORMS IN CULTURE

Under optimum conditions of culture, protoscoleces of *E. granulosus* and *E. multilocularis* develop into strobilate worms with apparently normal genitalia. However, under some culture conditions, at present not understood – as in early experiments – a variety of unusual forms develop. Thus, in cultures of *E. granulosus* and *E. multilocularis* remarkable 'monozoic' forms (Figs. 10.8(*c*), 10.9) have appeared, as well as worms with two or three proglottides (*797, 807*) (Figs. 10.8(*d*) and 10.9). These bizarre monozoic forms, which have an almost 'trematode-like' appearance, consist of organisms which contain a full, single set of genitalia but which fail to undergo segmentation. This clearly indicated that somatic growth had been inhibited independently of genital differentation, pointing to the existence of separate somatic and genital cell lines (*798, 806*). A characteristic of these monozoic forms is their asymmetrical appearance with a prominent bulge in the cirrus region.

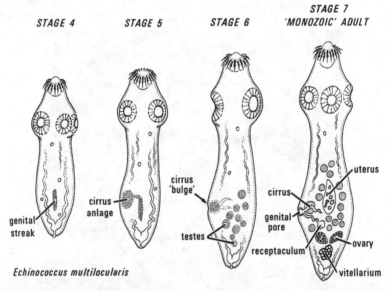

Fig. 10.9. Development of 'monozoic' forms of *Echinococcus multilocularis in vitro*. Such forms apparently develop only under abnormal conditions of culture; they fail to undergo strobilisation, but develop a complete set of male and female genitalia. Because somatic development is inhibited, the gentitalia swell within the organism to form a 'bulge'; see also Fig. 10.8(*c*). Similar forms have occasionally been found also in cultures of *E. granulosus*. (After Smyth & Barrett, 1979.)

MESOCESTOIDES CORTI

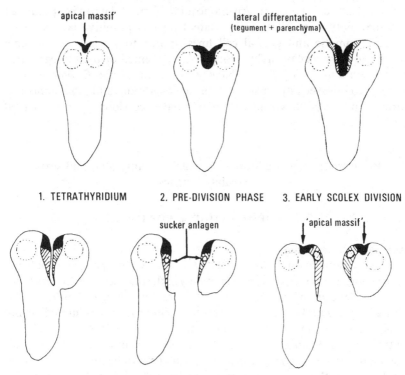

Fig. 10.10. Asexual multiplication of the tetrathyridium of *Mesocestoides corti*; the scolex in the adult worm appears to divide by the same mechanism. 1, In the non-dividing larva, a small area between the suckers is occupied by the 'apical massif', a polynucleated cell mass with cytomorphogenetic potential. During asexual multiplication it differentiates into tegumental syncytium, muscle, parenchymal cells and other cell types. 2, During the pre-division phase, the apical massif increases in size and its lateral areas (hatched zones) differentiate into parenchymal cells and tegument. 3, The scolex commences to divide longitudinally. 4, A tegumental constriction precedes the separation of a half-scolex and tegumental and parenchymal differentiation continues. The apical massif remains undifferentiated in the future apical centres of the two scoleces. 5, The scoleces separate asymmetrically leaving the larger fragment with a 'shoulder'. The anlage of a new sucker develops in each scolex. 6, Final regeneration and achievement of symmetry. An apical cleft (arrows) develops in each regenerating scolex. (Modified from Hess, 1980.)

Cultivation of cestodes in vitro

This unusual result has evolutionary overtones, for organisms in the subclass Cestodaria are similarly monozoic in form and it can be speculated that the ability to strobilate evolved later from this 'primitive' type. Growth of both somatic and genital cell lines appear to be capable of being suppressed, for, occasionally, elongated, segmented or pseudosegmented forms, containing no genitalia, have been seen in culture. *E. multilocularis* appears to be especially sensitive to culture conditions and other abnormal forms – such as adult worms with buds – have occasionally been observed (*807*).

Mesocestoides spp.: culture of asexual (tetrathyridia) and sexual (strobilar) stages

MESOCESTOIDES CORTI

Asexual differentiation
It has been pointed out (Chapter 9) that, uniquely, this species is capable of undergoing asexual reproduction not only in the intermediate (rodent) host but also in the gut of the definitive (carnivore) host (Fig. 9.12, p. 256). This species rapidly undergoes asexual reproduction (by longitudinal division) under relatively simple culture conditions (*606, 607, 907*). Asexual splitting of the scolex or budding of the scolex also occurred in sexually differentiating worms grown *in vitro* (see below). Asexual multiplication in this species appears to be related to the presence of a group of cells – the 'apical massif' – which occurs in the scolex (Fig. 10.10) (*322, 323*).

Sexual differentiation
Development of sexual forms in culture were first reported simultaneously by Barrett *et al.* (*47*) and Thompson *et al.* (*871*), essentially using media based on that used for *Echinococcus granulosus* (*801*). In these preliminary studies, fertilisation did not take place, as evidenced by the absence of oncospheres. The various stages of development are shown in Fig. 10.11. More critical examination of the culture conditions (*607*) showed that conditions favouring growth and sexual differentiation were: (*a*) pre-conditioning tetrathyridia by storing in saline for 24 h; (*b*) a liquid media; (*c*) a pH > 7.4; (*d*) anaerobic conditions + 5% CO_2 (Fig. 10.12). Although these conditions are unlikely to be optimal, in the most successful cultures insemination took place in some proglottides and a few hooked oncospheres were seen in the paruterine organ (Fig. 10.11). Since tetrathyridia of this species have been grown *in vitro* from oncospheres

278

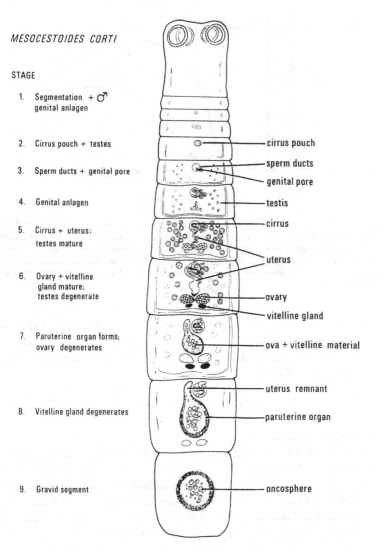

MESOCESTOIDES CORTI

STAGE

1. Segmentation + ♂ genital anlagen

2. Cirrus pouch + testes — cirrus pouch

3. Sperm ducts + genital pore — sperm ducts — genital pore

4. Genital anlagen — testis

5. Cirrus + uterus; testes mature — cirrus — uterus

6. Ovary + vitelline gland mature; testes degenerate — ovary — vitelline gland

7. Paruterine organ forms; ovary degenerates — ova + vitelline material

— uterus remnant

8. Vitelline gland degenerates — paruterine organ

9. Gravid segment — oncosphere

Fig. 10.11. *Mesocestoides corti*. Diagrammatic representation of the stages of sexual differentiation during development of a tetrathyridium to an adult worm *in vitro*. (Reprinted with permission from *International Journal for Parasitology*, **12**, Barrett, N. J., Smyth, J. D. & Ong, S. J., Spontaneous sexual differentiation of *Mesocestoides corti* tetrathyridia *in vitro*, © 1982, Pergamon Journals Ltd.)

279

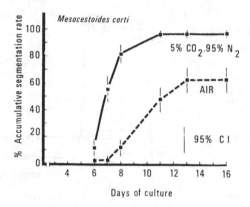

Fig. 10.12. *Mesocestoides corti*: effect of the gas phase on the induction of sexual differentiation of tetrathyridia *in vitro*. The anaerobic phase induced significantly higher segmentation rates than air. CI, confidence intervals. (Reprinted with permission from *International Journal for Parasitology*, **16**, Ong, S. J. & Smyth, J. D. Effects of some culture factors on sexual differentiation of *Mesocestoides corti* grown from tetrathyridia *in vitro*, © 1986, Pergamon Journals Ltd.)

(*909*), it should now be possible to complete the entire life cycle of *M. corti in vitro*. With improved techniques *in vitro*, this species is likely to become a valuable experimental model for studying the differentiation, biochemistry and physiology of cestodes.

MESOCESTOIDES LINEATUS

Sexual differentiation

This species has a life cycle similar to that of *M. corti* but, unlike it, does not undergo asexual reproduction in either the larval or adult stage. The adult of this species has also been grown *in vitro* from the tetrathyridium, but, in this case, the induction of sexual development appears to be related to the presence and activity of trypsin, for differentiation was inhibited or decreased by the presence of soybean trypsin inhibitor (*382*).

TAENIIDAE: CULTURE OF ASEXUAL AND SEXUAL STATES

Culture of cysticerci to adults

Although many different species of *Taenia* are of considerable medical or veterinary importance in man and domestic animals, attempts to culture them *in vitro* have been very unsatisfactory and, to date, no species has been grown from larva to adult worm with gravid proglottides containing fertile eggs.

The most successful species cultured to date has been *T. pisiformis*, using essentially the technique used for *E. granulosus* (*610, 801*). Although sexually mature adults with proglottides were obtained, these did not contain infective oncospheres.

A number of other species have been cultured to partial sexual maturity: *T. saginata* (*93*), *T. crassiceps* (*194*), *T. hydatigena* (*611*).

Culture of oncospheres to cysticerci

The following species of *Taenia* have been grown from oncospheres to active larvae with suckers: *T. hydatigena*, *T. ovis*, *T. pisiformis*, *T. serialis* and *E. granulosus* (*308, 314*). The techniques of Heath & Smyth (*314*), or modifications of them, have been used widely by other workers for the production and collection of E/S antigens of oncospheres in the preparation of potential vaccines (see Chapter 11, p. 302). Only limited success has been achieved with the culture of larval *T. saginata* (*310, 471*) and almost none with *T. solium*, and the *in vitro* culture of these important species remains a challenge.

Culture of cestode cells

In relation to the recent advances in molecular biology (Chapter 6), a major stimulus to this would be establishment of cestode cell lines. Early workers (*796*) using relatively crude *E. multilocularis* material were unable to establish cell cultures. Somewhat more successful results have been obtained using cells obtained from trypsinised germinal membranes of *Echinococcus granulosus* from liver hydatid cysts (from cattle) (*719*). As it is possible that such cultures could be contaminated with host liver cells, the *E. granulosus* cells were identified by karyotyping ($2n = 18$) and by direct immunofluorescence with a negative reaction to anti-bovine antibodies. Although the cells were reported as being 'grown for several months' with some mitosis occurring, there was no evidence that constant cell growth was achieved or a 'true' *Echinococcus* cell line obtained. This is clearly a problem which presents a major challenge to future research workers. Now that *Echinococcus*-specific DNA probes are available (Chapter 6, pp. 150–3), the precise nature of such cell lines, i.e. whether parasite, host or parasite/host hybrid, can be determined.

Cultivation of cestodes in vitro

Cryopreservation

The supply of living cestode material for *in vitro* and *in vivo* experiments presents a major barrier to progress in research. It is therefore worth drawing attention to the fact that recent work by Eckert & Ramp (*191*) have shown that cystic *Echinococcus mutilocularis* can be maintained successfully by cryopreservation without the proliferative capacity being lost. The establishment of this most useful technique should act as a great stimulus to future work. For technical details, the original paper (*191*) should be consulted.

11

Immunobiology of cestodes

General considerations

Host–parasite interaction

GENERAL COMMENTS

With rare exceptions (e.g. *Hymenolepis nana*) a cestode makes contact with the tissues of at least two different hosts during its life cycle. The degree of immunological response by each host is related to three main factors: (*a*) the nature of the tissue site invaded, (*b*) the intimacy of the host–parasite contact, and (*c*) the stage of development of the cestode, i.e. whether adult or larva. For example, in the case of *Taenia saginata*, the host–parasite contacts established during the life cycle are: (*a*) the scolex, when attached to the intestine of the definitive (human) host, (*b*) the oncosphere, during its penetration of the intestinal mucosa of the intermediate (bovine) host and subsequent migration, and finally (*c*) the developing larva established in its final tissue site.

This chapter will deal with the immune reactions of the host to adult and larval cestodes. Most work has been centred on species of *Taenia* and *Echinococcus* in man and domestic animals and on the Hymenolepididae in laboratory animals. In parallel with the remarkable developments in recombinant DNA technology (see Chapter 6) which holds out the tantalising possibilities of producing parasite vaccines – there has been an explosive interest in attempts to develop vaccines against cestodes of medical, veterinary or economic importance, so far without significant success (p. 301). It could, indeed, be argued that this effort, important as it is, has somewhat diverted attention from many basic problems of cestode biology, on whose solution the development of successful vaccines may ultimately depend. This chapter aims to present a brief overview of the basic immunological problems adult and larval cestodes face in their definitive or intermediate hosts; it does not attempt to review the whole immunobiology of the group.

The literature in cestode immunity is so voluminous that references in this chapter are largely confined to reviews.

Basic concepts of immunity

Because the basic concepts of immunity are now so well known, no attempt is made to deal with these here. For reference, however, the interrelationships between the main cells involved in immune processes are shown in Fig. 11.1 and the properties of the various immunoglobulins are summarised in Table 11.1. For more extensive coverage of theoretical and practical principles of immunology, reference should be made to recent texts or reviews, some of which are listed below.

Immunity in mammals including man
Fundenberg *et al.* (*241*), Jawetz *et al.* (*361*), Playfair (*644*), Roitt (*708*), Roitt *et al.* (*709*), Wakelin (*919*).

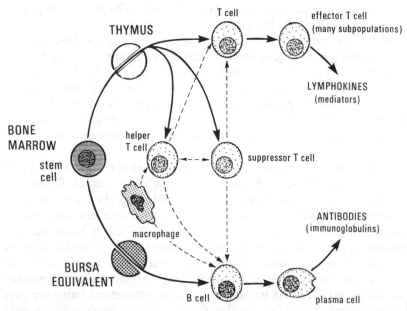

Fig. 11.1. Diagram of the major interactions of the immune system. B cells (B lymphocytes) react with soluble antigens and T cells (T lymphocytes) react with antigens occurring on cell surfaces. The population of T and B lymphocytes in individual organisms exhibit a very diverse antigen-binding repertoire due to the diversity of antigen receptors. However, individual B lymphocytes express only a single type of receptor; the same may be also true for T lymphocytes. (Reprinted with permission from Jawetz, E., Melnick, J. L. and Adelberg, E. A. *Review of Medical Microbiology*, 17th edn, © Appleton & Lange, 1987.)

Table 11.1. *Some characteristics of immunoglobulins. (Data from Jawetz et al., 1987)*

	IgG	IgM	IgA	IgD	IgE
Sedimentation coefficient	7 S	19 S	7 S or 11 S[a]	7 S	8 S
Molecular weight	150 000	900 000	170 000 or 400 000	180 000	190 000
Heavy chain symbol	γ	μ	α	δ	ε
Average concentration in normal serum (mg/dl)	1000–1500	60–180	100–400	3–5	0.03
Half-life in serum (days)	23	5	6	3	2.5
Prominent in external secretions	−	−	+ +	−	+
Percentage carbohydrate	4	15	10	18	18
Crosses placenta	+	−	−	−	−
Fixes complement	+	+	−	?	−
Examples of antibodies	Many Abs to toxins, bacteria, viruses; especially late in Ab response.	Many Abs to infectious agents, especially early in Ab response; antipolysaccharide Ab; cold agglutinins.	Important as secretory antibody on mucous membranes.	No proved Ab activity; main immunoglobulin on surface of β lymphocytes in newborn.	Binds to mast and basophil cells; raised in allergic and parasitic infections.

Notes:
−, no; +, + +, yes, to lesser or greater extent.
[a] 11 S, molecular weight 400 000 IgA in external secretions; 7 S molecular weight 170 000 IgA in serum.

Immunobiology

Immunity in lower vertebrates and invertebrates
Azzolina *et al.* (*31*), Marchalonis (*511*), van Muiswinkel & Cooper (*896*).

Immunity to helminths, including cestodes
Adams & Cobon (*1*), Befus *et al.* (*55*), Flisser *et al.* (*224, 226*), Heath (*309*), Ito & Smyth (*353*), Lloyd (*447*), Rickard (*686*), Rickard & Williams (*691*), Soulsby (*817*), Symons *et al.* (*848*), Thompson (*864*), Wakelin (*919–921*), Williams (*960*).

Immunity to adult cestodes

General comments

For many years, it was generally held that adult cestodes were non-immunogenic or poorly immunogenic (*353*). This view appeared to be based on the assumption that the scolex of most cestode species made only a loose, non-penetrative contact with the intestinal mucosa. It was recognised that an exceptional case was that of *H. nana*, which has a larval tissue phase in the villi of man and mice (Fig. 11.6) (which is strongly immunogenic) as well as an adult stage.

The concept that the scolex of adult cestodes is generally non-penetrative has been shown not to hold for species such as *Echinococcus granulosus* and *E. multilocularis*, where the scolex penetrates the crypts of Lieberkühn and, occasionally, even the lamina propria (Fig. 9.5) often resulting in a complete breakdown of the mucosal epithelium. These species can be regarded as both tissue and lumenal parasites. That the scolex contact is close is reflected in the fact that anti-*Echinococcus* antibodies appear in dog sera 14 days post-infection (p.i.) (*368*). It is likely, however, that in many cestodes the scolex contact is more superficial and breakdown of the mucosa may not occur, especially in those species which undergo diurnal migration – see Chapter 9.

Immunobiology of the intestinal mucosa

GENERAL COMMENTS

Before discussing the immunological reactions to adult cestodes, it is important to review our present knowledge of the immunobiology of the vertebrate intestine, about which many new data have become available in recent years (*53, 55, 80, 81*). The account below summarises briefly the local aspects of mucosal tissue responses (Fig. 11.2) which appear to be important in parasitic infections. An important concept which has emerged from these recent data is that local immunoregulatory events in mucosal infections may be distinct from those seen in systematic sites (*55*).

IMMUNOGLOBULIN A IN INTESTINAL INFECTIONS

Immunoglobulin A (Fig. 11.3) is the major immunoglobulin present in intestinal secretions. It is synthesised in the lamina propria by local plasma cells and is found there in the dimeric form $(IgA)_2$, which is transported into the intestinal secretions by two pathways. Firstly, it diffuses from the site of synthesis into the columnar cells of the gut and here it forms a remarkable

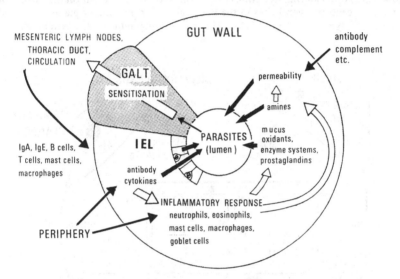

Fig. 11.2. Summary of the host responses to mucosal parasitic infections. Sensitisation occurs initially in gut-associated lymphoid tissue (GALT); various cells of mucosal or peripheral origin are recruited largely by thymus-dependent mechanisms; the inflammatory response so generated produces an environment hostile to the continued survival of the parasites. Heavy solid arrows denote possible sources of effector molecules; IEL, intraepithelial lymphocytes, some of which are granulated, and probably natural killer cells. (After Befus & Bienenstock, 1982, with permission from S. Karger AG, Basel.)

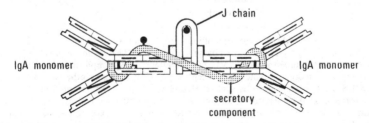

Fig. 11.3. Simplified diagram showing structure of human secretory IgA1 (sIgA1). The secretory component which is probably wound around the IgA dimer, as shown, is now known to be a proteolytic fragment of a receptor (poly(Ig)receptor); see also Fig. 11.4. The J chain is required for joining the two subunits. (Modified from Roitt *et al.*, 1986.)

complex with a protein originally known as the 'secretory component' (SC), which is a glycoprotein (M_r 60 000) synthesised in these cells. The secretory antibody (often referred to as sIgA1) forms the complex IgA_2 (($IgA)_2$SC), which is transported across the epithelial cells in vesicles and exocytosed into the intestinal lumen.

SC is now known to be a proteolytic fragment of a receptor (poly(Ig) receptor), synthesised in glandular epithelial cells, that mediates transport of polymeric IgA (and IgM) into external excretions (*512*). This receptor is present on the surface of epithelial cells at several sites in the body (gut, liver, mammary glands, etc.) and binds to polymeric IgA (and IgM). This receptor–ligand complex then undergoes

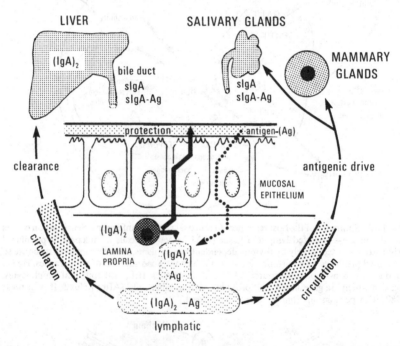

Fig. 11.4. Synthesis and transportation of immunoglobulin A in animals and (probably?) in man. Dimeric IgA (IgA_2) is synthesised by lymphoid cells in the intestinal lamina propria. It is transported into the lumen as secretory IgA (sIgA) after specific complexing with the secretory component, a proteolytic fragment of a receptor (poly(Ig)receptor) synthesised by these cells (see text). Some IgA (($IgA)_2$) diffuses into the lymphatics and the circulation, from which most of it is cleared as sIgA by hepatocytes and passed in the bile, but some is cleared and secreted by intestinal, lacrymal and salivary glands. Other cells with specific receptors for IgA, including monocytes, neutrophils and lymphocytes may also play a role in antigen clearance. (Redrawn with permission from Some thoughts on the biologic role of immunoglobulin A, by J. Bienenstock and A. D. Befus, *Gastroenterology*, **84**, 178–85. Copyright 1983 by The American Gastroenterological Association.)

endocytosis and is transported across the cytoplasm of the cell in vesicles to be excocytosed at the cell surface.

The second pathway of sIgA is via bile (Fig. 11.4); this has so far only been confirmed in laboratory animals, although it may occur in man. IgA receptors have been shown to occur in a range of cell types, including polymorphonuclear leukocytes, monocytes, lymphocytes, helper and suppressor cells; it is not known if IgA receptors occur on Küpffer cells or other cells of the stationary reticuloendothelial system. It is important to note that sIgA appears to be produced most effectively by *local* stimulation rather than by systemic infection or antigen administration (*361*).

In spite of the widespread occurrence of sIgA, not only in cestode infections, but in all other helminth intestinal infections, the role played by IgA remains unknown (*921*). In relation to this, it has been speculated (*53*) that 'IgA antibody is relevant in the control of the uptake of parasite antigens, their clearance from the circulation and in the modulation of the activities of various cells in the inflammatory events associated with infections; this may involve enhancement of cytotoxic functions or the minimization of potentially pathogenic reactions.'

MUCOSAL (INTESTINAL) MAST CELLS

Mast cells have long been recognised as cells which play a major role in the inflammatory responses by releasing mediators either when tissue is damaged or under the influence of antibody (IgE). By increasing vascular permeability, the resultant inflammation allows cells and complement to enter the tissues.

Within recent years, there has been increasing evidence that mucosal (intestinal) mast cells (MMC) differ in a number of fundamental respects from peritoneal (connective tissue) mast cells (PMC) (*55*). These differences are summarised briefly in Table 11.2.

Specific differences to note are a 10-fold lower level of histamine in the mucosal mast cell and the occurrence of distinct proteases and proteoglycans and (possibly) a different arachidonic acid metabolism for each type (*55*). These differences are sufficiently great to conclude that extrapolation of data on PMC (on which most research has been carried out) to MMC would be unwise.

The release of mediators (histamine, serotonin, bradykinin etc.) from mast cells is activated by IgE–antigen interactions and their action on smooth muscle, mucus glands and blood vessels probably plays an important role in the responses to intestinal infections. For example, mucus gland secretion can be stimulated by histamine and increased mucous secretion is a characteristic of several intestinal helminth infections (*80*).

289

Immunobiology

Table 11.2. *Comparative characteristics of mucosal (intestinal) mast cells and peritoneal (connective tissue) mast cells in the rat. (Data from Jarrett & Haig, 1984)*

Properties	Mucosal mast cells	Connective tissue mast cells
Morphology	Fewer variable-size granules/cell	Many uniform-size granules/cell
Size	9.7 μm	19.6 μm
Thymus-dependent proliferation in immune responses	Yes	No
Fixation and staining	Require special histological technique	Conventional technique
	Mature cell is Astra blue-positive	Mature cell stains with saffranin
Serine protease	RMCP II	RMCP I
Proteoglycan	Non-heparin	Heparin
Histamine	1.3 pg/cell	15 pg/cell
IgE	Cytoplasmic	Surface
Life span	<40 days	>6 months
Histamine release by 48/80	Resistant	Susceptible
Effects of anti-allergic compounds	Resistant to effects of theophylline and disodium chromoglycate	Susceptible

Notes: RMCP, rat mast cell protease.

INTRAEPITHELIAL LEUKOCYTES

In addition to mast cells, the intestinal epithelium contains a number of heterogeneous cell types referred to as *intraepithelial leukocytes (193)*. These cells are intimately associated with lumenal antigen and many types contain cytoplasmic granules. Many are lymphocyte-like (Fig. 11.2) and have been called 'granulated lymphocytes'; other cells present are globule leukocytes, which are believed to be (immature) mast cells, large granular leukocytes, non-granular lymphocytes and (occasionally) lymphocytes. Functionally, these cells constitute a population containing natural killer cells, cytotoxic T cells and natural cytotoxic cells (*193*). The role of many of these in intestinal immunoregulation appears to be largely unknown.

Immunobiology of specific species

GENERAL COMMENTS

Most research on immunity to adult cestodes has been carried out on the Hymenolepididae in laboratory animals, with comparatively few studies being made on the Taeniidae, which includes the important genera *Taenia* and *Echinococcus*. The account below deals briefly only with representative species from these cyclophyllidean families. Little appears to be known regarding immunity to adult pseudophyllidean cestodes, but some studies on responses to larvae in fish have been made. Immunity to adult cestodes in general has been reviewed comprehensively by Rickard (*686*) and Andreassen (*17*).

General comments
The immunobiology of this family has been reviewed by Hopkins (*332*) and Ito & Smyth (*353*), the species most studied being *Hymenolepis diminuta*, *H. nana*, *H. microstoma*, and *H. citelli*, in that order. Much of the experimental work has been carried out on what must be considered to be artificial host–parasite systems, i.e. those not occurring naturally. As discussed below, *H. diminuta* grows readily in rats (Fig. 11.5) but in the mouse host it develops to some extent and is then rejected without reaching maturity. In these

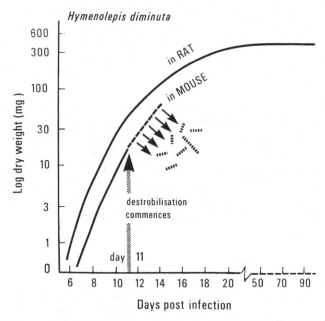

Fig. 11.5. *Hymenolepis diminuta*: comparative development in the rat (the natural definitive host) and the mouse (a non-permissive host). In most mice, growth ceases abruptly on day 10 + 1 and destrobilisation commences, leaving only a scolex and a neck 0.5–2 mm in length. (Modified from Hopkins *et al.*, 1972.)

291

infections mice are often referred to as 'abnormal' or 'unnatural' hosts. Both these terms are difficult to define precisely and the term non-permissive has been proposed as being more appropriate and is used in this text. A brief account of the immunological responses to *H. diminuta* and *H. nana* is given below.

Hymenolepis diminuta/rat system
H. diminuta develops readily in the rat (Fig. 11.5) and in the normal life span of the worm there is no evidence of worm expulsion or senescence – at least at infection levels of 10 or less. At levels above 10, there is some reduction in growth and egg production and expulsion may occur (Figs. 9.6 (p. 244) and 9.7 (p. 245)); there is controversy as to whether this is due to a non-immunological 'crowding effect' (see Chapter 10) or to a true immunological response. That an immunological response of some kind takes place, even in the permissive host, is reflected in the fact that IgG and IgE have been detected on the tegument of *H. diminuta* (*17*). Whether the scolex or the strobila is the chief source of antigen is still a matter of some dispute.

Hymenolepis diminuta/mouse system
In contrast to the rat, in the (non-permissive) mouse host, even a single worm infection of *H. diminuta* is rejected within 7–10 days (*353*). In most cases growth ceases abruptly at day 10 + 1, when worms are 14–30 cm (Fig. 11.5). In secondary infections, worm recovery is lower than in primary infections and worms are severely stunted and are more rapidly expelled (*332*).

Antibody responses in the *H. diminuta* mouse system have been reported from a number of workers and isotypes of IgA, IgG and IgM have been found on this cestode. Moreover, their titres increased coincidently with worm rejection and darkened areas suggested that these surface binding antibodies have a functional role in inducing morphological alterations. It is not known, however, whether the presence of these antibodies on the surface is due to specific or non-specific absorption (*353*). In this system, passive protection of mice by transfer of immune sera has not been demonstrated.

Hymenolepis nana/mouse/rat/human systems
The immunobiology of this parasite is complicated by the fact that hosts can be infected (*a*) *directly*, via egg infections, or (*b*) *indirectly*, via cysticercoids (in beetles). In the *direct* life cycle, eggs hatch in the duodenum and oncospheres immediately penetrate the villi and develop into cysticercoids (Fig. 11.6). When fully developed, these break out of the villi, attach and

Hymenolepis nana / MOUSE

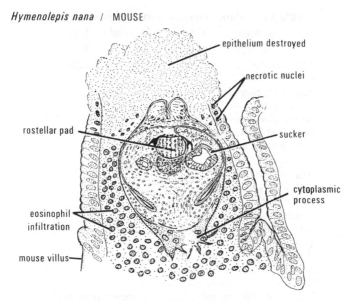

Fig. 11.6. *Hymenolepis nana*: section of a cysticercoid in the intestinal villus of a mouse; 96 h p.i. No fibrous capsule is formed, but marked eosinophilia develops.

grow to adult worms. There are thus both tissue and lumen phases in the cycle. In the *indirect* cycle, ingested cysticercoids evaginate and attach directly to the mucosa; i.e. there is only a lumen phase. The use of this species as a model thus allows immunogenesis of the tissue and lumen phases to be compared and these systems have been much investigated.

That the tissue phase of *H. nana* is highly immunogenic is witnessed by the fact that even a single oncosphere invading a villus can elicit complete protection (*353*). Moreover, this protection can develop within a few days. It must be stressed, however, that – as with most helminth parasites – the immune responses may vary substantially in different strains of the same host, a factor not always appreciated in these experiments. Immune responses known to be influenced by host genotype are listed in Table 11.3.

Egg-induced immunity appears to involve stage-specific immunogens against (*a*) the tissue phase of egg challenge (early response) and (*b*) the lumen phase of cysticercoid challenge (late response). This immunogenetic pattern is thus similar to the development of 'early' and 'late' immunity in larval taeniid cestodes (Fig. 11.7). The effector mechanism of the early response has been shown to be thymus dependent, X-irradiation sensitive, cell mediated and antibody mediated; the response is visualised by eosinophilia infiltration around the invading oncospheres (Fig. 11.6) (*353*).

Immunobiology

Table 11.3. *Parameters of anti-helminth immune responses known to be influenced by host genotype. (Data from Wakelin, 1986)*

Cellular
Antigen recognition by lymphocytes
Eosinophilia
Granulomatous hypersensitivity
Intestinal mastocytosis
Macrophage activation

Serological
Antibody isotype
Antibody specificity
Complement components
Hypergammaglobulinaemia
Rate of antibody production

Functional
Parasite reproduction
Survival of primary infection
Survival of secondary infection
Threshold levels of infections for immune response
Response to vaccination
Susceptibility to parasite-induced immunodepression

In most strains of mice, worms are lost within one to two months, but the mechanism of worm expulsion remains unknown. Although early workers believed that only the tissue phase and not the lumen phase induced immunity, recent work (*800*) using 'mouse derived' cysticercoids (i.e. extracted from villi) suggests that the lumen phase, too, appears to be immunogenic.

TAENIIDAE

Taenia and *Echinococcus*

Immunity to adult Taeniidae has been reviewed by Rickard (*687*) and Ito & Smyth (*353*). Due, undoubtedly, to the difficulty of obtaining suitable experimental material, relatively little work has been carried out on the immunobiology of adult cestodes of this family, which include many species of medical, veterinary or economic importance. Early workers were unable to demonstrate resistance to challenge infections with *Taenia taeniaeformis* in cats, or with *Multiceps glomeratus* or *T. hydatigena* in dogs. Some early experiments seemed to indicate that a certain degree of immunity developed against *Echinococcus granulosus* in dogs but, later, more critical work failed to support this conclusion (see Vaccination, below). This result is rather surprising, because, as shown earlier (Fig. 9.5, p. 240) the scolex of *E. granulosus* penetrates the crypts of Lieberkühn in the mucosa and may break down the lining epithelia and somatic and (probably) E/S antigens must certainly be taken into the circulation.

The development of *in vitro* culture techniques (Chapter 10) for oncospheres, larvae and adult worms has now made it possible to collect E/S antigens from all stages of the life cycle. The use of E/S scolex antigens with sera from dogs infected with *T. hydatigena, T. pisiformis* and *E. granulosus* have shown that a relatively strong antibody response develops within three weeks p.i. (*367*). Moreover, use of the E/S scolex or crude protoscolex antigens in enzyme-linked immunosorbent assay (ELISA) has recently led to the development of a specific immunological test for *E. granulosus* in dogs (*368*). Since this response disappears rapidly after purging and does not give cross-reactions with other dog Taeniidae, it is likely to prove to be a valuable serological test, particularly in control programmes.

Immunity to larval cestodes

General account

Most work has centred on cyclophyllidean species whose larvae develop in mammals. Few studies have been made on the other cestode orders, such as the Pseudophyllidea whose larvae develop in lower vertebrates, especially fish, amphibia and reptilia. Apart from the *H. nana*/rodent system, discussed earlier, the most studied species have been those which are either readily maintained in laboratory animals or are of medical, veterinary or economic importance, i.e. *Echinococcus granulosus, E. multilocularis, Mesocestoides corti, Taenia crassiceps, T. hydatigena, T. multiceps, T. ovis, T. saginata* and *T. solium*. The account given here has been restricted largely to these species.

Valuable reviews on various aspects of immunity to larval cestodes are those of: Flisser *et al.* (*225, 226*), Lloyd (*447*), Rickard (*686, 687*), Rickard & Lightowlers (*690*) and Rickard & Williams (*691*). Early work has been reviewed by Smyth (*796*).

Immunological responses to larval cestodes

GENERAL RESPONSES
Larval cestodes in mammals make intimate contact with host tissues at two sites (Fig. 11.7): (*a*) during the early phase of infection, when the recently hatched oncosphere penetrates the intestine; and (*b*) at the final encystment

Immunobiology

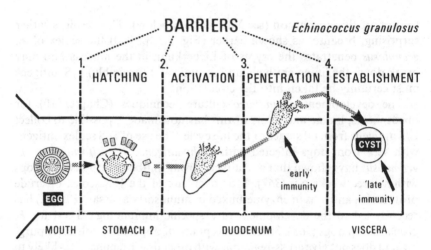

Fig. 11.7. Pattern of immune responses of an oncosphere of a taeniid cestode (e.g. *Echinococcus granulosus*) during initial penetration of the hatched oncosphere (initiating 'early immunity') and its subsequent migration to, and establishment in, its definitive site, where it is subject to 'late immunity'. The exact site of hatching (in man) is unknown; most eggs probably hatch in the upper duodenum.

site, commonly in viscera, muscles, nervous system or body cavity. It is not surprising, therefore, to find that a high degree of acquired resistance to infection is common – a situation which has greatly encouraged workers to attempt to develop vaccines against major cestode zoonoses such as hydatid disease and cysticercosis (p. 301). Both T- and B-cell responses have been shown to be produced in a number of experimental systems (e.g. *Echinococcus* spp./rodent) so that both humoral and cell-mediated mechanisms are initiated.

The immune response at the oncosphere penetration site is often referred to as 'early' or 'pre-encystment' immunity and that at the final establishment site as 'late', 'postoncospheral' or 'postencystment' immunity (Fig. 11.7). This division of larval cestode immunity into two areas, though convenient, may seldom represent the true immunological picture. This is because larvae of some species, such as *T. pisiformis*, undergo a migratory phase (through the liver) before emerging into the body cavity to undergo their final development and, during this migration, the host may be exposed to antigens involved in this migratory process. The importance of this is emphasised by the fact that the *T. pisiformis* larva, at 6 days p.i., develops an additional secretory gland (Fig. 8.23, p. 231) during its postoncospheral migration through the liver, and E/S antigens must clearly be released. The development of effective vaccines is therefore more likely to be successful if such vaccines are based on antigens produced by *all* phases of larval development, and not just those from the oncosphere and the established larva.

Secretory IgA in gut secretions and (where relevant) in colostrum probably plays an important role in attacking invading oncospheres, a situation related to the fact that this antibody may be resistant to intestinal enzymes. Thus, it has been shown that infection of mice with *T. taeniaeformis* results in the production of protective intestinal IgA antibodies against challenge infection and oral immunisation resulted in a protective immune response against this species (*447*). IgA antibody response has also been reported in *T. taeniaeformis* infection in rats and in *T. pisiformis* in rabbits (*687*).

Mast cells accumulate around both invading oncospheres and developing larvae and it is possible that IgE may react with antigen causing degranulation of these cells with a consequent increase in permeability; this would allow IgG antibodies to more easily reach the invading site. Regarding the cellular response, although eosinophils may accumulate around invading oncospheres, there is no evidence that these cells actually damage oncospheres or postoncospheral stages. The role of macrophages or neutrophils, which may accumulate at the site of dying oncospheres, remains unknown.

Attention has already been drawn to the fact that immunity in the *H. nana*/rodent system represents an unique situation in that the oncosphere develops in a villus after penetration, so that the penetration and final encystment sites are essentially the same. Passive protection against *H. nana* in mice can be induced by serum transfer, not only prior to the oncospheral invasion, but also after invasion has taken place. Oncospheral agglutination has also been shown to occur in *H. nana*-infected mouse serum but never in uninfected sera (*353*).

When newly hatched oncospheres reach their site of establishment, they become transformed from a stage of being highly susceptible to attack to one of almost complete insusceptibility. The possible basis of this insusceptibility is discussed further below (p. 299). Most studies on the responses of the postoncospheral stages have been carried out on larval species which can be passaged without using egg infection, e.g. *T. crassiceps*, *Echinococcus granulosus*, *E. multilocularis* and *M. corti*. It is important to emphasise the possible danger of extrapolating results from such studies, because the intermediate stages of infection – the oncosphere penetration of the gut and subsequent migration to the encystment site – have not been

Immunobiology

experienced by the host. In confirmation of this, it has been shown that in rats, antibody response from *T. taeniaeformis* strobilocerci implanted into rats was qualitatively different from that of an egg-induced infection (*686*).

Although the cellular and humoral response in experimental animals tends to be relatively uniform, it must be remembered that in man (and domestic animals) the immune responses can vary enormously. This is undoubtedly related to human genetic diversity – unlike the uniform genetic background of most experimental animals. These responses have been much studied in hydatid disease and (*T. solium*) cysticercosis. In the latter case, the frequency of different precipitation bands in serum immuno-electrophoresis (Fig. 11.8) and of the immunoglobulin classes (Table 11.4) show great variation between patients (*226*). Moreover, some patients show no humoral or cellular response whatsoever (*226*). Similarly, there is much variation in the immune responses to hydatid disease and, again, some patients show no detectable antibody (*734*).

Cellular responses to cestode infections have been much studied and – as in most helminth infections – eosinophils accumulate around establishing cestode larvae. It is known that sensitising parasite antigen stimulates sensitised lymphocytes to release an *eosinophil colony-stimulating factor* (*657*). The eosinophil granule is composed of an arginine-rich protein, major basic protein (MBP) and there is evidence that MBP may function in killing schistosomes *in vivo* and *in vitro*. Although eosinophils have been shown to adhere to and damage *T. taeniaeformis in vitro*, it is not clear if this reflects the *in vivo* situation, because mature strobilocerci survive transplantation in rats (*687*). Much work remains to be done to promote

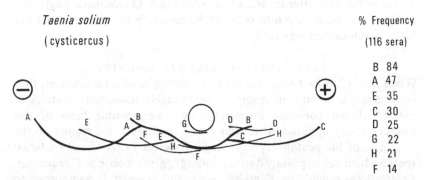

Fig. 11.8. Diagram of the precipitation bands in immunoelectrophoresis of sera from cases of human neurocystocercosis against an antigenic extract of *Taenia solium* cysticerci. Note the marked variation in the percentage frequency of different bands in different samples of sera. Some patients show no detectable immune response whatsoever. (After Flisser *et al.*, 1986.)

298

Table 11.4. *Correlation between the immune elements found on the surface of cysticerci of* Taenia solium *excised from human brain and antibodies found in the cerebrospinal fluid. (Data from Flisser et al., 1986)*

Cysticercus	Morphological appearance	Immune elements on the surface					Antibodies by ELISA in CSF			
		C3b	IgG	IgM	IgA	IgE	IgG	IgM	IgA	IgE
1	Hyalinised	−	+	−	+	−	+	−	−	−
2	Necrosed	−	+	+	+	−	+	−	−	−
3	Hyalinised	−	+	+	+	−	ND	ND	ND	ND
4	Hyalinised	+	+	−	−	−	+	+	−	−
5	Hyalinised	+	+	−	−	−	+	−	−	−
6	Viable[a]	+	+	+	+	−	+	−	−	−
	% Total	50	100	50	67	0	100	20	0	0

Notes:
ELISA, enzyme-linked immunosorbent assay; CSF, cerebrospinal fluid; ND, not determined; +, present, −, absent.
[a] Cysticercus of the racemose type, the rest were cellulosae.

our understanding of the host immune responses to established cestode larvae, a major problem being to distinguish between T-cell-dependent antibody-mediated and true cellular effects (*687*).

'EVASION' OF THE IMMUNE RESPONSE

Theoretical background
The ability of tissue parasites (e.g. larval cestodes) to survive in an immunologically hostile environment is clearly a necessary prerequisite for the completion of the life cycle and few topics in parasitology have aroused such interest and controversy. The phenomenon is referred to almost universally as an 'evasion' of the immune response. Use of the term evasion is unfortunate and a descriptive term such as insusceptibility would be more appropriate, as it avoids the teleological overtones of the word evasion, which suggests that the parasite recognises that it is being attacked and is taking 'evasive' action. Excellent summaries of the voluminous literature on this topic have been given by Parkhouse (*634*), Rickard (*687*) and Rickard & Williams (*691*). A brief account of the various mechanisms

which appear to block or moderate host immune attack is given below (reviewed: *687*).

Masking by host antigens

In trematodes (schistosomes), attachment of host cell surface antigens is believed to defend the parasites against immune attack, but this phenomenon (see Molecular mimicry, below) has not been demonstrated in cestodes. In larval cestodes, other host proteins have been demonstrated on the surface, including antibodies, these being referred to as 'blocking' antibodies. Thus, on the surface of the cysticerci of *T. solium*, all classes of immunoglobulin isotypes have been found (Table 11.4). In the case of *T. crassiceps* larvae, the predominating immunoglobulins appear to be IgM and IgG but, with this species, complete protection may not be provided.

Molecular mimicry

This well-known hypothesis has been substantially modified (*166, 167*), since its original conception (*165*), in the light of recent advances in molecular biology and immunology. The theory proposes that parasites disguise themselves as host (i.e. self) tissues by synthesising host-like antigens on their surface, with the result that they are not recognised as foreign (i.e. non-self) by the host. There has been much controversy as to whether the parasite is capable of synthesising such antigens or whether they have simply been absorbed from the host (as in the previous hypothesis) as there is some evidence that parasites may be capable of synthesising specific host receptors (e.g. parasite C3 and Fc-like receptors).

It was suggested originally (*165*) that host cell surface carbohydrate antigens appeared to be the best targets for parasites to mimic and evidence is accumulating in supporting the probable role of carbohydrate epitopes in this phenomenon (*167*). It should be noted, however, that in spite of evidence from a wide range of parasites, 'it is difficult to show any significant role for molecular mimicry in any host–parasite relationship' (*167*).

The possibility that parasites can synthesise host or host-like antigens raises the question of whether or not parasites can make use of mRNA to synthesise host protein (*799*) or whether putative molecular mimicry may actually represent cases of host 'capture' by parasites via natural transfection (perhaps by retroviruses?) – a hypothesis reviewed in detail by Howell (*338*). It has been pointed out earlier (Chapter 8) that there is evidence to suggest that this may have occurred in the sparganum of *Spirometra mansonoides*, which is capable of synthesising a growth factor with properties resembling those of a mammalian growth hormone (Tables 8.8 (p. 218) and 8.9 (p. 219)).

In taeniids too, the view that host-like antigens can be synthesised by

Immunisation (vaccination)

cestodes has recently received some support by experiments with *T. solium* in which IgG extracted from the cystericercus showed no specificity for the parasite. Moreover, that this protein is of parasite (rather than host) origin has been supported by the experiment (described in Chapter 6) in which *in vitro* translation of parasite-derived RNA produced a protein which was precipitated by anti-porcine IgG (*687*).

Alterations in antigenicity
Relatively little work has been done in this area, but there is evidence that the fact the postoncospheral stages rapidly outgrow their susceptibility to attack may be due to development of stage-specific antigens related to profound alterations in the surface structure of the developing larva (*687*). Thus, a strong immune response is generated against *T. taeniaeformis* in the liver, and larvae become insusceptible by 8 days – a time which coincides with the surface replacement of microvilli by microtriches.

Modulation of host defences
There is some evidence that some larvae of some species (e.g. *T. pisiformis*) can release substances which inhibit proteolytic enzymes such as trypsin and chymotrypsin (*591*) (see Chapter 4). Other host cellular activities reported to be modified by parasite secretions are: cell differentiation, macrophage activation, responsiveness to mitogens, cytotoxicity and complement activation (*687*).

Immunosuppression
There are a number of reports that some cestode larvae can induce immunosuppression in their hosts. For example, mice infected with *T. crassiceps* show some impaired ability to make antibodies to red blood cells (*687*).

Immunisation (vaccination)

Immunisation against adult cestodes

TAENIIDAE

Although promising results have been obtained in developing vaccination strategies against larval cestodes, attempts to develop vaccines against adult cestodes have been entirely unsuccessful, to date. The fact that the adult worms occupy an intestinal site, makes the problem of attacking them by immunological means especially difficult.

Obvious targets for vaccination are *Echinococcus granulosus* and

E. multilocularis, since successful vaccination of dogs (and foxes) would represent a major step forward in the control of hydatidosis. Although early attempts – by injecting crude antigens or feeding with irradiated protoscoleces – showed some promise, these early findings were not confirmed by subsequent work (*686*).

The development of *in vitro* culture techniques for adult worms (Chapter 10) appeared to hold out some promise for vaccine development by using E/S antigens isolated from the culture medium. The first experiments using such antigens by Herd *et al.* (*320*) were promising but, later, more critical experiments failed to support these earlier results and it was speculated that there may be an interaction between the vaccine and the innate resistance of the host (*319*). Other attempts to vaccinate dogs using somatic antigens from oncospheres of a heterologous cestode, *T. hydatigena*, also proved to be unsuccessful (*686*). Some success in vaccinating dogs against *E. granulosus* has been achieved using irradiated protoscoleces. However, this approach is dangerous and has not been followed up as in some dogs, X-irradiated worms regenerated and grew to produce infective eggs.

The development of vaccines against the adult stage of all pathogenic cestode species thus remains a major challenge.

Immunisation against larval cestodes

BASIC PROBLEMS

The development of successful vaccines against the highly pathogenic larvae of *E. granulosus* and *E. multilocularis* (in man and domestic animals), *T. solium* (in man and pigs) and those of economic importance, such as *T. saginata* (in cattle) and *T. ovis* and *T. multiceps* (both in sheep) would clearly revolutionise the control of these important parasites, and much effort has been put into this research area (reviewed: *137, 447, 687, 691, 784, 961*).

ANTIGENS UTILISED

A basic problem of immunisation against larvae has been to identify and characterise (in each species) those antigens which can be regarded as *functional* antigens, i.e. those antigens which, if their function is impaired or (ideally) blocked by antibodies or by other immune mechanisms, as a result of immunisation, would result in the death of the parasite. Unfortunately, few functional antigens have been identified for helminths and these largely in nematodes (*447*). Regarding cestodes, a functional antigen has been isolated and partly characterised in the strobilicercus of *T. taeniaeformis*, the antigen proving to have molecular weight 140 000 (*447*). Further candidate molecules have been identified in *T. taeniaeformis* (*90*) and *T. saginata*

Immunisation (vaccination)

(*302, 303*). In *E. granulosus* hydatid cyst fluid, five major subunit antigens (Fig. 6.2, p. 118) have been characterised, using radio-iodination, immunoprecipitation, SDS/polyacrylamide gel electrophoresis and immunoblotting (*754*). The antigens were shown to have an M_r of 12 000, 16 000, 20 000 (two), and 38 000 when reduced. The 12 000 and 16 000 M_r molecules are specific to *E. granulosus* and appear to be E/S antigens of all strains of the parasite and all human isolates examined, although not all hydatid patients produce antibodies to them. These molecules may prove to be useful for specific immunodiagnosis, but they have not yet been considered as possible vaccine candidates. In most cases, however, the antigens used in immunisation have remained uncharacterised. The following have been utilised:

(i) Homogenised fresh, dried or formalin-killed adults, eggs or larval stages, or fluids or extracts of these; both homologous and heterologous species have been used.
(ii) Whole living stages of homologous or heterologous species.
(iii) E/S antigens from eggs, larvae or adults.
(iv) Attenuated eggs or larval stages.

PASSIVE IMMUNISATION
Early work (*796*) on *T. taeniaeformis* and *T. pisiformis* showed that the high degree of immunity developed against larvae of these species could be transmitted by passive transfer of immune serum. Results with other species, such as *E. granulosus* and *T. solium* have not been so successful and not as promising as procedures involving active immunisation (see below).

ACTIVE IMMUNISATION
Functional antigens (although not characterised) can be extracted from the somatic tissues of several *Taenia* spp. and in some cases can provide complete protection against challenge infection (*447*). More successful, however, have been the use of E/S antigens collected from oncospheres cultured *in vitro*, and highly effective immunisation of cattle against *T. saginata* and of sheep against *T. ovis* has been achieved (*688, 689*). Later work, however, has shown that it is not necessary to culture oncospheres (which is both difficult and expensive) and that equally effective results can be obtained by using frozen/thawed or sonicated oncospheres (*687*).

A major difficulty in vaccine development is the supply of material, especially the eggs of *T. solium*, *T. saginata*, *E. granulosus* and *E. multilocularis* which are both difficult to obtain and (except *T. saginata*) highly dangerous to handle. Progress in the characterisation of antigens, and the identification of which of them are 'functional' antigens, clearly now holds out some promise for vaccine development. If the appropriate cells producing such antigens can be identified, recombinant DNA techniques or *in vitro* culture of the appropriate cell lines should allow the mass production of these and the major problem of shortage of material would be overcome. The use of biotechnology and recombinant DNA technology has thus the potential of providing a powerful tool for the development of vaccines against parasitic helminths. Progress in this field, which is considered further in Chapter 6, has been reviewed by Gamble & Zarlenga (*249*), Leng *et al.* (*438*), Smithers (*784*) and Williams (*961*).

Immunodiagnosis

It is beyond the scope of this text to consider the immunodiagnosis of adult and larval cestode infections. Various aspects of this topic have been extensively reviewed by Flisser *et al.* (*226*), Fox *et al.* (*228*), Draper & Lillywhite (*183*), Houba (*337*), Gottstein *et al.* (*261*), Rickard & Lightowlers (*690*) and Walls & Schantz (*928*). A major breakthrough in diagnosis has been the development of an anti-oncospheral monoclonal antibody for the unequivocal identification of eggs of *Echinococcus* spp. by immunofluorescence (*154*).

References

Wherever possible, the relevant number in *Helminthological Abstracts* (series A) for papers in the less-well known foreign languages has been given, e.g. [HA/48/367]. This will enable readers to consult an abstract in English.

1. **Adams, D. B. & Cobon, G.** (1984). Basis for the development of vaccines for control of diseases produced by metazoan parasites. In *Biotechnology and recombinant DNA technology in animal production industries*, ed. J. C. Nugent & M. O'Connor, pp. 67–74. *Reviews in Rural Science*, no. 6. CSIRO Division of Animal Health: Armidale, New South Wales.
2. **Adibi, S. A. & Kim, Y. S.** (1981). Peptide absorption and hydrolysis. In *Physiology of the gastrointestinal tract*, ed. L. R. Johnson, pp. 1073–122. Raven Press: New York.
3. **Agosin, M. & Aravena, L.** (1959). Studies on the metabolism of *Echinococcus granulosus*. III. Glycolysis, with special reference to hexokinases and related glycolytic enzymes. *Biochimica et Biophysica Acta*, **34**: 90–102.
4. — (1960). Studies on the metabolism of *Echinococcus granulosus*. IV. Enzymes of the pentose phosphate pathway. *Experimental Parasitology*, **10**: 23–38.
5. **Agosin, M. & Naquira, C.** (1978). Translation of *Taenia crassiceps* mRNA in cell-free heterologous systems. *Comparative Biochemistry and Physiology*, **60B**: 183–7.
6. **Agosin, M. & Repetto, Y.** (1961). Studies on the metabolism of *Echinococcus granulosus*. VI. Pathways of glucose ^{14}C metabolism in *E. granulosus* scolices. *Biologica*, **23**: 33–8.
7. — (1967). Studies on the metabolism of *Echinococcus granulosus* protoscolices. *Experimental Parasitology*, **21**: 195–208.
8. **Agosin, M., Repetto, Y. & Dicowsky, L.** (1971). Ribonucleic acid of *Echinococcus granulosus* protoscoleces. *Experimental Parasitology*, **30**: 233–43.
9. **Alberts, B., Bray, D., Lewis, J., Raff, M., Roberts, K. & Watson, J. D.** (1983). *Molecular biology of the cell*. Garland Publishing Inc.: New York & London.
10. **Allen, A., Hutton, D. A., Pearson, J. P. & Sellers, L. A.** (1984). Mucus

305

References

glycoprotein structure, gel formation and gastrointestinal mucus function. In *Mucus and mucosa*, ed. J. C. Nugent, & M. O'Connor, pp. 137–156. Ciba Foundation Symposium, no. 109. Pitman: London.

11. **Allen, P. C.** (1973). Helminths: Comparison of their rhodoquinones. *Experimental Parasitology*, **34**: 211–19.
12. **Anderson, K. I.** (1975a). Ultrastructure studies on *Diphyllobothrium ditremum* and *D. dendriticum* (Cestoda, Pseudophyllidea), with emphasis on the scolex tegument and the tegument in the area around the gential atrium. *Zeitschrift für Parasitenkunde*, **46**: 253–64.
13. — (1975b). Comparison of surface topography of three species of *Diphyllobothrium* (Cestoda: Pseudophyllidea) by scanning electron microscopy. *International Journal for Parasitology*, **5**: 293–300.
14. **Andersen, K. I. & Lysfjord, S.** (1982). The functional morphology of the scolex of two *Tetrabothrius* species (Cestoda; Tetrabothriidae) from penguins. *Zeitschrift für Parasitenkunde*, **67**: 299–307.
15. **Anderson, R. M.** (1978). The regulation of host population growth by parasitic species. *Parasitology*, **76**: 119–57.
16. **Anderson, R. M. & Lethbridge, R. C.** (1975). An experimental study of the survival characteristics, activity and energy reserves of the hexacanths of *Hymenolepis diminuta*. *Parasitology*, **71**: 137–51.
17. **Andreassen, J.** (1981). Immunity to adult cestodes. *Parasitology* (EMOP Proceedings, 4): **82**: 153–9.
18. **Anikieva, L. V. & Lutta, A. S.** (1977). [Stored nutrients at different stages of the development of *Proteocephalus exiguus* La Rue, 1911 and *Diphyllobothrium latum* (L., 1758)]. In Russian. In *Comparative biochemistry of fish and their helminths. Lipids, enzymes, proteins. (Collected works)*, ed. V. S. Sidorov, pp. 116–27. USSR Akademii Nauk, Institut Biologii Karel' skiĭ Filial: USSR. [HA/50/3943].
19. **Arai, H. P.** (ed.). (1980a). Biology of the tapeworm *Hymenolepis diminuta*. Academic Press: New York.
20. — (1980b). Migratory activity and related phenomena in *Hymenolepis diminuta*. In *Biology of the tapeworm* Hymenolepis diminuta, ed. H. P. Arai, pp. 615–37. Academic Press: New York.
21. **Arfin, M. & Nizami, W. A.** (1986). Chemical nature and mode of stabilization of eggshell/capsule of some cyclophyllidean cestodes. *Journal of Helminthology*, **60**: 105–12.
22. **Arme, C.** (1975). Tapeworm–host interactions. *Symposia of the Society for Experimental Biology*, **29**: 505–32.
23. — (1982). Nutrition. In *Modern parasitology*, ed. F. E. G. Cox, pp. 148–172. Blackwell Scientific Publications: Oxford.
24. — (1987). Cestoda. In *In vitro methods for parasite cultivation*, 3rd edn, ed. A. E. R. Taylor & J. R. Baker, pp. 282–317. Academic Press: London.
25. **Arme, C., Bridges, J. F. & Hoole, D.** (1983). Pathology of cestode infections in the vertebrate host. In *Biology of the Eucestoda*, vol. 2, ed. C. Arme & P. W. Pappas, pp. 499–538. Academic Press: London.
26. **Arme, C., Griffiths, D. V. & Sumpter, J. P.** (1982). Evidence against the hypothesis that the plerocercoid larva of *Ligula intestinalis*

References

(Cestoda: Pseudophyllidea) produces a sex steroid that interferes with host reproduction. *Journal of Parasitology*, **68**: 169–71.
27. **Arme, C. & Pappas, P. W.** (eds.) (1983*a*). *Biology of the Eucestoda*, vol. 1. Academic Press: London.
28. — (eds.) (1983*b*). *Biology of the Eucestoda*, vol. 2. Academic Press: London.
29. **Asanov, S.** (1979). [Study of the amino acid composition of *Echinococcus granulosus* (Batch, 1786).] In Russian. *Trudy Uzbekskogo Nauchno-Issledovatel' skogo Veterinarnogo Instituta (Bolezni sel'skokhozyaĭstvennykh zhivotnykh)*, **29**: 28–30. [HA/51/3190]
30. **Atkinson, B. G. & Podesta, R. B.** (1982). Two-dimensional electrophoretic separation and fluorographic analysis of the gene products synthesised by *Hymenolepis diminuta* with particular reference to the parasite-specific polypeptides in the brush border membrane. *Molecular and Biochemical Parasitology*, **6**: 33–43.
31. **Azzolina, L. S., Tridente, G. & Cooper, E. L.** (1980). *Comparative immunology*. Pergamon Press: New York.
32. **Baer, J. G. & Joyeux, C.** (1961). Platyhelminthes, Mésozaires, Acanthocéphales, Némertiens. In: *Traité de zoologie*, ed. P. Grassé, vol. IV, part I, pp. 561–692. Masson et Cie: Paris.
33. **Bahr, J. M., Frayha, G. J. & Hajjar, J. J.** (1979). Mechanism of cholesterol absorption by the hydatid cysts of *Echinococus granulosus* (Cestoda). *Comparative Biochemistry and Physiology*, **62A**: 485–9.
34. **Balasubramanian, M. P., Nellaiappan, K. & Ramalingam, K.** (1982). Characterization of esterase isoenzymes of *Raillietina tetragona* (Molin, 1858) (Cestoda). *Veterinary Parasitology*, **10**: 313–16.
35. — (1983). Studies on non-specific esterases activity in *Raillietina tetragona*. *Helminthologia*, **20**: 45–52.
36. **Baldwin, J. L., Berntzen, A. K. & Brown, B. W.** (1978). *Mesocestoides corti*: cation concentration in calcareous corpuscles of tetrathyridia grown *in vitro*. *Experimental Parasitology*, **44**: 190–6.
37. **Bankov, I. & Barrett, J.** (1985). Sphingomyelin synthesis in *Hymenolepis diminuta* (Cestoda). *Molecular and Biochemical Parasitology*, **15**: 341–7.
38. **Barker, I. K.** (1970). The penetration of oncospheres of *Taenia pisiformis* into the intestine of the rabbit. *Canadian Journal of Zoology*, **48**: 1329–32.
39. **Barrett, J.** (1981). *Biochemistry of parasitic helminths*. Macmillian Publishers Ltd: London.
40. — (1983). Lipid metabolism. In *Biology of the Eucestoda*, vol. 2, ed. C. Arme & P. W. Pappas, pp. 391–419. Academic Press: London.
41. — (1984). The anaerobic end-products of helminths. *Parasitology*, **88**: 179–98.
42. **Barrett, J. & Beis, I.** (1973). Nicotinamide and adenosine nucleotide levels in *Ascaris lumbricoides*, *Hymenolepis diminuta* and *Fasciola hepatica*. *International Journal for Parasitology*, **3**: 271–3.
43. **Barrett, J. & Körting, W.** (1977). Lipid catabolism in the plerocercoids

References

of *Schistocephalus solidus* (Cestoda: Pseudophyllidea). *International Journal for Parasitology*, 7: 419–22.

44. **Barrett, J. & Lloyd, G. M.** (1981). A novel phosphagen phosphotransferase in the plerocercoids of *Schistocephalus solidus* (Cestoda: Pseudophyllidea). *Parasitology*, **82**: 11–16.
45. **Barrett, N. J.** (1984). Developmental biology of *Echinococcus multilocularis in vivo* and *in vitro*. Ph.D. thesis. University of London (Imperial College).
46. **Barrett, N. J. & Smyth, J. D.** (1983). Observations on the structure and ultrastructure of sperm development in *Echinococcus multilocularis*, both *in vitro* and *in vivo*. *Parasitology*, **87**: li.
47. **Barrett, N. J., Smyth, J. D. & Ong, S. J.** (1982). Spontaneous sexual differentiation of *Mesocestoides corti* tetrathyridia *in vitro*. *International Journal for Parasitology*, **12**: 315–22.
48. **Baugh, S. C. & Sing, J. P.** (1979). Glycogen distribution in *Raillietina* (*Raillietina*) *echinobothrida* (Megnin, 1880). *Revista Ibérica de Parasitología*, **39**: 223–32.
49. **Beach, D. H., Mueller, J. F. & Holz, G. G., Jr** (1980a). Lipids of stages in the life-cycle of the cestode *Spirometra mansonoides*. *Molecular and Biochemical Parasitology*, **1**: 249–68.
50. — (1980b). Benzoquinones in stages of the life cycle of the cestode *Spirometra mansonoides*. *Molecular and Biochemical Parasitology*, **1**: 269–78.
51. **Beach, D. H., Sherman, I. W. & Holz, G. G., Jr** (1973). Incorporation of docosahexaenoic fatty acid into the lipids of a cestode of marine elasmobranchs. *Journal of Parasitology*, **59**: 655–66.
52. **Befus, A. D.** (1977). *Hymenolepis diminuta* and *H. microstoma*: mouse immunoglobulins binding to the tegumental surface. *Experimental Parasitology*, **41**: 242–51.
53. **Befus, A. D. & Bienenstock, J.** (1982). Factors involved in symbiosis and host resistance at the mucosa–parasite interface. *Progress in Allergy*, **31**: 76–177.
54. — (1984). Induction and expression of mucosal immune responses and inflammation to parasitic infections. In *Immunobiolology of parasites and parasitic infections. Contemporary topics in immunology*, vol. 12, ed. J. J. Marchalonis, pp. 71–108. Plenum Press: New York.
55. **Befus, A. D., Lee, T., Ernst, P., Egwang, T., McElroy, P., Gauldie, J. & Bienenstock, J.** (1986). Unique characteristics of local responses in host resistance to mucosal parasitic infections. *Veterinary Parasitology*, **20**: 175–94.
56. **Befus, A. D. & Podesta, R. B.** (1976). Intestine. In *Ecological aspects of parasitology*. ed. C. R. Kennedy, pp. 303–25. North-Holland Publishing Co.: Amsterdam.
57. — (1984). Platyhelminth host–parasite interface. In *Biology of the integument*. I. *Invertebrates*, ed. J. Bereiter–Hahn, A. G. Matolosky & K. S. Richards, pp. 192–204. Springer-Verlag: Berlin.
58. **Béguin, F. (1966).** Étude au microscope électronique de la cuticle et de ses structures associées chez quelques cestodes. Essai d'histologie

References

comparée. *Zeitschrift für Zellforschung und microscopie Anatomie*, **72**: 30–46.

59. **Behm, C. A. & Bryant, C.** (1975a). Studies of regulatory metabolism in *Moniezia expansa*: general conditions. *International Journal for Parasitology*, **5**: 209–17.

60. — (1975b). Studies of regulatory metabolism in *Moniezia expansa*: the role of phosphofructokinase (with a note on pyruvate kinase). *International Journal for Parasitology*, **5**: 339–46.

61. — (1975c). Studies of regulatory metabolism in *Moniezia expansa*: the role of phosphoenolpyruvate carboxykinase. *International Journal for Parasitology*, **5**: 347–54.

62. **Behm, C. A., Bryant, C. & Jones, A. J.** (1987). Studies of glucose metabolism in *Hymenolepis diminuta* using ^{13}C nuclear magnetic resonance. *International Journal for Parasitology*, **17**: 1333–41.

63. **Beis, I. & Barrett, J.** (1979). The contents of adenine nucleotides and glycolytic and tricarboxylic acid cycle intermediates in activated and non-activated plerocercoids of *Schistocephalus solidus* (Cestoda: Pseudophyllidea). *International Journal for Parasitology*, **9**: 465–8.

64. — (1980). Oxidative enzymes in the plerococercoids of *Schistocephalus solidus* (Cestoda: Pseudophyllidea). *International Journal for Parasitology*, **10**: 151–3.

65. **Beis, I. & Theophilidis, G.** (1982). Phosphofructokinase in the plerocercoids of *Schistocephalus solidus* (Cestoda: Pseudophyllidea). *International Journal for Parasitology*, **12**: 389–93.

66. **Belton, C. M.** (1977). Freeze-fracture study of the tegument of larval *Taenia crassiceps*. *Journal of Parasitology*, **63**: 306–13.

67. **Benito, A. A.** (1984). [Hydatidosis of sheep. II. Levels of glucose, lipids and proteins in hydatid liquid.] In Spanish. *Archivos de Zootecnia*, **33**: 163–9.

68. **Benito, A. A., Gómez, F. M. & Gaitán, A. U.** (1983). [Ovine hydatidosis. I. The pH, sodium content and potassium content of hydatid fluid.] In Spanish. *Archivos de Zootecnia*, **32**: 279–84.

69. **Bern, H. A.** (1962). The properties of neurosecretory cells. In *General and comparative endocrinology Supplement*, **1**: 117–32.

70. **Berntzen, A. K.** (1961). The *in vitro* cultivation of tapeworms. I. Growth of *Hymenolepis diminuta* (Cestoda: Cyclophyllidea). *Journal of Parasitology*, **47**: 351–5.

71. — (1962). *In vitro* cultivation of tapeworms. II. Growth and maintenance of *Hymenolepis nana* (Cestoda: Cyclophyllidea). *Journal of Parasitology*, **48**: 785–97.

72. **Berntzen, A. K. & Mueller, J. F.** (1964). *In vitro* cultivation of *Spirometra mansonoides* (Cestoda) from the plerocercoid to the early adult. *Journal of Parasitology*, **5**: 705–11.

73. — (1972). *In vitro* cultivation of *Spirometra* spp. (Cestoda) from the plerocercoid to the gravid adult. *Journal of Parasitology*, **58**: 750–2.

74. **Berntzen, A. K. & Voge, M.** (1965). *In vitro* hatching of oncospheres of four hymenolepidid cestodes. *Journal of Parasitology*, **51**: 235–42.

75. **Bessonov, A. S. & Arkhipova, N. S.** (1983). [A model for screening

References

anthelmintics against *Cysticercus*.] In Russian. *Veterinariya, Moscow*, **1**: 38–9. [HA/52/2689]

76. **Beveridge, I., Rickard, M. D., Gregory, G. G. & Munday, B. L.** (1975). Studies on *Anoplotaenia dasyuri* Beddard, 1911 (Cestoda: Taeniidae), a parasite of the Tasmanian devil: observations on the egg and metacestode. *International Journal for Parasitology*, **5**: 257–67.

77. **Bhalya, A., Seth, A., Malhotra, S. K. & Capoor, V. N.** (1983*a*). Chemotaxonomic differentiation of 3 subgenera *Parionella*, *Raillietina* and *Skrjabinia* of cestode genus *Raillietina*. *Advances in Biosciences*, **2**: 93–6.

78. — (1983*b*). Amino-acids of *Amoebotaenia cuneata* (Cestoda: Dilepidoidea Wardle, McLeod & Radinovsky, 1974). *Journal of Helminthology*, **57**: 9–10.

79. — (1985). Amino acids of *Hymenolepis palmarum* (Johri, 1956) and chemotaxonomic studies on hymenolepidid cestodes. *Journal of Helminthology*, **59**: 39–42.

80. **Bienenstock, J. & Befus, A. D.** (1980). Mucosal immunology. *Immunology*, **41**: 249–70.

81. — (1983). Some thoughts on the biologic role of immunoglobulin A. *Gastroenterology*, **84**: 178–85.

82. **Black, M. I., Scarpino, P. V., O'Donnell, C. J., Meyer, K. B., Jones, J. V. & Kaneshiro, E. S.** (1982). Survival rates of parasite eggs in sludge during aerobic and anaerobic digestion. *Applied and Environmental Microbiology*, **44**: 1138–43.

83. **Blackburn, B. J., Hutton, H. M., Novak, M. & Evans, W. S.** (1986). *Hymenolepis diminuta*: nuclear magnetic resonance analysis of the excretory products resulting from the metabolism of D-[$^{13}C_6$]glucose. *Experimental Parasitology*, **62**: 381–8.

84. **Blažek, K. & Schramlová, J.** (1981). The organ reaction during the localization of *C. bovis* in the internal organs of cattle. *Veterinárni Medicina Praha*, **26**: 37–47.

85. **Blitz, N. M. & Smyth, J. D.** (1973). Tegumental ultrastructure of *Raillietina cesticillus* during the larval–adult transformation, with emphasis on the rostellum. *International Journal for Parasitology*, **3**: 561–70.

86. **Bloom, F. E.** (1984). The functional significance of neurotransmitter diversity. *American Journal of Physiology*, **246**: C184–95.

87. — (1985). Neurotransmitter diversity and its functional significance. *Journal of the Royal Society of Medicine*, **78**: 189–92.

88. **Boddington, M. J. & Mettrick, D. F.** (1981). Production and reproduction in *Hymenolepis diminuta* (Platyhelminthes: Cestoda). *Canadian Journal of Zoology*, **59**: 1962–72.

89. **Bonsdorff, B. von** (1977). *Diphyllobothriasis in man*. Academic Press: New York.

90. **Bowtell, D. D. L., Mitchell, G. F., Anders, R. F., Lightowlers, M. W. & Rickard, M. D.** (1983). *Taenia taeniaeformis*: immunoprecipitation analysis of the protein antigens of oncospheres and larvae. *Experimental Parasitology*, **56**: 416–27.

91. **Bowtell, D. D. L., Saint, R. B., Rickard, M. D. & Mitchell, G. F.** (1984).

References

Expression of *Taenia taeniaeformis* antigens in *Escherichia coli*. *Molecular and Biochemical Parasitology*, **13**: 173–85.

92. — (1986). Immunochemical analysis of *Taenia taeniaeformis* antigens expressed in *Escherichia coli*. *Parasitology*, **93**: 599–610.

93. **Brandt, J. R. A. & Sewell, M. M. H.** (1979/80). Preliminary observations on the *in vitro* culture of metacestodes of *Taenia saginata*. *Veterinary Science Communications*, **3**: 317–24.

94. **Branford-White, C. J., Hipkiss, J. B. & Peters, T. J.** (1984). Evidence for a Ca^{2+}-dependent activator in the rat tapeworm *Hymenolepis diminuta*. *Molecular and Biochemical Parasitology*, **13**: 201–11.

95. **Bråten, T. & Hopkins, C. A.** (1969). The migration of *Hymenolepis diminuta* in the rat's intestine during normal development and following surgical transplantation. *Parasitology*, **59**: 891–905.

96. **Brunet, P. C. J.** (1967). Sclerotins. *Endeavour*, **26**: 68–74.

97. **Bryant, C.** (1970). Electron transport in parasitic helminths and protozoa. *Advances in Parasitology*, **8**: 139–72.

98. — (1972). Metabolic regulation in *Moniezia expansa* (Cestoda): the role of pyruvate kinase. *International Journal for Parasitology*, **2**: 333–40.

99. — (1975). Carbon dioxide utilisation, and the regulation of respiratory metabolic pathways in parasitic helminths. *Advances in Parasitology*, **13**: 35–69.

100. — (1978). The regulation of respiratory metabolism in parasitic helminths. *Advances in Parasitology*, **16**: 311–31.

101. — (1982). Biochemistry. In *Modern parasitology. A textbook of parasitology*, ed. F. E. G. Cox, pp. 84–115. Blackwell Scientific Publications: Oxford.

102. — (1983). Intraspecific variations of energy metabolism in parasitic helminths. *International Journal for Parasitology*, **13**: 327–32.

103. **Bryant, C. & Behm, C. A.** (1976). Regulation of respiratory metabolism in *Moniezia expansa* under aerobic and anaerobic conditions. In *Biochemistry of parasites and host–parasite relationships*, ed. H. Van den Bossche, pp. 89–94. North-Holland Publishing Co.: Amsterdam.

104. **Bryant, C. & Flockart, H. A.** (1986). Biochemical strain variation in parasitic helminths. *Advances in Parasitology*, **25**: 276–319.

105. **Bryant, C. & Morseth, D. J.** (1968). The metabolism of radioactive fumaric acid and some other substrates by whole adult *Echinococcus granulosus* (Cestoda). *Comparative Biochemistry and Physiology*, **25**: 541–6.

106. **Bürger, H.-J.** (1984). Survival of *Taenia saginata* eggs in sewage and on pasture. *C.E.C. (Agriculture). Parasitological Symposium; Lyons, 1983*, pp. 155–65. Official Publication of the European Community, Luxembourg.

107 **Burenina, E. A.** (1977). [The carbohydrate metabolism of *Bothriocephalus scorpii*.] In Russian. *Trudy Biologo-Pochovennogo Instituta (Paraziticheskie i svobodnozhivushchie chervi fauny Dal'nego Vostoka), Novaya Seriya*, **47** 106–8. [HA/48/367]

108. **Burke, W. F., Gracy, R. W. & Harris, B. G.** (1972). Studies on enzymes from parasitic helminths. III. Purification and properties of lactate

References

dehydrogenase from the tapeworm, *Hymenolepis diminuta. Comparative Biochemistry and Physiology*, **43B**: 345–59.

109. **Bursey, C. C., McKenzie, J. A. & Burt, M. D. B.** (1980). Polyacrylamide gel electrophoresis in the differentiation of *Taenia* (Cestoda) by total protein. *International Journal for Parasitology*, **10**: 167–74.

110. **Buteau, G. H., Jr, Simmons, J. E. & Fairbairn, D.** (1969). Lipid metabolism in helminth parasites. IX. Fatty acid composition of shark tapeworms and of their hosts. *Experimental Parasitology*, **26**: 209–13.

111 **Cai, X. P. & Jin, J. S.** (1984). [Study on the life cycle of *moniezia benedeni*.] In Chinese. *Journal of Veterinary Science & Technology (Shouyi Keji Zazhi)*. **12** 26–30. [HA/54/3966]

112. **Cain, G. D., Johnson, W. J. & Oaks, J. A.** (1977). Lipids from subcellar fractions of the tegument of *Hymenolepis diminuta. Journal of Parasitology*, **63**: 486–91.

113. **Campbell, W. C.** (1986). The chemotherapy of parasitic infections. *Journal of Parasitology*, **72**: 45–61.

114. **Campbell, W. C. & Rew, R. S.** (eds.). (1986). *Chemotherapy of parasitic diseases*. Plenum Press: New York.

115. **Cannon, C. E. & Mettrick, D. F.** (1970). Changes in the distribution of *Hymenolepis diminuta* (Cestoda: Cyclophyllidea) within the rat intestine during prepatent development. *Canadian Journal of Zoology*, **48**: 761–9.

116. **Carlstedt, I. & Sheehan, J. K.** (1984). Macromolecular properties and polymeric structure of mucus glycoproteins. In *Mucus and mucosa*, ed. J. C. Nugent & M. O'Connor, pp. 157–72. Ciba Foundation Symposium, no. 109. Pitman Press: London.

117. **Carter, C. E. & Fairbairn, D.** (1975). Multienzymic nature of pyruvate kinase during development of *Hymenolepis diminuta* (Cestoda). *Journal of Experimental Zoology*, **194**: 439–48.

118. **Carter, C. E., Wells, J. R. & Macinnis, A. J.** (1972). DNA from anaerobic adult *Ascaris lumbricoides* and *Hymenolepis diminuta* mitochondria isolated by zonal centrifugation. *Biochimica et Biophysica Acta*, **262**: 135–44.

119. **Chandra, K. J., Hanumantha Rao, K. & Shyamasundari, K.** (1985). Observations on the histology and histochemistry of *Penetrocephalus* plerocercoid (Pseudophyllidea: Cestoda). *Proceedings of the Indian Academy of Sciences, Animal Sciences*, **94**: 11–19.

120. **Charles, G. H.** (1971). The ultrastructure of the developing pseudophyllid tegument (epidermis) with reference to the larval stages of *Schistocephalus solidus* and *Ligula intestinalis. Journal of Parasitology, ICOPA II Proceedings* (special vol. 59, part 4), pp. 38–9.

121. **Cheah, K. S.** (1967*a*). The oxidase systems of *Moniezia expansa* (Cestoda). *Comparative Biochemistry and Physiology*, **23**: 277–302.

122. — (1967*b*). Spectrophotometric studies on the succinate oxidase system of *Taenia hydatigena. Comparative Biochemistry and Physiology*, **20**: 867–75.

123. — (1968). The respiratory components of *Moniezia expansa* (Cestoda). *Biochimica et Biophysica Acta*, **153**: 718–20.

References

124. — (1970). Evidence for the existence of an α-glycerophosphate oxidase system with three phosphorylation sites and sensitive to rotenone and piericidin A. *FEBS Letters*, **10**: 109–12.
125. — (1971). Oxidative phosphorylation in *Moniezia* muscle mitochondria. *Biochimica et Biophysica Acta*, **253**: 1–11.
126. — (1972*a*). Cytochromes in *Ascaris* and *Moniezia*. In *Comparative biochemistry of parasites*, ed. H. Van den Bossche, pp. 417–32. Academic Press: New York.
127. — (1972*b*). Oxidative phosphorylation in *Moniezia* mitochondria. In *Comparative biochemistry of parasites*, ed. H. Van den Bossche, pp. 479–90. Academic Press: New York.
128. — (1975). Purification and properties of *Moniezia* cytochrome c_{550}. *Comparative Biochemistry and Physiology*, **51B**: 41–5.
129. — (1983). Electron transport systems. In *Biology of the Eucestoda*, vol. 2, ed. C. Arme & P. W. Pappas, pp. 421–40. Academic Press: London & New York.
130. **Cheah, K. S. & Bryant, C.** (1966). Studies on the electron transport system of *Moniezia expansa* (Cestoda). *Comparative Biochemistry and Physiology*, **19**: 197–223.
131. **Chew, M. W. K.** (1983). *Taenia crassiceps*: ultrastructural observations on the oncosphere and associated structures. *Journal of Helminthology*, **57**: 101–13.
132. **Cho, C. H. & Mettrick, D. F.** (1982). Circadian variation in the distribution of *Hymenolepis diminuta* (Cestoda) and 5-HT (serotonin) levels in the gastro-intestinal tract of the laboratory rat. *Parasitology*, **84**: 431–41.
133. **Chopra, A. K.** (1981). Glycogen content and its distribution in three cyclophyllidean cestodes of sheep. *Comparative Physiology and Ecology*, **6**: 173–6.
134. **Chowdhury, N. & De Rycke, P. H.** (1976). Qualitative distribution of neutral lipids and phospholipids in *Hymenolepis microstoma* from the cysticercoid to the egg producing adult. *Zeitschrift für Parasitenkunde*, **50**: 151–60.
135. — (1977). Structure, formation, and functions of calcareous corpuscles in *Hymenolepis microstoma*. *Zeitschrift für Parasitenkunde*, **53**: 159–69.
136. **Chowdhury, N. & Kinger, S. & Ahuja, S. P.** (1986). The chemical composition of secondary hydatid cysts of buffalo origin. *Annals of Tropical Medicine and Parasitology*, **80**: 469–71.
137. **Clegg, J. A. & Smith, M. A.** (1978). Prospects for the development of dead vaccines against helminths. *Advances in Parasitology*, **16**: 165–218.
138. **Clegg, J. A. & Smyth, J. D.** (1968). Growth, development, and culture methods: parasitic platyhelminths. In *Chemical Zoology*, vol. 2, ed. M. Florkin & B. T. Scheer, pp. 395–446. Academic Press: New York.
139. **Coggins, J. R.** (1980). Tegument and apical end organ fine structure in the metacestode and adult *Proteocephalus ambloplitis*. *International Journal for Parasitology*, **10**: 409–18.
140. **Coil, W. H.** (1970). Studies on the biology of the tapeworm *Shipleya inermis* Fuhrmann, 1908. *Zeitschrift für Parasitenkunde*, **35**: 40–54.

References

141. — (1975). The histochemistry and fine structure of the embryophore of *Shipleya inermis* (Cesoda). *Zeitschrift für Parasitenkunde*, **48**: 9–14.
142. — (1977). Studies on the embryogenesis of the tapeworm *Shipleya inermis* Fuhrmann, 1908, using transmission and scanning electron microscopy. *Zeitschrift für Parasitenkunde*, **52**: 311–18.
143. — (1979). Studies on the embryogenesis of the tapeworm *Cittotaenia variabilis* (Stiles, 1895) using transmission and scanning microscopy. *Zeitschrift für Parasitenkunde*, **59**: 151–9.
144. — (1984). Studies on the development of *Dioecocestus acotylus* (Cestoda) with emphasis on the scanning electron microscopy of embryogenesis. *Proceedings of the Helminthological Society of Washington*, **51**: 113–20.
145. **Coles, G. C. & Simpkin, K. G.** (1977). Metabolic gradients in *Hymenolepis diminuta* under aerobic conditions. *International Journal for Parasitology*, **7**: 127–8.
146. **Collin, W. K.** (1968). Electron microscope studies of the muscle and hook systems of hatched oncospheres of *Hymenolepis citelli* McLeod, 1933 (Cestoda: Cyclophyllidea). *Journal of Parasitology*, **54**: 74–88.
147. **Conder, G. A., Marchiondo, A. A., Williams, J. F. & Andersen, F. L.** (1983). Freeze-etch characterization of the teguments of three metacestodes: *Echinococcus granulosus, Taenia crassiceps* and *Taenia taeniaeformis*. *Journal of Parasitology*, **69**: 539–48.
148. **Conn, D. B.** (1985a). Fine structure of the embryonic envelopes of *Oochoristica anolis* (Cestoda: Linstowiidae). *Zeitschrift für Parasitenkunde*, **71**: 639–48.
149. — (1985b). Scanning electron microscopy and histochemistry of embryonic envelopes of the porcupine tapeworm *Monoecocestus americanus* (Cyclophyllidea: Anoplocephalidae). *Canadian Journal of Zoology*, **63**: 1194–5.
150. **Conn, D. B. & Etges, F. J.** (1984). Fine structure and histochemistry of the parenchyma and uterine egg capsules of *Oochoristica anolis* (Cestoda: Linstowiidae). *Zeitschrift für Parasitenkunde*, **70**: 769–79.
151. **Conn, D. B., Etges, F. J. & Sidner, R. A.** (1984). Fine structure of the gravid paruterine organ and embryonic envelopes of *Mesocestoides lineatus* (Cestoda). *Journal of Parasitology*, **70**: 68–77.
152. **Conway-Jones, P. B. & Rothman, A. H.** (1978). *Hymenolepis microstoma*: ultrastructural localization of adenyl cyclase in the tegument. *Experimental Parasitology*, **46**: 152–6.
153. **Cornish, R. A. & Bryant, C.** (1975). Studies of regulatory metabolism in *Moniezia expansa*: glutamate, and the absence of the γ-aminobutyrate pathway. *International Journal for Parasitology*, **5**: 355–62.
154. **Craig, P. S., Macpherson, C. N. L. & Nelson, G. S.** (1986). The identification of eggs of *Echinococcus* by immunofluorescence using a specific anti-oncospheral monoclonal antibody. *American Journal of Tropical Medicine and Hygiene*, **35**: 152–8.
155. **Crane, R. K.** (1960). Intestinal absorption of sugars. *Physiological Review*, **40**: 789–825.
156. — (1977). The gradient hypothesis and other models of carrier-

References

mediated active transport. *Reviews of Physiology, Biochemistry and Pharmacology*, **78**: 99–159.
157. **Crompton, D. W. T.** (1973). The sites occupied by some parasitic helminths in the alimentary tract of vertebrates. *Biological Reviews*, **48**: 27–83.
158. **Crompton, D. W. T. & Nesheim, M. C.** (1969). Amino acid patterns during digestion in the small intestine of ducks. *Journal of Nutrition*, **99**: 43–50.
159. **Crompton, D. W. T., Shrimpton, D. H. & Silver, I.A.** (1965). Measurements of the oxygen tension in the lumen of the small intestine of the domestic duck. *Journal of Experimental Biology*, **43**: 473–8.
160. **Crompton, D. W. T. & Whitfield, P. J.** (1969). A hypothesis to account for the anterior migrations of adult *Hymenolepis diminuta* (Cestoda) and *Moniliformis dubius* (Acanthocephala) in the intestine of rats. *Parasitology*, **58**: 227–9.
161. **Csáky, T. Z.** (ed.) (1975). *Intestinal absorption and malabsorption*. Raven Press: New York.
162. **Cuesta-Bandera, C., McManus, D. P. & Rishi, A. K.** (1988). Characterization of *Echinococcus granulosus* of Spanish origin by DNA restriction endonuclease analysis and Southern blot hybridisation. *International Journal for Parasitology*, **18**: 137–41.
163. **Cyr, D., Gruner, S. & Mettrick, D. F.** (1983). Uptake of 5-hydroxytryptamine (serotonin), glucose and changes in worm glycogen levels. *Canadian Journal of Zoology*, **61**: 1469–74.
164. **Dabrowski K. R.** (1980). Amino-acid composition of *Ligula intestinalis* (L.) (Cestoda) plerocercoids and of the host parasitized by these cestodes. *Acta Parasitologica Polonica*, **27**: 45–8.
165. **Damian, R. T.** (1964). Molecular mimicry: antigen sharing by parasite and host and its consequences. *American Naturalist*, **98**: 129–49.
166. — (1987a). The exploitation of host immune response by parasites. *Journal of Parasitology*, **73**: 1–13.
167. — (1987b). Molecular mimicry revisited. *Parasitology Today*, **3**: 263–6.
168. **Darwish, A. A., Sangster, N. C. & Prichard, R. K.** (1983). Ribonucleotide levels in six nematodes, a cestode and a trematode. *Molecular and Biochemical Parasitology*, **8**: 109–17.
169. **Davey, K. G. & Breckenridge, W. R.** (1967). Neurosecretory cells in a cestode, *Hymenolepis diminuta. Science*, **158**: 931–2.
170. **Davis, R. E. & Roberts, L. S.** (1983a). Platyhelminths – Eucestoda. In *Reproductive biology of invertebrates*, vol. 1: *Oogenesis, oviposition, and oosorption*, ed. K. G. & R. G. Adiyodi, pp. 109–33. John Wiley & Sons: London.
171. — (1983b). Platyhelminths – Eucestoda. In *Reproductive biology of invertebrates*, vol. II: *Spermiogenesis and sperm function*, ed. K. G. Adiyodi & R. G. Adiyodi, pp. 131–49. John Wiley & Sons: London.
172. **Davydov, V. G.** (1981). [The penetration of the plerocercoids of some cestodes into the host tissues.] In Russian. *Biologiya Vnutrennykh Vod. Informatsionnyi Byulleten*, **52**: 57–62. [HA/53/778]
173. **De Leon, L. D., Arcos, L. & Willms, K.** (1982). The use of cell-free

315

References

systems of *Cysticercus cellulosae* antigens. In *Cysticercosis: present state of knowledge and perspectives*, ed. A. Flisser, K. Willms, J. P. Laclette, C. Larralde, C. Ridaura, F. Beltrán & M. W. Vogt, pp. 465–75. Academic Press: New York.

174. **De Rycke, P. H. & Berntzen, A. K.** (1967). Maintenance of and growth of *Hymenolepis microstoma* (Cestoda: Cyclophyllidea) *in vitro. Journal of Parasitology*, **53**: 352–4.

175. **Demshin, N. I.** (1981*a*). [The development of the metacestode *Polycercus paradoxa* (Rud. 1802) Spasskaya et Spasski, 1970 (Dilepididae) in Oligochaeta.] In Russian. *AN SSR: Dal'nevostochnyi Nauchnyi Tsentr*, USSR: 12–23. [HA/52/2202]

176. — (1981*b*). [Post-embryonic development of the cestode *Aploparaksis filum* (Goeze, 1782) (Hymenolepididae).] In Russian. *Yaroslavskii Gosudarstvennyi Universitet*: 18–23. [HA/51/6462]

177. **Dhandayuthapani, S., Nellaiappan, K. & Ramalingam, K.** (1983). Quinones in *Penetrocephalus ganaptii* (Cestoda: Pseudophyllidea). *Journal of Parasitology*, **69**: 996–8.

178. **Dick, T. A. & Pool, B. C.** (1985). Identification of *Diphyllobothrium dendriticum* and *Diphyllobothrium latum* from some freshwater fishes of central Canada. *Canadian Journal of Zoology*, **63**: 196–201.

179. **Dixon, B. R. & Arai, H. P.** (1985). Isoelectric focusing of soluble proteins in the characterization of three species of *Hymenolepis* (Cestoda). *Canadian Journal of Zoology*, **63**: 1720–3.

180. **Douch, P. G. C.** (1976*a*). Azo- and nitro-reductases of the cestode *Moniezia expansa*. Substrate specificity, reaction products and the effects of flavins and other compounds. *Xenobiotica*, **6**: 339–404.

181. — (1976*b*). Azo- and nitro-reductase activities and cytochromes of *Ascaris lumbricoides* var. *suum* and *Moniezia expansa*. *Xenobiotica*, **6**: 531–6.

182. — (1978). L-Leucyl-β-naphthylamidases of the cestode *Moniezia expansa*, and the nematode *Ascaris suum*. *Comparative Biochemistry and Physiology*, **60B**: 63–6.

183. **Draper, C. C. & Lillywhite, J. E.** (1984). Immunodiagnosis of tropical parasitic infections. *Recent Advances in Tropical Medicine*, **1**: 267–88.

184. **Dubinina, M. N.** (1966). *Cestoda: Ligulidae*. Publishing House of Science: Moscow & Leningrad.

185. **Dubovskaya, A. Ya** (1979). [A study of the arginase activity in cestode plerocercoids and adults.] In Russian. *Trudy Gel'mintologicheskoĭ Laboratorii (Gel'minty zhivotnykh i rasteniĕ)*, **29**: 46–9. [HA/49/2010]

186. — (1982*a*). [Study of the properties of arginase in adult and larval cestodes, parasitic, in freshwater fish.] In Russian. In *Gel'minty v presnovodnykh biotsenozakh*, pp. 85–90. Nauka: Moscow, USSR. [HA/52/2611]

187. — (1982*b*). [Temperature dependence of arginase activity in *Triaenophorus nodulosus* and *Diphyllobothrium latum*.] In Russian. *Parazitologiya*, **16**: 494–7. [HA/52/2028]

188. **Dunn, J. & Threadgold, L. T.** (1984). *Taenia crassiceps*: temperature, polycations, and cysticercal endocytosis. *Experimental Parasitology*, **58**: 110–24.

References

189. **Dupont, F. & Gabrion, C.** (1987). The concept of specificity in the procercoid–copepod system: *Bothriocephalus claviceps* (Cestoda) a parasite of the eel (*Anguilla anguilla*). *Parasitological Research*, **73**: 151–8.
190. **Eckert, J.** (1986). Prospects for treatment of the metacestode stage of *Echinococcus*. In *The biology of* Echinococcus *and hydatid disease*, ed. R. C. A. Thompson, pp. 250–284. George Allen & Unwin: London.
191. **Eckert, J. & Ramp, T.** (1985). Cryopreservation of *Echinococcus multilocularis* metacestodes and subsequent proliferation in rodents (*Meriones*). *Zeitschrift für Parasitenkunde*, **71**: 777–87.
192. **Eckert, J., von Brand, T. & Voge, M.** (1969). Asexual multiplication of *Mesocestoides corti* (Cestoda) in the intestine of dogs. *Journal of Parasitology*, **55**: 241–9.
193. **Ernst, P. B., Befus, A. D. & Bienenstock, J.** (1985). Leukocytes in the intestinal epithelium: an unusual immunological compartment. *Immunology Today*, **6**: 50–5.
194. **Esch, G. W. & Smyth, J. D.** (1976). Studies on the *in vitro* culture of *Taenia crassiceps*. *International Journal for Parasitology*, **6**: 143–9.
195. **Etges, D. J. & Bogitsh, B. J.** (1985). The effect of colchicine on translocation of incorporated [³H]proline in *Hymenolepis diminuta*. *Journal of Parasitology*, **71**: 290–6.
196. **Euzet, L., Swiderski, Z. & Mokhtar-Maamouri, F.** (1981). Ultrastructure comparée du spermatozoïde des cestodes. Relations avec la phylogénèse. *Annals de Parasitologie Humaine et Comparée*, **56**: 247–59.
197. **Evans, W. S.** (1970). The *in vitro* cultivation of *Hymenolepis microstoma* from cysticercoid to egg-producing adult. *Canadian Journal of Zoology*, **48**: 1135–7.
198. — (1980). The cultivation of *Hymenolepis in vitro*. In *Biology of the tapeworm* Hymenolepis diminuta, ed. H. P. Arai, pp. 425–48. Academic Press: New York.
199. **Evans, W. S. & Novak, M.** (1983). Growth and development of *Hymenolepis microstoma* in mice acclimated to different environmental temperatures. *Canadian Journal of Zoology*, **61**: 2899–903.
200. **Fairbairn, D., Wertheim, G., Harpur, R. P. & Schiller, E. L.** (1961). Biochemistry of normal and irradiated strains of *Hymenolepis diminuta*. *Experimental Parasitology*, **11**: 248–63.
201. **Fairweather, I.** (1976). Neurosecretion in hymenolepidid tapeworms. *Parasitology*, **73**: XXXIV.
202. — (1978). Studies on the nervous system of *Hymenolepis nana* (Cestoda: Cyclophyllidea). Ph.D. thesis, University of London.
203. — (1979). *Hymenolepis nana*: the fine structure of the oncosphere. *Parasitology*, **79**: XVI.
204. **Fairweather, I. & Threadgold, L. T.** (1981*a*). *Hymenolepis nana*: the fine structure of the embryonic envelopes. *Parasitology*, **82**: 429–43.
205. — (1981*b*). *Hymenolepis nana*: the fine structure of the 'penetration' gland and nerve cells within the oncosphere. *Parasitology*, **82**: 445–58.
206. — (1983). *Hymenolepis nana*: the fine structure of the adult nervous system. *Parasitology*, **86**: 89–103.

References

207. **Falkmer, S., Gustafsson, M. K. S. & Sundler, F.** (1985). Phylogenetic aspects of the neuroendocrine system. *Nordisk Psykiatrsik Tidsskrift* **39** (suppl. 11): 21–30.
208. **Farland, W. H. & Macinnis, A. J.** (1978). *In vitro* thymidine kinase activity: present in *Hymenolepis diminuta* (Cestoda) and *Moniliformis dubius* (Acanthocephala), but apparently lacking in *Ascaris lumbricoides* (Nematoda). *Journal of Parasitology*, **64**: 564–5.
209. **Farooq, R. & Farooqi, H. U.** (1983). Histochemical localization of esterases in *Avitellina lahorea* Woodland, 1927 (Cestoda: Anoplocephalida). *Journal of Helminthology*, **57**: 39–41.
210. — (1984). Histochemical localization of phosphomonoesterases in *Avitellina lahorea* Woodland, 1927 (Cestoda: Anoplocephalida). *Journal of Helminthology*, **58**: 169–73.
211. **Featherston, D. W.** (1969). *Taenia hydatigena*. I. Growth and development of adult stage in the dog. *Experimental Parasitology*, **25**: 329–38.
212. — (1971). *Taenia hydatigena*. III. Light and electron microscope study of spermatogenesis. *Zeitschrift für Parasitenkunde*, **37**: 148–68.
213. **Fielding, L. P.** (1980). *Gastro-intestinal mucosal blood flow*. Churchill Livingston: London & New York.
214. **Fioravanti, C. F.** (1981). Coupling of mitochondrial NADPH:NAD transhydrogenase with electron transport in adult *Hymenolepis diminuta*. *Journal of Parasitology*, **67**: 823–31.
215. — (1982*a*). Mitochondrial NADH oxidase activity of adult *Hymenolepis diminuta* (Cestoda). *Comparative Biochemistry and Physiology*, **72B**: 591–6.
216. — (1982*b*). Mitochondrial malate dehydrogenase, decarboxylating (malic enzyme) and transhydrogenase activities of adult *Hymenolepis microstoma* (Cestoda). *Journal of Parasitology*, **68**: 213–20.
217. **Fioravanti, C. F. & Kim, Y.** (1983). Phospholipid dependence of the *Hymenolepis diminuta* mitochondrial NADH:NAD transhydrogenase. *Journal of Parasitology*, **69**: 1048–54.
218. **Fioravanti, C. F. & MacInnis, A. J.** (1977). The identification and characterization of a prenoid constituent (farnesol) of *Hymenolepis diminuta* (Cestoda). *Comparative Biochemistry and Physiology*, **57B**: 227–33.
219. **Fioravanti, C. F. & Saz, H. J.** (1976). Pyridine nucleotide transhydrogenases of parasitic helminths. *Archives of Biochemistry and Biophysics*, **175**: 21–30.
220. — (1978). 'Malic' enzyme, fumarate reductase and transhydrogenase systems in the mitochondria of adult *Spirometra mansonoides* (Cestoda). *Journal of Experimental Zoology*, **206**: 167–77.
221. — (1980). Energy metabolism of adult *Hymenolepis diminuta*. In *Biology of the tapeworm* Hymenolepis diminuta, ed. H. P. Arai, pp. 463–504. Academic Press: London & New York.
222. **Flemström, G. & Garner, A.** (1984). Some characteristics of duodenal epithelium. In *Mucus and mucosa*, ed. J. C. Nugent & M. O'Connor, pp. 94–108. Ciba Foundation Symposium, no. 109. Pitman: London.
223. **Fletcher, T. C., White, A. & Baldo, B. A.** (1980). Isolation of a

phosphorylcholine-containing component from the turbot tapeworm, *Bothriocephalus scorpii* (Müller), and its reaction with C-reactive protein. *Parasite Immunology*, **2**: 237–48.

224. **Flisser, A., Espinoza, B., Tovar, A., Plancarte, A. & Correa, D.** (1986). Host–parasite relationship in cysticercosis: immunologic study in different compartments of the host. *Veterinary Parasitology*, **20**: 95–102.

225. **Flisser, A., Pérez-Monfort, R. & Larralde, C.** (1979). The immunology of human and animal cysticercosis: a review. *Bulletin of the World Health Organization*, **57**: 839–56.

226. **Flisser, A., Willms, K., Laclette, J. P., Larralde, C., Ridaura, C., Beltrán, F. & Vogt, M. W.** (1982). *Cysticercosis: present state of knowledge and perspectives.* Academic Press: New York.

227. **Flockart, H. A.** (1979). *Ligula intestinalis* (L.1758) – *in vitro* and *in vivo* studies. Ph.D. thesis, University of London.

228. **Fox, J. C., Jordan, H. E., Kocan, K. M., George, T. J., Mullins, S. T., Barnett, C. E. & Glenn, B. L.** (1986). An overview of serological tests currently available for laboratory diagnosis of parasitic infection. *Veterinary Parasitology*, **20**: 13–29.

229. **Frayha, G. J.** (1971). Comparative metabolism of acetate in the taeniid tapeworms *Echinococcus granulosus*, *E. multilocularis* and *Taenia hydatigena*. *Comparative Biochemistry and Physiology*, **39B**: 167–70.

230. — (1974). Synthesis of certain cholesterol precursors by hydatid protoscoleces of *Echinococcus granulosus* and cysticerci of *Taenia hydatigena*. *Comparative Biochemistry and Physiology*, **49B**: 93–8.

231. **Frayha, G. J., Bahr, G. M. & Haddad, R.** (1980). The lipids and phospholipids of hydatid protoscolices of *Echinococcus granulosus* (Cestoda). *International Journal for Parasitology*, **10**: 213–6.

232. **Frayha, G. J. & Haddad, R.** (1980). Comparative chemical composition of protoscolices and hydatid cyst fluid of *Echinococcus granulosus* (Cestoda). *International Journal for Parasitology*, **10**: 359–64.

233. **Frayha, G. J. & Smyth, J. D.** (1983). Lipid metabolism in parasitic helminths. *Advances in Parasitology*, **22**: 309–87.

234. **Freeman, R. S.** (1973). Ontogeny of cestodes and its bearing on their phylogeny and systematics. *Advances in Parasitology*, **11**: 481–557.

235. **Freze, V. I., Dalin, M. V. & Sergeeva, E. G.** (1985). [Agglutination reaction in pseudophyllidea coracidia and its use in cestodological studies]. In Russian. *Parazitologiya*, **19**: 257–63. [HA/54/4662]

236. **Freze, V. I., Sidorov, V. S. & Gur'yanova, S. D.** (1974). [Amino-acids of the plerocercoids from the genus *Diphyllobothrium* (Cestoidea: Pseudophyllidea) and of their hosts.] In Russian. *Helminthologia, Bratislava*, **15**: 487–94. [HA/47/3672]

237. **Friedman, P. A., Weinstein, P. P., Davidson, L. A. & Mueller, J. F.** (1982). *Spirometra mansonoides*: lectin analysis of tegumental glycopeptides. *Experimental Parasitology*, **54**: 93–103.

238. **Friedman, P. A., Weinstein, P. P., Mueller, J. F. & Allen, R. H.** (1983).

References

Characterization of cobalamin receptor sites in brush-border plasma membranes of the tapeworm *Spirometra mansonoides*. *Journal of Biological Chemistry*, **258**: 4261–5.

239. **Fukase, T., Matsuda, Y., Akihama, S. & Itagaki, H.** (1985). Purification and some properties of cysteine protease of *Spirometra erinacei* plerocercoid (Cestoda: Diphyllobothriidae). *Japanese Journal of Parasitology*, **34**: 351–60.

240. **Fukumoto, S.** (1985). Pyruvate kinase isoenzymes and phosphoenolpyruvate carboxykinase during development of *Spirometra erinacei*. *Yonago Acta Medica*, **28**: 89–105.

241. **Fundenberg, H. H., Stites, D. P., Caldwell, J. L. & Wells, J. V.** (1980). *Basic and clinical immunology*. Lange Medical Publications: Los Altos, CA.

242. **Furukawa, T., Miyazato, T., Okamoto, K. & Nakai, Y.** (1977). The fine structure of the hatched oncospheres of *Hymenolepis nana*. *Japanese Journal of Parasitology*, **26**: 49–62.

243. **Gabrion, C.** (1975). Étude experimentale du développement larvaire d'*Anomotaenia constricta* (Molin, 1858) Cohn, 1900 chez un Coléoptère *Pimpelia sulcata* Geoffr. *Zeitschrift für Parasitenkunde*, **47**: 249–62.

244. — (1982). Origine du tégument définitive chez les Cestodes Cyclophyllides. *Bulletin de la Société Zoologique de France*, **107**: 565–9.

245. **Gabrion, C. & Gabrion, J.** (1976). Étude ultrastructural de la larve d'*Anomotaenia constricta* (Cestoda, Cyclophyllidea). *Zeitschrift für Parasitenkunde*, **49**: 161–77.

246. **Gaitanaki, C. & Beis, I.** (1985). Enzymes of adenosine metabolism in *Hymenolepis diminuta* (Cestoda). *International Journal for Parasitology*, **15**: 651–4.

247. **Gallie, G. J. & Sewell, M. M. H.** (1970). A technique for hatching *Taenia saginata* eggs. *Veterinary Record*, **86**: 749.

248. **Gamble, H. R. & Pappas, P. W.** (1981). Adenosine deaminase (E.C. 3.5.4.4.) from *Hymenolepis diminuta*. *Journal of Parasitology*, **67**: 759–60.

249. **Gamble, H. R. & Zarlenga, D. S.** (1986). Biotechnology in the development of vaccines for animal parasites. *Veterinary Parasitology*, **20**: 237–250.

250. **Gaur, A. S. & Agarwal, S. M.** (1981). Non-specific phosphomonoesterases in three species of caryophyllids from *Clarias batrachus* (Linn.). *Proceedings of the Indian Academy of Parasitology*, **2**: 71–5.

251. **Gemmell, M. A.** (1977). Taeniidae: modification of the life span of the egg and the regulation of tapeworm populations. *Experimental Parasitology*, **41**: 314–28.

252. **Gemmell, M. A. & Johnstone, P. D.** (1977). Experimental epidemiology of hydatidosis and cysticercosis. *Advances in Parasitology*, **15**: 311–69.

253. **Gemmell, M. A. & Lawson, J. R.** (1982). Ovine cysticercosis: an epidemiological model for the cysticercoses. I. The free-living egg phase. In *Cysticercosis: present state of knowledge and perspectives*, ed.

320

References

A. Flisser, K. Willms, J. P. Laclette, C. Larralde, C. Ridaura, F. Beltrán & M. W. Vogt, pp. 87–98. Academic Press: New York.
254. **Ghazal, A. M. & Avery, R. A.** (1974). Population dynamics of *Hymenolepis nana* in mice: fecundity and the 'crowding effect'. *Parasitology*, **69**: 403–15.
255. **Giles, N.** (1983). Behavioural effects of the parasite *Schistocephalus solidus* (Cestoda) on an intermediate host, the three-spined stickleback *Gasterosteus aculeatus* L. *Animal Behaviour*, **31**: 1192–4.
256. **Goddeeris, B.** (1980a). The role of insects in dispersing eggs of tapeworms, in particular *Taeniarhynchus saginatum*. 1. Review of the literature. *Annales de la Société Belge de Médecine Tropicale*, **60**: 195–201.
257. — (1980b). The role of insects in dispersing eggs of tapeworms. II. Personal investigations. *Annales de la Société Belge de Médecine Tropicale*, **60**: 277–83.
258. **Godfrey, D. G.** (1984). Molecular biochemical characterisation of human parasites. *Recent Advances in Tropical Medicine*, **1**: 289–319.
259. **Golitsyna, N. B., Kharin, V. N. & Anikieva, L. V.** (1981). [Study of the effect of infection intensity on the morphology of *Diphyllobothrium latum* (Cestoda: Diphyllobothriidae), with an analysis of the relationship.] In Russian. *Parazitologiya*, **15**: 313–17. [HA/51/1873]
260. **Golubev, A. I. & Kashapova, L. A.** (1975). [Some ultrastructural pecularities of neurons in the cestode *Pelichnobothrium speciosum* (Monticelli, 1889) (Cestoda: Tetraphyllidea).] In Russian. *Parazitologiya*, **9**: 439–42. [HA/46/134]
261. **Gottstein, B., Schantz, P. M., Todorov, T., Saimot, A. G. & Jaquier, P.** (1986). An international study on the serological differential diagnosis of human and cystic alveolar echonococcosis. *Bulletin of the World Health Organization*, **64**: 101–5.
262. **Grabiec, S., Guttowa, A., Malzahn, E. & Michajłow, W.** (1969). Investigations on the oxidation–reduction activity in embryos and coracidia of *Triaenophorus nodulosus* (Pall.) by the chemiluminescence method. *Bulletin de l'Académie Polonaise des Sciences*, Cl. II. **17**: 609–12.
263. **Grabiec, S., Guttowa, A. & Michajłow, W.** (1963). Effect of light stimulus on hatching of coracidia of *Diphyllobothrium latum* (L.). *Acta Parasitologica Polonica*, **11**: 229–38.
264. — (1964). Investigation on the respiratory metabolism of eggs and coracidia of *Diphyllobothrium latum* (L.) Cestoda. *Bulletin de l'Académie Polonaise des Sciences. Cl.II, Série des Sciences Biologiques*, **12**: 29–34.
265. **Grammeltvedt, A.-F.** (1973). Differentiation of the tegument and associated structures in *Diphyllobothrium dendriticum* Nitsch (1824) (Cestoda: Pseudophyllidea). An electron microscopical study. *International Journal for Parasitology*, **3**: 321–7.
266. **Granath, W. O. & Esch, G. W.** (1983). Temperature and other factors that regulate the composition and infrapopulation densities of *Bothriocephalus acheilognathi* (Cestoda) in *Gambusia affinis* (Pisces). *Journal of Parasitology*, **69**: 1116–24.

321

References

267. **Granath, W. O., Lewis, J. C. & Esch, G. W.** (1983). An ultrastructural examination of the scolex and tegument of *Bothriocephalus acheilognathi* (Cestoda: Pseudophyllidea). *Transactions of the American Microscopical Society*, **102**: 240–50.
268. **Gruner, S. & Mettrick, D. F.** (1984). The effect of 5-hydroxytryptamine on glucose absorption by *Hymenolepis diminuta* (Cestoda) and by the mucosa of the rat small intestine. *Canadian Journal of Zoology*, **62**: 798–803.
269. **Guillou, F.** (1982). La ligulose pisciaire. Ph.D. thesis, École National Vétérinaire d'Alfort, France.
270. **Gurbanov, F. Sh. & Simonov, A. P.** (1982). [Action of chemicals on the eggs of dog taeniids in the soil.] In Russian. *Byulleten Vsesoyuznogo Instituta Gel'mintologii im. K. I. Skryabina.* **28**: 85–9. [HA/51/5875]
271. **Gur'yanova, S. D.** (1977). [The amino-acid composition of the total proteins of plerocercoids of the genus *Diphyllobothrium* and their hosts.] In Russian. In [*Comparative biochemistry of fish and their helminths. Lipids, enzymes, proteins. (Collected works)*.] ed. V. S. Sidorov, pp. 99–109. USSR Akademii Nauk, Institut Biologi, Karel'skiĭ Filial. USSR.
272. — (1983). [Comparative study of the lipids of some cestodes and of their freshwater fish hosts.] In Russian. *Sravnitel'naya biokhimiya vodnykh zhivotnykh,* ed. V. S. Sidorov & R. U. Vysotskaya, pp. 159–66 Petrozavodsk, USSR: Karel'skiĭ Filial AN SSSR. [HA/55/197]
273. **Gur'yanova, S. D. & Freze, V. I.** (1978). [Lipid content in different eco- and host-forms of *Diphyllobothrium* plerocercoids.] In Russian. In *Ékologicheskaya biokhimiya zhivotnykh,* ed. V. S. Sidorov, pp. 24–33. Petrozavodsk, USSR: Institut Biologii, Karel'skiĭ Filial AN SSSR. [HA/51/4333]
274. — (1981). [Lipid composition of larval and adult forms of some cestodes.] In Russian. In *Sravitel'nye aspeckty biokhimii ryb in hekotorykh drugikh zhivotnykh,* ed. V. S. Sidorov, pp. 116–21. Petrozavodsk, USSR: Institut Biologii, Karel'skiĭ Filial AN SSSR. [HA/53/1927]
275. **Gur'yanova, S. D., Sidorov, V. S. & Freze, V. I.** (1976). [The amino-acid composition of the plerocercoids of *Diphyllobothrium*, related to ecological factors.] In Russian. In *Parazitologicheskie issledovaniya Karel'skoi ASSR i Murmanskoĭ oblasti,* pp. 222–230. Petrozavodsk; Karel'skoĭ filial Akademii Nauk SSSR, Institut Biologii. [HA/46/4919]
276. **Gustafsson, M. K. S.** (1976). Observations on the histogenesis of nervous tissue in *Diphyllobothrium dendriticum* Nitzsch, 1824 (Cestoda, Pseudophyllidea). *Zeitschrift für Parasitenkunde,* **50**: 313–21.
277. — (1984). Synapses in *Diphyllobothrium dendriticum* (Cestoda). An electron microscopical study. *Annales Zoologici Fennici,* **21**: 167–75.
278. — (1985). Cestode neurotransmitters. *Parasitology Today,* **1**: 72–5.
279. **Gustafsson, M. K. S. & Vaihela, B.** (1981). Two types of frontal glands in *Diphyllobothrium dendriticum* (Cestoda, Pseudophyllidea) and their fate during the maturation of the worm. *Zeitschrift für Parasitenkunde,* **66**: 145–54.

References

280. **Gustafsson, M. K. S. & Wikgren, M. C.** (1981a). Peptidergic and aminergic neurons in adult *Diphyllobothrium dendriticum* Nitzsch, 1824 (Cestoda, Pseudophyllidea). *Zeitschrift für Parasitenkunde*, **64**: 121–34.

281. — (1981b). Activation of the peptidergic neurosecretory system in *Diphyllobothrium dendriticum* (Cestoda: Pseudophyllidea). *Parasitology*, **83**: 243–7.

282. — (1981c). Release of neurosecretory material by protrusions of bounding membranes extending through the axolemma, in *Diphyllobothrium dendriticum* (Cestoda). *Cell and Tissue Research*, **220**: 473–9.

283. **Gustafsson, M. K. S., Wikgren, M. C., Karhi, T. J. & Schot, L. P. C.** (1985). Immunocytochemical demonstration of neuropeptides and serotonin in the tapeworm *Diphyllobothrium dendriticum*. *Cell and Tissue Research*, **240**: 255–60.

284. **Guttowa, A.** (1958). Further research on the effect of temperature on the development of the cestode *Triaenophorus lucii* (Müll) embryos in eggs, and on the invadability of their oncospheres. *Acta Parasitologica Polonica*, **6**: 367–81.

285. **Guttowa, A. & Moczoń, T.** (1974). Oxidoreductase histochemistry in larval stages of pseudophyllidean cestodes. I. Coracidium. *Acta Parasitologica Polonica*, **22**: 1–7.

286. **Hadžiosmanović, A. & Kravica, S.** (1982). [The activity of phosphoenolpyruvate carboxykinase (PEPCK) and pyruvate kinase (PK) in some parasitic helminths.] In Russian. *Veterinarski Arkiv*, **52**: 55–63. [HA/52/1493]

287. **Halawa, B. & Jakacka, B.** (1977a). [Proteolytic activity of *Taenia saginata* oncospheres examined by the isotope method.] In Polish. *Wiadomości Parazytologiczne*, **23**: 363–7. [HA/48/3531]

288. — (1977b). [Proteolytic activity of *Cysticercus bovis*.] In Polish. *Wiadomości Parazytologiczne*, **23**: 369–73. [HA/48/2942]

289. **Halton, D. W.** (1982). Morphology and ultrastructure of parasitic helminths. In *Parasites – their world and ours*. Proceedings of the 5th International Congress of Parasitology, Toronto, ed. D. F. Mettrick & S. S. J. Desser, pp. 60–9. Elsevier Biomedical Press: Amsterdam.

290. **Halverson, O. & Andersen, K. I.** (1974). Some effects of population density in infections of *Diphyllobothrium dendriticum* (Nitzsch) in golden hamster (*Mesocricetus auratus* Waterhouse) and common gull (*Larus larus* L.). *Parasitology*, **69**: 149–60.

291. **Hamilton, J. D., Dawson, A. M. & Webb, J. P. W.** (1968). Observations on the small gut 'mucosal' pO_2 and pCO_2 in anesthetized dogs. *Gastroenterology*, **55**: 52–60.

292. **Hammerberg, B., Dangler, C. & Williams, J. F.** (1980). *Taenia taeniaeformis*: chemical composition of parasite factors affecting coagulation and complement cascades. *Journal of Parasitology*, **66**: 569–76.

293. **Hammerberg, B. & Williams, J. F.** (1978a). Interaction between *Taenia taeniaeformis* and the complement system. *Journal of Immunology*, **120**: 1033–8.

294. — (1978b). Physicochemical characterization of complement

References

interacting factors from *Taenia taeniaeformis*. *Journal of Immunology*, **120**: 1039–45.
295. **Hariri, M.** (1974*a*). Quantitative measurements of endogenous levels of acetylcholine and choline in tetrathyridia of *Mesocestoides corti* (Cestoda) by means of combined gas chromatograph–mass spectrometry. *Journal of Parasitology*, **60**: 227–30.
296. — (1974*b*). Occurrence and concentration of biogenic amines in *Mesocestoides corti* (Cestoda). *Journal of Parasitology*, **60**: 737–43.
297. — (1975). Uptake of 5-hydroxytryptamine by *Mesocestoides corti* (Cestoda). *Journal of Parasitology*, **61**: 440–8.
298. **Harlow, D. R. & Byram, J. E.** (1971). Isolation and morphology of the mitochondrion of the cestode *Hymenolepis diminuta*. *Journal of Parasitology*, **57**: 559–65.
299. **Harris, A., Heath, D. D., Lawrence, S. B. & Shaw, R. J.** (1987). Ultrastructure of changes at the surface during the early development phases of *Taenia ovis* cysticerci *in vitro*. *International Journal for Parasitology*, **17**: 903–10.
300. **Harris, B. G.** (1983). Protein metabolism. In *Biology of the Eucestoda*, vol. 2, ed. C. Arme & P. W. Pappas, pp. 335–41. Academic Press: London.
301. **Harrison, L. J. S., Le Riche, P. D. & Sewell, M. M. H.** (1986). Variations in *Echinococcus granulosus* of bovine origin identified by enzyme electrophoresis. *Tropical Animal Health and Production*, **18**: 48–50.
302. **Harrison, L. J. S. & Parkhouse, R. M. E.** (1986). Identification of protective antigens in *Taenia saginata* cysticercosis. *Coopers Animal Health Symposium on the immunobiology and molecular biology of cestode infections.* Abstract A2. University of Melbourne, Veterinary School: Melbourne.
303. **Harrison, L. J. S., Parkhouse, R. M. E. & Sewell, M. M. H.** (1984). Variation in 'target' antigens between appropriate and inappropriate hosts of *Taenia saginata* metacestodes. *Parasitology*, **88**: 659–63.
304. **Hart, J. L.** (1967). Studies on the nervous system of tetrathyridia (Cestoda: *Mesocestoides*). *Journal of Parasitology*, **53**: 1032–9.
305. **Haslewood, G. A. D.** (1967). *Bile salts.* Methuen: London.
306. **He, Q. L.** (1983). [Experiment on artificial infestation of dogs with *Taenia hydatigena*.] In Chinese. *Chinese Journal of Veterinary Medicine*, **9**: 8–10. [HA/53/1]
307. **Heath, D. D.** (1971). The migration of the oncospheres of *Taenia pisiformis*, *T. serialis*, and *Echinococcus granulosus* within the intermediate host. *International Journal for Parasitology*, **1**: 145–52.
308. — (1973). An improved technique for the *in vitro* culture of taeniid larvae. *International Journal for Parasitology*, **3**: 481–4.
309. — (1986). Immunology of *Echinococcus* infections. In *The biology of Echinococcus and hydatid disease*, ed. R. C. A. Thompson, pp. 164–88. George Allen & Unwin: London.
310. **Heath, D. D. & Elsdon-Dew, R.** (1972). The *in vitro* culture of *Taenia saginata* and *Taenia taeniaeformis* larvae from the oncosphere with

References

observations on the role of serum for *in vitro* culture of larval cestodes. *International Journal for Parasitology*, **2**: 119–30.

311. **Heath, D. D. & Lawrence, S. B.** (1976). *Echinococcus granulosus*: development *in vitro* from oncosphere to immature hydatid cyst. *Parasitology*, **73**: 417–23.

312. — (1980). Prepatent period of *Taenia ovis* in dogs (correspondence). *New Zealand Veterinary Journal*, **28**: 193–4.

313. **Heath, D. D. & Osborn, P. J.** (1976). Formation of *Echinococcus granulosus* laminated membrane in a defined medium. *International Journal for Parasitology*, **6**: 467–71.

314. **Heath, D. D. & Smyth, J. D.** (1970). *In vitro* culture of *Echinococcus granulosus*, *Taenia hydatigena*, *T. ovis*, *T. pisiformis* and *T. serialis* from oncosphere to cystic larva. *Parasitology*, **61**: 329–43.

315. **Heath, R. L. & Hart, J. L.** (1970). Biosynthesis *de novo* of purines and pyrimidines in *Mesocestoides* (Cestoda). II. *Journal of Parasitology*, **56**: 340–5.

316. **Heaton, K. W.** (1972). *Bile salts in health and disease*. Churchill Livingstone: London.

317. **Henderson, D.** (1977). The effect of worm age, weight and number in the infection on the absorption of glucose by *Hymenolepis diminuta*. *Parasitology*, **75**: 277–84.

318. **Henry, K.** (1982). Ultrastructure of the small intestine. In *Gasterology*, vol. 2: *Small intestine*, ed. V. S. Chadwick & S. Phillips, pp. 19–39. Butterworth Scientific: London.

319. **Herd, R. P.** (1977). Resistance of dogs to *Echinococcus granulosus*. *International Journal for Parasitology*, **7**: 135–8.

320. **Herd, R. P., Chappel, R. J. & Biddell, D.** (1975). Immunization of dogs against *Echinococcus granulosus* using worm secretory antigens. *International Journal for Parasitology*, **5**: 395–9.

321. **Hermoso, R., Valero, A. & Monteoliva, M.** (1982). [A study of isoenzymes and soluble proteins in *Moniezia expansa* and *Avitellina centripunctata* by electrophoresis and electrofocusing.] In Spanish. *Revista Ibérica de Parasitologia*, **42**: 109–16. [HA/52/941]

322. **Hess, E.** (1980). Ultrastructural study of the tetrathyridium of *Mesocestoides corti* Hoeppli, 1925 (Cestoda): tegument and parenchyma. *Zeitschrift für Parasitenkunde*, **61**: 135–59.

323. — (1981). Ultrastructural study of the tetrathyridium of *Mesocestoides corti* Hoeppli, 1925 (Cestoda): pool of germinative cells and suckers. *Revue Suisse de Zoologie*, **88**: 661–74.

324. **Hess, E. & Guggenheim, R.** (1977). A study of the microtriches and sensory processes of the tetrathyridium of *Mesocestoides corti* Hoeppli, 1925, by transmission and scanning electron microscopy. *Zeitschrift für Parasitenkunde*, **53**: 189–99.

325. **Hesselberg, C. A. & Andreassen, J.** (1975). Some influences of population density on *Hymenolepis diminuta* in rats. *Parasitology*, **71**: 517–523.

326. **Hill, B., Kilsby, J., Rogerson, G. W., McIntosh, R. T. & Ginger, C. D.** (1981). The enzymes of pyrimidine biosynthesis in a range of parasitic

References

protozoa and helminths. *Molecular and Biochemical Parasitology*, **2**: 123–34.

327. **Hilliard, D. K.** (1972). Studies on the helminth fauna of Alaska. LI. Observations on egg-shell formation in some diphyllobothriid cestodes. *Canadian Journal of Zoology*, **50**: 585–92.

328. **Hipkiss, J. B., Skinner, A. & Branford White, C. J.** (1987). Biochemical and structural investigation of the effect of Stelazine (trifluoperazine) on *Hymenolepis diminuta* (Cestoda). *Parasitology*, **94**: 135–49.

329. **Hirai, K., Nishida, H., Shiwaku, K. & Okuda, H.** (1978). Studies on the plerocercoid growth factor of *Spirometra erinacei* (Rudolphi, 1819) with special reference to the effect on lipid mobilization *in vitro*. *Japanese Journal of Parasitology*, **27**: 527–33.

330. **Hirai, K., Shiwaku, K., Tsuboi, T., Torii, M., Nishida, H. & Yamane, Y.** (1983). Biological effects of *Spirometra erinacei* plerocercoids in several species of rodents. *Zeitschrift für Parasitenkunde*, **69**: 489–99.

331. **Holmes, S. D. & Fairweather, I.** (1982). *Hymenolepis diminuta*: the mechanism of egg hatching. *Parasitology*, **85**: 237–50.

332. **Hopkins, C. A.** (1980). Immunity and *Hymenolepis diminuta*. In *Biology of the tapeworm* Hymenolepis diminuta, ed. H. P. Arai, pp. 551–614. Academic Press: New York.

333. **Hopkins, C. A. & Allen, L. M.** (1979). *Hymenolepis diminuta*: the role of the tail in determining the position of the worm in the intestine of the rat. *Parasitology*, **79**: 401–9.

334. **Hopkins, C. A. & Barr, I. F.** (1982). The source of antigen in the adult tapeworm. *International Journal for Parasitology*, **12**: 327–33.

335. **Hopkins, C. A., Law, L. M. & Threadgold, L. T.** (1978). *Schistocephalus solidus*: pinocytosis by the plerocercoid tegument. *Experimental Parasitology*, **44**: 161–72.

336. **Hopkins, C. A., Subramanian, G. & Stallard, H.** (1972). The development of *Hymenolepis diminuta* in primary and secondary infections in mice. *Parasitology*, **64**: 401–12.

337. **Houba, V.** (1980). *Immunological investigation of tropical parasitic diseases. Practical methods in clinical immunology*, vol. II. Churchill Livingstone: London.

338. **Howell, M. J.** (1985). Gene exchange between hosts and parasites. *International Journal for Parasitology*, **15**: 597–600.

339. — (1986). Cultivation of *Echinococcus* species *in vitro*. In *The biology of* Echinococcus *and hydatid disease*, ed. R. C. A. Thompson, pp. 143–63. George Allen & Unwin: London.

340. **Howells, R. E. & Erasmus, D. A.** (1969). Histochemical observations on the tegumental epithelium and interproglottid glands of *Moniezia expansa* (Rud., 1805) (Cestoda, Cyclophyllidea). *Parasitology*, **59**: 505–18.

341. **Hrženjak, T. Kljaić, K. & Častek, A.** (1983). Identification of glyceroglycolipids in cytolipin P isolated from hydatid fluid of *Echinococcus granulosus*. *Iugoslavica Physiologica et Pharmacologica Acta*, **19**: 445–51.

342. **Hulinska, D.** (1981). Scanning electron microscopy of the surface of

References

the adult *Multiceps endothoracicus* and a comparison of its larval and adult scoleces. *Folia Parasitologica*, **28**: 349–52.
343. **Hulinska, D. & Lavrov, V. I.** (1981). Morphology and cytochemistry of the adult tapeworm *Multiceps endothoracicus* with emphasis on the scolex. *Folia Parasitologica*, **28**: 227–34.
344. **Hustead, S. T. & Williams, J. F.** (1977). Permeability studies on taeniid metacestodes. I. Uptake of protein by larval stages of *Taenia taeniaeformis*, *T. crassiceps*, *Echinococcus granulosus*. *Journal of Parasitology*, **63**: 314–21.
345. **Hyman, L. H.** (1951). *The invertebrates*, vol. II. McGraw-Hill Book Co. Inc.: New York.
346. **Il'yasov, I. N.** (1978). [Amino acid composition of *Raillietina tetragona* (Molin, 1858) and *R. echinobothrida* (Megnin, 1881).] In Russian. *Trudy Nauchno-Issledovatel' skogo Veterinarnogo Instituta Tadzhikskoĭ SSR*, **8**: 75–9. [HA/51/5048]
347. **Insler, G. D.** (1981). Population and developmental changes in thymidine uptake kinetics of *Hymenolepis diminuta* (Cestoda: Cyclophyllidea). *Comparative Biochemistry and Physiology*, **70B**: 697–702.
348. **Insler, G. D. & Roberts, L. S.** (1980*a*). Developmental physiology of cestodes. XV. A system for testing possible crowding factors *in vitro*. *Journal of Experimental Zoology*, **211**: 45–54.
349. — (1980*b*). Developmental physiology of cestodes. XVI. Effects of certain excretory products on incorporation of [³H]thymidine into DNA of *Hymenolepis diminuta*. *Journal of Experimental Zoology*, **211**: 55–61.
350. **Isavkov, S. I.** (1982). [Development of *Echinococcus granulosus* in the animal host.] In Russian. *Nauchno-Technicheskiĭ Byulleten' VASKhNIL, Sibirskoe Otdelenie*, **9**: 27–9. [HA/54/635]
351. **Ishii, A. I.** (1984). Fe-rich corpuscles in *Diplogonoporus grandis* detected using X-ray microanalysis. *Zeitschrift für Parasitenkunde*, **70**: 199–202.
352. **Ito, A.** (1982). Induction of adult formation from cysticercoid of *Hymenolepis nana* established in Fischer (F344) rat with cortisone acetate. *Japanese Journal of Parasitology*, **31**: 141–6.
353. **Ito, A. & Smyth, J. D.** (1987). Immunology of the lumen-dwelling cestode infections. In *Immune responses in parasitic infections: immunology, immunopathology and immunoprophylaxis*, vol. II *Trematodes and cestodes*, ed. E. J. L. Soulsby, pp. 115–63. CRC Press Inc.: Boca Raton, FL.
354. **Jacobsen, N. S. & Fairbairn, D.** (1967). Lipid metabolism in helminth parasites. III. Biosynthesis and interconversion of fatty acids by *Hymenolepis diminuta* (Cestoda). *Journal of Parasitology*, **53**: 355–61.
355. **Jakutowicz, K. & Korpaczewska, W.** (1976). Some trace elements in plerocercoid and adult forms of *Ligula intestinalis* (L., 1758). (Cestoda: Diphyllobothriidae). *Bulletin de l'Académie Polonaise des Sciences (Sciences Biologiques)*, **24**: 525–7.
356. — (1977*a*). Abundances of Mn, Zn, Co, Ag, U and Ba in males and females of *Dioecocestus asper* (Mehlis, 1831) (Cestoda:

References

Dioecocestidae). *Bulletin de l'Académie Polonaise des Sciences* (*Sciences Biologiques*), **24**: 757–8.

357. — (1977*b*). Comparison of the levels of some trace elements detected by the non-destructive neutron activation analysis in five tapeworms – parasites of birds. *Bulletin de l'Académie Polonaise des Sciences* (*Sciences Biologiques*), **25**: 49–54.

358. — (1978). Levels of some trace elements in a canine tapeworm, *Taenia pisiformis* (Bloch, 1780). *Bulletin de l'Académie Polonaise des Sciences* (*Sciences Biologiques*), **26**: 186–6.

359. **Jarrett, E. E. E. & Haig, D. M.** (1984). Mucosal mast cells *in vivo* and *in vitro*. *Immunology Today*, **5**: 115–19.

360. **Jarroll, E. L. Jr** (1980). Population dynamics of *Bothriocephalus rarus* (Cestoda) in *Notophthalmus viridescens*. *American Midland Naturalist*, **103**: 360–6.

361. **Jawetz, E., Melnick, J. L. & Adelberg, E. A.** (1987). *Review of medical microbiology*, 17th edn. Appleton & Lange: Norwalk, CT.

362. **Jayabaskaran, C. & Ramalingam, K.** (1985). Characterization of phenol oxidase in a pseudophyllidean cestode *Penetrocephalus ganapatii*. *Parasitology*, **90**: 433–40.

363. **Jeffs, S. A. & Arme, C.** (1984). *Hymenolepis diminuta*: protein synthesis in cysticercoids. *Parasitology*, **88**: 351–7.

364. — (1985*a*). *Hymenolepis diminuta*: characterization of the neutral amino acid transport loci of the metacestode. *Comparative Biochemistry and Physiology*, **81A**: 387–90.

365. — (1985*b*). *Hymenolepis diminuta* (Cestoda): uptake of cycloleucine by metacestodes. *Comparative Physiology and Biochemistry*, **81A**: 495–9.

366. — (1986). *Echinococcus granulosus*: absorption of cycloleucine and α-aminoisobutyric acid by protoscoleces. *Parasitology*, **92**: 153–63.

367. **Jenkins, D. J. & Rickard, M. D.** (1985). Specific antibody responses to *Taenia hydatigena*, *T. pisiformis* and *Echinococcus granulosus* in dogs. *Australian Veterinary Journal*, **62**: 72–8.

368. — (1986). Specific antibody responses in dogs experimentally infected with *Echinococcus granulosus*. *American Journal of Tropical Medicine and Hygiene*, **35**: 345–49.

369. **Jilek, R. & Crites, J. L.** (1979). Scanning electron microscopic examination of the scolex and external tegumental surface of *Proteocephalus ambloplitis*. *Journal of Microscopy*, **118**: 443–6.

370. **Johnson, L. R.** (ed). (1981). *Physiology of the gastointestinal tract*. Raven Press: New York.

371. **Johnson, W. J. & Cain, G. D.** (1985). Biosynthesis of polyisoprenoid lipids in the rat tapeworm *Hymenolepis diminuta*. *Comparative Biochemistry and Physiology*, **82**: 487–95.

372. **Jones, A.** (1975). The morphology of *Bothriocephalus scorpii* (Müller) (Pseudophyllidea, *Bothriocephalidae*) from littoral fishes in Britain. *Journal of Helminthology*, **49**: 251–61.

373. **Jones, A. W. & Tan, B. D.** (1971). Effect of crowding upon growth and fecundity in the mouse bile tapeworm, *Hymenolepis microstoma*. *Journal of Parasitology*, **57**: 88–93.

References

374. **Jones, B. R., Smith, B. F. & Leflore, W. B.** (1979). The ultrastructural localization of alkaline phosphatase activity in the tegument of the cysticercus of *Hydatigera taeniaeformis*. *Cytobios*, **24**: 195–209.
375. — (1980). Surface topography features of the scolex and strobila of the cysticercus of *Hydatigera taeniaeformis*. *IRCS Medical Science: Microbiology, Parasitology and Infectious Diseases*, **8**: 28–9.
376. **Juhász, S.** (1979). Studies on the nature of the protease inhibitor of *Ligula intestinalis*. *Helminthologia*, **16**: 293–8.
377. — (1980). [Studies on the protease inhibitor of *Ligula intestinalis*.] [In Hungarian.] *Magyar Állatorvosok Lapja*, **35**: 548–9. [HA/50/828]
378. **Kamiya, M., Ooi, H.-K., Oku, Y., Yagi, K. & Ohbayashi, M.** (1985). Growth and development of *Echinococcus multilocularis* in experimentally infected cats. *Japanese Journal of Veterinary Research*, **33**: 135–40.
379. **Kamo, H., Yazaki, S., Fukumoto, S., Maejima, J. & Kawasaki, H.** (1986). [Evidence of egg-discharging periodicity in experimental human infection with *Diphyllobothrium latum* Linné, 1758]. In Japanese. *Japanese Journal of Parasitology*, **35**: 53–7. [HA/55/2698]
380. **Kashin, V. A. & Pluzhnikov, L. T.** (1983). [Ultrastructure of mature ova of the cestode *Fimbriaria fasciolaris* (Cestoidea, Hymenolepididae). In Russian. *Parazitologiya*, **17**: 430–5. [HA/53/1964]
381. **Kassis, A. I. & Tanner, C. E.** (1976). The role of complement in hydatid disease: *in vitro* studies. *International Journal for Parasitology*, **6**: 25–35.
382. **Kawamoto, F., Fujioka, H. & Kumada, N.** (1986). Studies on the post-larval development of cestodes of the genus *Mesocestoides*: trypsin-induced development of *M. lineatus in vitro*. *International Journal for Parasitology*, **16**: 333–40.
383. **Kawamoto, F., Fujioka, H., Mizuno, S., Kumada, N. & Voge, M.** (1986). Studies on the post-larval development of the genus *Mesocestoides*: shedding and further development of *Mesocestoides lineatus* and *M. corti* tetrathyridia *in vivo*. *International Journal for Parasitology*, **16**: 323–31.
384. **Keenan, L., Koopowitz, H. & Solon, M. H.** (1984). Primitive nervous systems: electrical activity in the nerve cords of the parasitic flatworm, *Gyrocotyle fimbriata*. *Journal of Parasitology*, **70**: 131–8.
385. **Kegley, L. M., Baldwin, J., Brown, B. W. & Berntzen, A. K.** (1970). *Mesocestoides corti*: environmental cation concentration in calcareous corpuscles. *Experimental Parasitology*, **27**: 88–94.
386. **Kegley, L. M., Brown, B. W. & Berntzen, A. K.** (1969). *Mesocestoides corti*: inorganic components in calcareous corpuscles. *Experimental Parasitology*, **25**: 85–92.
387. **Kelsoe, G. H., Ubelaker, J. E. & Allison, V. F.** (1977). The fine structure of spermatogenesis in *Hymenolepis diminuta* (Cestoda) with a description of the mature spermatozoon. *Zeitschrift für Parasitenkunde*, **54**: 175–87.
388. **Kennedy, C. R.** (1976). *Ecological aspects of parasitology*. North-Holland: Amsterdam & Oxford.
389. — (1983). General ecology. In *Biology of the Eucestoda*, vol. 1, ed. C.

References

Arme & P. W. Pappas, pp. 27–80. Academic Press: London.
390. **Kerr, T.** (1948). The pituitary in normal and parasitised roach (*Leuciscus rutilus* Flem). *Quarterly Journal of Microscopical Science*, **89**: 129–37.
391. **Keymer, A. E.** (1980). The influence of *Hymenolepis diminuta* on the survival and fecundity of the intermediate host *Tribolium confusum*. *Parasitology*, **81**: 405–21.
392. **Keymer, A. E., Crompton, D. W. T. & Singhvi, A.** (1983). Mannose and the 'crowding effect' of *Hymenolepis* in rats. *International Journal for Parasitology*, **13**: 561–70.
393. **Khan, Z. I. & De Rycke, P. H.** (1976). Studies on *Hymenolepis microstoma in vitro*. I. Effect of heme compounds on growth and reproduction. *Zeitschrift für Parasitenkunde*, **49**: 253–61.
394. **Kilejian, A. & Macinnis, A. J.** (1976). Density distribution of DNA from parasitic helminths with special reference to *Ascaris lumbricoides*. *Rice University Studies*, **62**: 161–74.
395. **Kilejian, A. & Schwabe, C. W.** (1971). Studies on the polysaccharides of the *Echinococcus granulosus* cyst, with observations on a possible mechanism for laminated membrane formation. *Comparative Biochemistry and Physiology*, **40B**: 25–36.
396. **Kim, Y. & Fioravanti, C. F.** (1985). Reduction and oxidation of cytochrome *c* by *Hymenolepis diminuta* (Cestoda) mitochondria. *Comparative Biochemistry and Physiology*, **81B**: 335–9.
397. **Knowles, W. J. & Oaks, J. A.** (1979). Isolation and partial biochemical characterization of the brush border plasma membrane from the cestode *Hymenolepis diminuta*. *Journal of Parasitology*, **65**: 715–31.
398. **Köhler, P.** (1985). The strategies of energy conservation in helminths. *Molecular and Biochemical Parasitology*, **17**: 1–18.
399. **Köhler, P. & Hanselmann, K.** (1974). Anaerobic and aerobic metabolism in the larvae (tetrathyridia) of *Mesocestoides corti*. *Experimental Parasitology*, **36**: 178–88.
400. **Kohlhagen, S., Behm, C. A. & Bryant, C.** (1985). Strain variation in *Hymenolepis diminuta*: enzyme profiles. *International Journal for Parasitology*, **15**: 479–83.
401. **Komuniecki, R. W. & Roberts, L. S.** (1977*a*). Hexokinase from the rat tapeworm *Hymenolepis diminuta*. *Comparative Biochemistry and Physiology*, **57B**: 45–9.
402. — (1977*b*). Galactose utilization by the rat tapeworm *Hymenolepis diminuta*. *Comparative Biochemistry and Physiology*, **57B**: 329–33.
403. — (1977*c*). Enzymes of galactose utilization in the rat tapeworm *Hymenolepis diminuta*. *Comparative Biochemistry and Physiology*, **58B**: 35–8.
404. **Koopowitz, H.** (1973). Organization of primitive nervous system. Neuromuscular physiology of *Gyrocotyle urna*, a parasitic flatworm. *Biological Bulletin*, **144**: 489–502.
405. **Körting, W.** (1976). Metabolism in parasitic helminths of freshwater fish. In *Biochemistry of parasites and host–parasite relationships*, ed. H. Van den Bossche, pp. 95–100. North-Holland Publishing Co: Amsterdam.

References

406. **Körting, W. & Barrett, J.** (1977). Carbohydrate catabolism in the plerocercoids of *Schistocephalus solidus* (Cestoda: Pseudophyllidea). *International Journal for Parasitology,* **7**: 411–17.
407. — (1978). Studies on beta oxidation in the plerocercoids of *Ligula intestinalis* (Cestoda: Pseudophyllidea). *Zeitschrift für Parasitenkunde,* **57**: 243–6.
408. **Kralj, N.** (1967). Morphologic and histochemical studies on the nervous system of tapeworms revealed by the cholinesterase method (*Taenia hydatigena, Dipylidium caninum* and *Moniezia expansa*). *Veterinarski Arhiv,* **37**: 277–86.
409. **Krasnoshchekov, G. P., Kashin, V. A. & Kontrimavichus, V. L.** (1978). [Dynamics of lipid distribution in the developing larvae of *Aploparaxis polystictae* (Cestoda: Hymenolepididae).] In Russian. *Doklady Akademii Nauk SSSR,* **241**: 1481–4. [HA/51/1371]
410. **Krasnoshchekov, G. P. & Pluzhnikov, L. T.** (1981). [The rostellar glands of *Taenia crassiceps* (Cestoda: Taeniidae) larvae.] In Russian. *Parazitologiya,* **15**: 519–24. [HA/51/3605]
411. **Krishna, G. V. R. & Simha, S. S.** (1980). A study of the nerve arrangement and cholinesterase distribution in *Oochoristica sigmoides* and *Raillietina tetragona* (Cestoda). *Journal of Animal Morphology & Physiology,* **27**: 10–13.
412. **Kuei, Li., Chang, Y. S. & Wei, H. C.** (1982). [The life history and taxonomy of *Pseudanoplocephala crawfordi*]. In Chinese. *Acta Veterinaria et Zootechnica Sinica,* **13**: 173–180. [HA/52/1107]
413. **Kumaratilake, L. M. & Thompson, R. C. A.** (1979). A standardized technique for the comparison of tapeworm soluble proteins by thin-layer isoelectric focusing in polyacrylamide gels, with particular reference to *Echinococcus granulosus*. *Science Tools,* **26**: 21–4.
414. — (1984). Biochemical characterization of Australian strains of *Echinococcus granulosus* by isoelectric focusing of soluble proteins. *International Journal for Parasitology,* **14**: 581–6.
415. **Kumaratilake, L. M., Thompson, R. C. A. & Dunsmore, J. D.** (1983). Comparative strobilar development of *Echinococcus granulosus* of sheep origin from different geographical areas of Australia *in vivo* and *in vitro*. *International Journal for Parasitology,* **13**: 151–6.
416. **Kumaratilake, L. M., Thompson, R. C. A., Eckert, J. & Alessandro, A. D.** (1986). Sperm transfer in *Echinococcus* (Cestoda: Taeniidae). *Zeitschrift für Parasitenkunde,* **72**: 265–9.
417. **Kumazawa, H. & Suzuki, N.** (1983). Kinetics of proglottid formation, maturation and shedding during development of *Hymenolepis nana*. *Parasitology,* **86**: 275–89.
418. **Kuperman, B. I.** (1973). [Tapeworms of the genus *Triaenophorus*, parasites of fishes. Experimental, taxonomy, ecology.] In Russian. Izdatel'stvo 'Nauka': Leningrad. [Translation: British Library No: RTS 10341–10347]
419. — (1980). [Ultrastructure of the cestode integument and its importance in systematics.] In Russian. *Parazitologicheskiĭ Sbornik, Leningrad,* **29**: 84–95. [HA/50/3142]
420. — (1981). [Ultrastructure of the tegument and glandular apparatus of

References

cestodes in ontogenesis.] In Russian. *Biologiya Vnutrennikh Vod. Informatsionnyi Byulleten*, **51**: 29–36. [HA/54/2064]

421. **Kuperman, B. I. & Davydov, V. G.** (1982*a*). The fine structure of glands in oncospheres, procercoids and plerocercoids of Pseudophyllidea (Cestoidea). *International Journal for Parasitology*, **12**: 135–44.

422. — (1982*b*). The fine structure of frontal glands in adult cestodes. *International Journal for Parasitology*, **12**: 285–93.

423. **Kurelec, B. & Rijavec, M.** (1976). Occurrence of γ-glutamyl cycle in some parasitic helminths. In *Biochemistry of parasites and host–parasite relationships*, ed. H. Van den Bossche, pp. 101–7. North-Holland Publishing Co.: The Netherlands.

424. **Kwa, B. H.** (1972*a*). Studies on the sparganum of *Spirometra erinacei*. I. The histology and histochemistry of the scolex. *International Journal for Parasitology*, **2**: 23–8.

425. — (1972*b*). Studies on the sparganum of *Spirometra erinacei*. II. Proteolytic enzyme(s) in the scolex. *International Journal for Parasitology*, **2**: 29–34.

426. — (1972*c*). Studies on the sparganum of *Spirometra erinacei*. III. The fine structure of the tegument of the scolex. *International Journal for Parasitology*, **2**: 35–43.

427. **Lackie, A. M.** (1975). The activation of infective stages of endoparasites of vertebrates. *Biological Reviews*, **50**: 285–323.

428. **Laclette, J. P., Ornelas, Y., Merchant, M. T. & Willms, K.** (1982). Ultrastructure of the surrounding envelopes of *Taenia solium* eggs. In *Cysticercosis: present state of knowledge and perspectives*, ed. A. Flisser, K. Willms, J. P. Laclette, C. Larralde, C. Ridaura, F. Beltrán & M. W. Vogt, pp. 375–87. Academic Press: New York.

429. **Laurie, J. S.** (1957). The *in vitro* fermentation of carbohydrates by two species of cestodes and one species of Acanthocephala. *Experimental Parasitology*, **6**: 245–60.

430. **Laws, G. F.** (1968). The hatching of taeniid eggs. *Experimental Parasitology*, **23**: 1–10.

431. **Le Riche, P. D., Dwinger, R. H. & Kühne, G. I.** (1982). Bovine echinococcosis in the north-west of Argentina. *Tropical Animal Health and Production*, **14**: 205–6.

432. **Le Riche, P. D. & Sewell, M. M.** (1977). Differentiation of *Taenia saginata* and *Taenia solium* by enzyme electrophoresis. *Transactions of the Royal Society of Tropical Medicine and Hygiene*, **71**: 327–8.

433. — (1978*a*). Differentiation of taeniid cestodes by enzyme electrophoresis. *International Journal for Parasitology*, **8**: 479–83.

434. — (1978*b*). Identification of *Echinococcus* strains by enzyme electrophoresis. *Research in Veterinary Science*, **25**: 247–8.

435. **Lee, M. B., Beuding, E. & Schiller, E. L.** (1978). The occurrence and distribution of 5-hydroxytryptamine in *Hymenolepis diminuta* and *H. nana*. *Journal of Parasitology*, **64**: 257–64.

436. **Leid, R. W. & McConnell, L. A.** (1983). PGE$_2$ generation and release by the larval stage of the cestode *Taenia taeniaeformis*. *Prostaglandins, Leukotrienes and Medicine*, **11**: 317–23.

437. **Leid, R. W. & Suquet, C. M.** (1986). A superoxide dismutase of

References

metacestodes of *Taenia taeniaeformis*. *Molecular and Biochemical Parasitology*, **18**: 301–11.

438. **Leng, R. A., Barker, J. S. F., Adams, D. B. & Hutchinson, K. J.** (1985). Biotechnology and recombinant DNA technology in the animal productions industries. *Reviews in Rural Science*, **6**: 67–74.
439. **Lethbridge, R. C.** (1971*a*). The hatching of *Hymenolepis diminuta* eggs and penetration of the hexacanths in *Tenebrio molitor* beetles. *Parasitology*, **62**: 445–56.
440. — (1971*b*). The chemical composition and some properties of the egg layers in *Hymenolepis diminuta* eggs. *Parasitology*, **63**: 275–88.
441. — (1972). The *in vitro* hatching of *Hymenolepis diminuta* eggs in *Tenebrio molitor* extracts and in defined enzyme preparations. *Parasitology*, **64**: 389–400.
442. — (1980). The biology of the oncosphere of cyclophyllidean cestodes. *Helminthological Abstracts*, series A, **49**: 59–72.
443. **Lethbridge, R. C. & Gijsbers, M. F.** (1974). Penetration gland secretion by hexacanths of *Hymenolepis diminuta*. *Parasitology*, **68**: 303–11.
444. **Levitt, M. D., Bond, J. H. & Levitt, D. G.** (1981). Gastrointestinal gas. In *Physiology of the gastrointestinal tract*, ed. L. R. Johnson, pp. 1301–16. Raven Press: New York.
445. **Li, T., Gracy, R. W. & Harris, B. G.** (1972). Studies on enzymes from parasitic platyhelminthes. II. Purification and properties of malic enzyme from the tapeworm *Hymenolepis diminuta*. *Archives of Biochemistry and Biophysics*, **150**: 397–406.
446. **Litchford, R. G.** (1963). Observations on *Hymenolepis microstoma* in three laboratory hosts: *Mesocricetus auratus*, *Mus musculus*, and *Rattus norvegicus*. *Journal of Parasitology*, **49**: 403–10.
447. **Lloyd, S.** (1981). Progress in immunization against parasitic helminths. *Parasitology*, **83**: 225–42.
448. **Logan, J., Ubelaker, J. E. & Vrijenhoek, R. C.** (1977). Isozymes of L(+) LDH in *Hymenolepis diminuta*. *Comparative Biochemistry and Physiology*, **57B**: 51–3.
449. **Lonc, E.** (1980*a*). The possible role of the soil fauna in the epizootiology of cysticercosis in cattle. I. Earthworms – the biotic factor in a transmission of *Taenia saginata* eggs. *Angewandte Parasitologie*, **21**: 133–39.
450. — (1980*b*). The possible role of the soil fauna in the epizootiology of cysticercosis in cattle. II. Dung beetles – a biotic factor in the transmission of *Taenia saginata* eggs. *Angewandte Parasitologie*, **21**: 139–44.
451. **Löser, E.** (1965*a*). Der Feinbau des Oogenotop bei cestoden. *Zeitschrift für Parasitenkunde*, **25**: 413–58.
452. — (1965*b*). Die Eibildung bei cestoden. *Zeitschrift für Parasitenkunde*, **25**: 556–80.
453. — (1965*c*). Die postembryonale Entwicklung des Oogenotop bei cestoden. *Zeitschrift für Parasitenkunde*, **25**: 581–96.
454. **Lumsden, R. D.** (1965). Macromolecular structure of glycogen in some cyclophyllidean and trypanorhynch cestodes. *Journal of Parasitology*, **51**: 501–15.

References

455. — (1967). Ultrastructure of mitochondria in a cestode *Lacistorhynchus tenuis* (V. Beneden, 1858). *Journal of Parasitology*, **53**: 65–77.
456. — (1975*a*). The tapeworm tegument: a model system for studies of membrane structure and function in host–parasite relationships. *Transactions of the American Microscopical Society*, **94**: 501–7.
457. — (1975*b*). Surface ultrastructure and cytochemistry of parasitic helminths. *Experimental Parasitology*, **37**: 267–339.
458. **Lumsden, R. D. & Hildreth, M. B.** (1983). The fine structure of adult tapeworms. In *Biology of the Eucestoda*, vol. 1, ed. C. Arme & P. W. Pappas, pp. 177–233. Academic Press: London.
459. **Lumsden, R. D. & Murphy, W. A.** (1980). Morphological and functional aspects of the cestode surface. In *Cellular interactions in symbiosis and parasitism*, ed. C. B. Cook, P. W. Pappas & E. D. Rudolph, pp. 95–130. Ohio State University Press: Ohio.
460. — (1984). A comparison of phlorizin and phloretin absorption by the tapeworm *Hymenolepis diminuta*. *Comparative Biochemistry and Physiology*, **79A**: 137–41.
461. **Lumsden, R. D. Oaks, J. A. & Mueller, J. F.** (1974). Brush border development of the tegument of the tapeworm *Spirometra mansonoides*. *Journal of Parasitology*, **60**: 209–26.
462. **Lumsden, R. D. & Specian, R. D.** (1980). The morphology, histology, and fine structure of the adult stage of the cyclophyllidean tapeworm *Hymenolepis diminuta*. In *The biology of* Hymenolepis diminuta, ed. H. P. Arai, pp. 157–280. Academic Press: New York.
463. **Lumsden, R. D., Threadgold, L. T., Oaks, J. A. & Arme, C.** (1970). On the permeability of cestodes to colloids: an evaluation of the transmembranosis hypothesis. *Parasitology*, **60**: 185–93.
464. **Lumsden, R. D., Voge, M. & Sogandares-Bernal, F.** (1982). The metacestode tegument: fine structure, development, topochemistry, and interactions with the host. In *Cysticercosis: present state of knowledge and perspectives*, ed. A. Flisser, K. Willms, J. P. Laclette, C. Larralde, C. Ridaura, F. Beltrán & M. W. Vogt, pp. 307–61. Academic Press: New York.
465. **Lussier, P. E., Podesta, R. B. & Mettrick, D. F.** (1979). *Hymenolepis diminuta*: Na$^+$-dependent and Na$^+$-independent components of neutral amino acid transport. *Journal of Parasitology*, **65**: 842–8.
466. — (1982). *Hymenolepis diminuta*: the non-saturable component of methionine uptake. *International Journal for Parasitology*, **12**: 265–70.
467. **McCaig, M. L. O. & Hopkins, C. A.** (1963). Studies on *Schistocephalus solidus*. II. Establishment and longevity in the definitive host. *Experimental Parasitology*, **13**: 273–83.
468. **McCracken, R. O. & Taylor, D. D.** (1983*a*). Biochemical effects of thiabendazole and cambendazole on *Hymenolepis diminuta* (Cestoda) *in vivo*. *Journal of Parasitology*, **69**: 295–301.
469. — (1983*b*). Biochemical effects of fenbendazole on *Hymenolepis diminuta in vivo*. *International Journal for Parasitology*, **13**: 267–72.
470. **Machnicka, B. & Grzybowski, J.** (1986). Host serum proteins in

References

Taenia saginata metacestode fluid. *Veterinary Parasitology*, **19**: 47–54.

471. **Machnicka, B. & Smyth, J. D.** (1985). The early development of larval *Taenia saginata in vitro* and *in vivo*. *Acta Parasitological Polonica*, **30**: 47–52.

472. **MacInnis, A. J.** (ed.) (1987). *Molecular paradigms for eradicating helminthic parasites*. UCLA Symposia on Molecular and Cellular Biology, new series, vol. 60. Alan R. Liss, Inc.: New York.

473. **MacInnis, A. J. & Carter, C.** (1980). Nucleic acids from hymenolepidids. In *Biology of the tapeworm* Hymenolepis diminuta, ed. H. P. Arai, pp. 449–61. Academic Press: New York.

474. **MacInnis, A. J. & Voge, M.** (eds.) (1970). *Experiments and techniques in parasitology*. W. H. Freeman: San Francisco.

475. **McIntyre, P., Coppell, R. L., Smith, H. D., Stahl, L. M., Corcoran, C. J., Langford, C. J., Favaloro, J. M., Crewther, P. E., Brown, G. V., Mitchell, G. F., Anders, R. F. & Kemp, D. J.** (1987). Expression of parasite antigens in *Escherichia coli*. *International Journal for Parasitology*, **17**: 59–67.

476. **McKelvey, J. R. & Fioravanti, C. F.** (1984). Coupling of 'malic' enzyme and NADPH:NAD transhydrogenase in the energetics of *Hymenolepis diminuta* (Cestoda). *Comparative Biochemistry and Physiology*, **77B**: 737–42.

477. — (1985). Intramitochondrial localization of fumarate reductase, NADPH→NAD transhydrogenase, 'malic' enzyme and fumarase in adult *Hymenolepis diminuta*. *Molecular and Biochemical Parasitology*, **17**: 253–63.

478. — (1986). Localization of cytochrome *c* oxidase and cytochrome *c* peroxidase in mitochondria of *Hymenolepis diminuta* (Cestoda). *Comparative Biochemistry and Physiology*, **85B**: 333–5.

479. **Mackiewicz, J. S.** (1981). Caryophyllidea (Cestoidea): evolution and classification. *Advances in Parasitology*, **19**: 139–206.

480. — (1984). Cercomer theory: significance of sperm morphology, oncosphere metamorphosis, polarity reversal, and the cercomer to evolutionary relationships of Monogenea to Cestoidea. *Acta Parasitologica Polonica*, **29**: 11–21.

481. **Mackinnon, A. D. & Featherston, D. W.** (1982). Location and means of attachment of *Bothriocephalus scorpii* (Müller) (Cestoda: Pseudophyllidea) in red cod, *Pseudophycis bacchus* (Forster in Bloch & Schneider), from New Zealand waters. *Australian Journal of Marine & Freshwater Research*, **33**: 595–8.

482. **Mackinnon, B. M. & Burt, M. D. B.** (1982). The comparative ultrastructure of sperm from *Bothrimonus sturionis* (Pseudophyllidea), *Pseudanthobothrium* sp. (Tetraphyllidea) and *Monoecocestus americanus* (Cyclophyllidea). *Parasitology*, **85**: xxxi.

483. — (1984). The development of the tegument and cercomer of the polycephalic larvae (cercoscolices) of *Paricterotaenia paradoxa* (Rudolphi, 1802) (Cestoda: Dilepididae) at the ultrastructural level. *Parasitology*, **88**: 117–30.

484. — (1985). The comparative ultrastructure of the plerocercoid and

References

adult primary scolex of *Haplobothrium globuliforme* Cooper, 1914 (Cestoda: Haplobothrioidea). *Canadian Journal of Zoology*, **63**: 1488–95.

485. **Mackinnon, B. M., Jarecka, L. & Burt, M. D. B.** (1985). Ultrastructure of the tegument and penetration glands of developing procercoids of *Haplobothrium globuliforme* Cooper, 1914 (Cestoda: Haplobothrioidea). *Canadian Journal of Zoology*, **63**: 1470–7.

486. **McManus, D. P.** (1975a). Tricarboxylic acid cycle enzymes in the plerocercoid of *Ligula intestinalis* (Cestoda: Pseudophyllidea). *Zeitschrift für Parasitenkunde*, **45**: 319–22.

487. — (1975b). Pyruvate kinase in the plerocercoid of *Ligula intestinalis* (Cestoda: Pseudophyllidea). *International Journal of Biochemistry*, **6**: 79–84.

488. — (1981). A biochemical study of adult and cystic stages of *Echinococcus granulosus* of human and animal origin from Kenya. *Journal of Helminthology*, **55**: 21–7.

489. —(1985). Enzyme analyses of natural populations of *Schistocephalus solidus* and *Ligula intestinalis*. *Journal of Helminthology*, **59**: 323–32.

490. — (1987). Intermediary metabolism in parasitic helminths. *International Journal for Parasitology*, **17**: 79–95.

491. **McManus, D. P. & Barrett, N. J.** (1985). Isolation, fractionation and partial characterization of the tegumental surface from protoscoleces of the hydatid organism, *Echinococcus granulosus*. *Parasitology*, **90**: 111–29.

492. **McManus, D. P. & Bryant, C.** (1986). Biochemistry and physiology of *Echinococcus*. In *The biology of* Echinococcus *and hydatid disease*, ed. R. C. A. Thompson, pp. 114–42. George Allen & Unwin: London.

493. **McManus, D. P., Knight, M. & Simpson, A. J. G.** (1985). Isolation and characterisation of nucleic acids from the hydatid organisms *Echinococcus* spp. (Cestoda). *Molecular and Biochemical Parasitology*, **16**: 251–66.

494. **McManus, D. P., McLaren, D. J., Clark, N. W. T. & Parkhouse, R. M. E.** (1987). A comparison of two procedures for labelling the surface of the hydatid disease organism, *Echinococcus granulosus*, with [125]I. *Journal of Helminthology*, **61**: 47–52.

495. **McManus, D. P. & Macpherson, C. N. L.** (1984). Strain characterization in the hydatid organism *Echinococcus granulosus*: current status and new perspectives. *Annals of Tropical Medicine and Parasitology*, **78**: 193–8.

496. **McManus, D. P. & Simpson, A. J. G.** (1985). Identification of the *Echinococcus* (hydatid disease) organisms using cloned DNA markers. *Molecular and Biochemical Parasitology*, **17**: 171–8.

497. **McManus, D. P., Simpson, A. J. G. & Rishi, A. K.** (1987). Characterization of the hydatid disease organism *Echinococcus granulosus*, from Kenya using cloned DNA markers. In *Helminth zoonoses*, ed. S. Geerts, V. Kumar & J. Brandt, pp. 29–36. Martinus Nijhoff Publishers: Dordrecht.

498. **McManus, D. P. & Smyth, J. D.** (1978). Differences in the chemical

References

composition and carbohydrate metabolism of *Echinococcus granulosus* (horse and sheep strains) and *E. multilocularis*. *Parasitology*, **77**: 103–9.

499. — (1979). Isoelectric focusing of some enzymes from *Echinococcus granulosus* (horse and sheep strains) and *E. multilocularis*. *Transactions of the Royal Society of Tropical Medicine and Hygiene*, **73**: 259–65.

500. — (1982). Intermediary carbohydrate metabolism in protoscoleces of *Echinococcus granulosus* (horse and sheep strains) and *E. multilocularis*. *Parasitology*, **84**: 351–66.

501. — (1986). Hydatiodosis: changing concepts in epidemiology and speciation. *Parasitology Today*, **2**: 163–8.

502. **McManus, D. P. & Sterry, P. R.** (1982). *Ligula intestinalis*: intermediary carbohydrate metabolism in plerocercoids and adults. *Zeitschrift für Parasitenkunde*, **67**: 73–85.

503. **McPhail, J. D. & Peacock, S. D.** (1983). Some effects of the cestode (*Schistocephalus solidus*) on reproduction in the three-spined stickleback (*Gasterosteus aculeatus*). *Canadian Journal of Zoology*, **61**: 901–8.

504. **Macpherson, C. N. L. & McManus, D. P.** (1982). A comparative study of *Echinococcus granulosus* from human and animal hosts in Kenya using isoelectric focusing and isoenzyme analysis. *International Journal for Parasitology*, **12**: 515–21.

505. **Macpherson, C. N. L. & Smyth, J. D.** (1985). *In vitro* culture of the strobilar stage of *Echinococcus granulosus* from protoscoleces of human, camel, cattle, sheep and goat origin from Kenya and buffalo origin from India. *International Journal for Parasitology*, **15**: 137–40.

506. **McVicar, A. H.** (1972). The ultrastructure of the parasite–host interface of three tetraphyllidean tapeworms of the elasmobranch *Raja naevus*. *Parasitology*, **65**: 77–88.

507. **Madge, D. S.** (1975). *The mammalian alimentary system: a functional approach*. Edward Arnold: London.

508. **Malhotra, S. K.** (1982). Studies on amino acids in *Raillietina* (*Raillietina*) *saharanpurensis* (Malhotra & Capoor, 1981) with a note on biochemical variations in cyclophyllidean cestodes. *Comparative Physiology and Ecology*, **7**: 207–10.

509. **Maniatis, T., Fritsch, E. F. & Sambrook, J.** (1982). *Molecular cloning: A laboratory manual*, Cold Spring Harbor Publications: Cold Spring Harbor, NY.

510. **Mansour, T. E.** (1984). Serotonin receptors in parasitic worms. *Advances in Parasitology*, **23**: 1–26.

511. **Marchalonis, J. J.** (1976). *Comparative immunology*. Blackwell Scientific Publications: London.

512. **Mastov, K. E., Friedlander, M. & Blobel, G.** (1984). The receptor for transepithelial transport of IgA and IgM contains multiple immunoglobulin-like domains. *Nature*, **308**: 37–43.

513. **Matskási, I. & Hajdú, É.** (1983). Studies on the lipase activity of parasitic platyhelminths. *Parasitologia Hungarica*, **16**: 53–7.

References

514. **Matskási, I. & Juhász, S.** (1977). *Ligula intestinalis* (L. 1785): investigation of plerocercoids and adults for protease and protease inhibitor activity. *Parasitologia Hungarica*, **10**: 51–60.

515. **Matskási, I. & Németh, I.** (1979). *Ligula intestinalis* (Cestoda: Pseudophyllidea): studies on the properties of proteolytic and protease inhibitor activities of plerocercoid larvae. *International Journal for Parasitology*, **9**: 221–7.

516. — (1980). [Characterization of proteolytic and protease inhibitor activities of the plerocercoid larvae of *Ligula intestinalis*.] In Hungarian: *Magyar Állatorvosok Lapja*, **35**: 550–2. [HA/50/829]

517. **May, R. M.** (1977). Dynamical aspects of host–parasite associations: Crofton's model revisited. *Parasitology*, **75**: 259–76.

518. **Meakins, R. H.** (1974). A quantitative approach to the effects of the plerocercoid of *Schistocephalus solidus* Müller 1776 on the ovarian maturation of the three-spined stickleback *Gasterosteus aculeatus*. *Zeitschrift für Parasitenkunde*, **44**: 73–9.

519. **Mehlorn, H., Becker, B., Andrews, P. & Thomas, H.** (1981). On the nature of the proglottids of cestodes: and light and electron microscopic study of *Taenia*, *Hymenolepis* and *Echinococcus*. *Zeitschrift für Parasitenkunde*, **65**: 243–59.

520. **Mendis, A. H. W., Rees, H. H. & Goodwin, T.** (1984). The occurrence of ecdysteroids in the cestode, *Moniezia expansa*. *Molecular and Biochemical Parasitology*, **10**: 123–38.

521. **Mercer, J. G.** (1985). Developmental hormones in parasitic helminths. *Parasitology Today*, **1**: 96–100.

522. **Mercer, J. G., Munn, A. E. & Rees, H. H.** (1987). *Echinococcus granulosus*: occurrence of ecdysteroids in protoscoleces and hydatid cyst fluid. *Molecular and Biochemical Parasitology*, **24**: 203–14.

523. **Mettrick, D. F.** (1970). Protein nitrogen, amino acid and carbohydrate gradients in the rat intestine. *Comparative Physiology and Biochemistry*, **37**: 517–41.

524. — (1971). Effect of host dietary constituents on intestinal pH and the migratory behaviour of the rat tapeworm, *Hymenolepis diminuta*. *Canadian Journal of Zoology*, **49**: 1513–25.

525. — (1972). Changes in the distribution and chemical composition of *Hymenolepis diminuta*, and in the intestinal nutritional gradients of uninfected and parasitized rats following a glucose meal. *Journal of Helminthology*, **46**: 407–29.

526. — (1973). Competition for ingested nutrients between the tapeworm *Hymenolepis diminuta* and the rat host. *Canadian Journal of Public Health*, **64**: 70–82.

527. — (1980). The intestine as an environment for *Hymenolepis diminuta*. In *Biology of the tapeworm* Hymenolepis diminuta, ed. H. P. Arai, pp. 281–356. Academic Press: London & New York.

528. **Mettrick, D. F. & Cho, C. H.** (1981). Effect of electrical vagal stimulation on migration of *Hymenolepis diminuta*. *Journal of Parasitology*, **67**: 386–90.

529. — (1982). Changes in the tissue and intestinal serotonin (5-HT) levels in the laboratory rat following feeding and the effect of 5-HT inhibi-

References

tors on the migratory response of *Hymenolepis diminuta* (Cestoda). *Canadian Journal of Zoology*, **60**: 790–7.

530. **Mettrick, D. F. & Podesta, R. B.** (1974). Ecological and physiological aspects of helminth–host interactions in the mammalian gastrointestinal canal. *Advances in Parasitology*, **12**: 183–278.
531. **Mettrick, D. F. & Rahman, M. S.** (1984). Effects of parasite strain and intermediate host species on carbohydrate intermediary metabolism in the rat tapeworm *Hymenolepis diminuta*. *Canadian Journal of Zoology*, **62**: 355–61.
532. **Meyer, F., Kimura, S. & Mueller, J. L.** (1966). Lipid metabolism in the larval and adult forms of the tapeworm *Spirometra mansonoides*. *Journal of Biological Chemistry*, **241**: 4224–32.
533. **Meyer, H., Mueller, J. & Meyer, F.** (1978). Isolation of an acyl-CoA carboxylase from the tapeworm *Spirometra mansonoides*. *Biochemical and Biophysical Research Communications*, **82**: 834–9.
534. **Mied, P. A. & Bueding, E.** (1979*a*). Glycogen synthase of *Hymenolepis diminuta*. I. Allosteric activation and inhibition. *Journal of Parasitology*, **65**: 14–24.
535. — (1979*b*). Glycogen synthase of *Hymenolepis diminuta*. II. Nutritional state, interconversion of forms, and primer glycogen molecular weight as control factors. *Journal of Parasitology*, **65**: 25–30.
536. **Mikulikova, L.** (1979). Identification of helminth species by means of disc electrophoresis. *Folia Parasitologica*, **26**: 111–22.
537. **Mills, G. L., Coley, S. C. & Williams, J. F.** (1983). Chemical composition of lipid droplets isolated from larvae of *Taenia taeniaeformis*. *Journal of Parasitology*, **69**: 850–6.
538. — (1984). Lipid and protein composition of the surface tegument from larvae of *Taenia taeniaeformis*. *Journal of Parasitology*, **70**: 197–207.
539. **Mills, G. L., Taylor, D. C. & Williams, J. F.** (1981*a*). Lipid composition of metacestodes of *Taenia taeniaeformis* and lipid changes during growth. *Molecular and Biochemical Parasitology*, **3**: 301–18.
540. — (1981*b*). Lipid composition of the helminth parasite *Taenia crassiceps*. *Comparative Biochemistry and Physiology*, **69B**: 553–7.
541. **Moczoń, T.** (1972). Histochemistry of oncospheral envelopes of *Hymenolepis diminuta* (Rudolphi 1819) (Cestoda, Hymenolepididae). *Acta Parasitologica Polonica*, **20**: 517–32.
542. — (1974). Histochemical studies on the enzymes of *Hymenolepis diminuta* (Rud., 1819) (Cestoda). III. Some oxidoreductases in the mature parasite. *Acta Parasitologica Polonica*, **22**: 191–202.
543. — (1975*a*). Glycogen distribution and accumulation of radioactive compounds in tissues of mature specimens of *Hymenolepis diminuta* (Cestoda) after incubation in glucose – $^{14}C_{1-6}$. *Acta Parasitologica Polonica*, **23**: 135–45.
544. — (1975*b*). Histochemical studies on the enzymes of *Hymenolepis diminuta* (Rud., 1819) (Cestoda). V. Some enzymes of the synthesis and phosphorolytic degradation of glycogen in mature cestodes. *Acta Parasitologica Polonica*, **23**: 569–92.
545. — (1977*a*). Glycogen distribution and accumulation of radioactive

References

compounds in oncospheres and cysticercoids of *Hymenolepis diminuta* (Cestoda) after incubation in glucose – $^{14}C_{1-6}$. *Acta Parasitologica Polonica*, **24**: 269–74.

546. — (1977*b*). Histochemical studies on the enzymes of *Hymenolepis diminuta* (Rud., 1819) (Cestoda). VI. Some enzymes of the synthesis and phosphorolytic degradation of glycogen in oncospheres and cysticercoids. *Acta Parasitologica Polonica*, **24**: 275–82.

547. — (1977*c*). Histochemical studies on the enzymes of *Hymenolepis diminuta* (Rud., 1819) (Cestoda). VII. Enzymes of the synthesis and hydrolytic degradation of glycogen in three developmental stages. *Acta Parasitologica Polonica*, **25**: 45–54.

548. — (1977*d*). Negative results of biochemical tests for collagenase and hyaluronidase activities in extracts of *Hymenolepis diminuta* (Cestoda) oncospheres and *Nippostrongylus muris* (Nematoda) invasive larvae. *Bulletin de l'Académie Polonaise des Sciences*, Class II, **25**: 479–81.

549. — (1977*e*). Penetration of *Hymenolepis diminuta* oncospheres across the intestinal tissues of *Tenebrio molitor* beetles. *Bulletin de l'Académie Polonaise des Sciences*, Class II. **25**: 531–5.

550. — (1977*f*). Penetration glands of oncospheres of *Hymenolepis diminuta* (Cestoda). Histochemical studies. *Bulletin de l'Académie Polonaise des Sciences*, Class II, **25**: 619–22.

551. — (1978*a*). Histochemical studies on the enzymes of *Hymenolepis diminuta* (Rud., 1819) (Cestoda). VIII. Induction of the synthesis of cytochrome oxidase in oncospheres after invading *Tribolium castaneum* beetle. *Acta Parasitologica Polonica*, **25**: 163–8.

552. — (1978*b*). Oxygen consumption by oncospheres of *Hymenolepis diminuta* (Cestoda). *Acta Parasitologica Polonica*, **26**: 19–23.

553. — (1980*a*). Histochemical studies on the enzymes of *Hymenolepis diminuta* (Rud., 1819) (Cestoda). IX. Isocitrate oxidation in three developmental stages. *Acta Parasitologica Polonica*, **27**: 49–52.

554. — (1980*b*). Histochemical studies on the enzymes of *Hymenolepis diminuta* (Rud., 1819) (Cestoda). X. Other oxidoreductases in three developmental stages of the parasite. *Acta Parasitologica Polonica*, **27**: 471–5.

555. — (1980*c*). Histochemical studies on the enzymes of *Hymenolepis diminuta* (Rud., 1819) (Cestoda). XI. Some hydrolases of oligo-saccharides in mature tapeworms. *Acta Parasitologica Polonica*, **27**: 477–86.

556. — (1980*d*). Oxidoreductase histochemistry in the mature stage of *Triaenophorus nodulosus* (Pallas, 1781), and some remarks on the regulation of the respiratory metabolism in the ontogeny of pseudophyllidean cestodes. *Acta Parasitologica Polonica*, **27**: 359–66.

557. — (1980*e*). Oxidoreductase histochemistry in larval stages of pseudophyllidean cestodes. III. The plerocercoid. *Acta Parasitoloica Polonica*, **27**: 367–71.

558. — (1981*a*). Histochemical studies on the enzymes of *Hymenolepis diminuta* (Rud., 1819) (Cestoda). XIV. Uridine diphosphoglucose

References

(UDPG) dehydrogenase, UDPG epimerase and UDP-glucuronate epimerase. *Acta Parasitologica Polonica*, **28**: 187–96.
559. — (1981*b*). Histochemical studies on the enzymes of *Hymenolepis diminuta* (Rud., 1819) (Cestoda). XII. Activities of some glycosidases of the mature parasite in different nutritional conditions. *Acta Parasitologica Polonica*, **28**: 97–102.
560. — (1981*c*). Histochemical studies on the enzymes of *Hymenolepis diminuta* (Rud., 1819) (Cestoda). XIII. Some hydrolases of oligosaccharides in oncospheres and cysticercoids. *Acta Parasitologica Polonica*, **28**: 103–8.
561. **Mokhtar-Maamouri, F.** (1979). Étude en microscopie électronique de la spermiogénese et du spermatozoïde de *Phyllobothrium gracile* Weld 1855 (Cestoda, Tetraphyllidea, Phyllobothriidae). *Zeitschrift für Parasitenkunde* **59**: 245–8.
562. — (1980). Particularités des processus de la fécondation chez *Acanthobothrium filicolle* Zschokke, 1888 (Cestoda: Tetraphyllidea, Onchobothriidae). *Archives de l'Institut Pasteur de Tunis*, **57**: 191–205.
563. — (1982). Étude ultrastructurale de la spermiogénèse de *Acanthobothrium filicolle* var. *filicolle* Zschokke, 1888. (Cestoda, Tetraphyllidea, Onchobothriidae). *Annales de Parasitologie Humaine et Comparée*, **57**: 429–42.
564. **Mokhtar-Maamouri, F. & Swiderski, Z.** (1975). Étude en microscopie électronique de la spermatogénèse de deux Cestodes *Acanthobothrium filicolle benedenii* Loennberg, 1889 et *Onchobotrium uncinatum* (Rud., 1819) (Tetraphyllidea, Onchobothriidae). *Zeitschrift für Parasitenkunde*, **47**: 269–81.
565. — (1976*a*). Vitellogénèse chez *Echeneibothrium beauchampi* Euzet, 1959 (Cestoda: Tetraphyllidea, Phyllobothriidae). *Zeitschrift für Parasitenkunde*, **50**: 293–302.
566. — (1976*b*). Ultrastructure du spermatozoïde d'un cestode Tetraphyllidea Phyllobothriidae, *Echeneibothrium beauchampi*, Euzet, 1959. *Annales de Parasitologie Humaine et Comparée*, **51**: 673–4.
567. **Moon, T. W., Hulbert, W. C., Mustafa, T. & Mettrick, D. F.** (1977). A study of lactate dehydrogenase and malate dehydrogenase in adult *Hymenolepis diminuta* (Cestoda). *Comparative Biochemistry and Physiology*, **56B**: 249–254.
568. **Moon, T. W., Mustafa, T., Hulbert, W. C., Podesta, R. B. & Mettrick, D. F.** (1977). The phosphoenolpyruvate branchpoint in adult *Hymenolepis diminuta* (Cestoda): a study of pyruvate kinase and phosphoenolpyruvate carboxykinase. *Journal of Experimental Zoology*, **200**: 325–36.
569. **Moreno, M. S. & Barrett, J.** (1979). Monoamine oxidase in adult *Hymenolepis diminuta* (Cestoda). *Parasitology*, **78**: 1–5.
570. **Morseth, D. J.** (1967). Observations on the fine structure of the nervous system of *Echinococcus granulosus*. *Journal of Parasitology*, **53**: 492–500.

References

571. — (1969). Sperm tail fine structure of *Echinococcus granulosus* and *Dicrocoelium dendriticum. Experimental Parasitology*, **24**: 47–53.
572. **Movsesyan, S. O., Chubaryan, F. A. & Kurbet, A. V.** (1981). [Biological and morphological characteristics of the cestodes *Taenia pisiformis* (Bloch, 1780) and *Hydatigera taeniaeformis* (Batsch, 1786).] In Russian. In *Raboty po gel'mintologii*, pp. 128–37. Nauka: Moscow. [HA/51/3978]
573. **Mueller, J. F.** (1959). The laboratory propagation of *Spirometra mansonoides* as an experimental tool. I. Collecting, incubation and hatching of the eggs. *Journal of Parasitology*, **45**: 353–61.
574. — (1963). Parasite-induced weight gain in mice. *Annals of the New York Academy of Sciences*, **113**: 217–33.
575. — (1966). Host–parasite relationships as illustrated by the cestode *Spirometra mansonoides*. In *Host–parasite relationships*, ed. J. E. McCauley, pp. 11–58. Oregon State University Press: Oregon.
576. — (1974). The biology of *Spirometra. Journal of Parasitology*, **60**: 3–14.
577. **Munck, B. G.** (1981). Intestinal absorption of amino acids. In *Physiology of the gastrointestinal tract*, ed. L. R. Johnson, pp. 1097–122. Raven Press: New York.
578. **Muravskiĭ, N. N., Mar'enko, M. A., Grebenshchikov, V. M. & Tsitko, A. A.** (1976). [A study of the content of some elements in *Hydatigera taeniaeformis* and *Dipylidium caninum* and in their cat hosts in town and country areas of the same biogeochemical province.] In Russian. In *Problemy éksperimental'noĭ, morfofiziologii i genetiki*, pp. 170–3. Kemerovo, USSR; Kemerovskiĭ Gosudarstvennyĭ Universitet. [No named editor]. [HA/46/3829]
579. **Murphy, W. A. & Lumsden, R. D.** (1984a). Phloretin inhibition of glucose transport by the tapeworm *Hymenolepis diminuta*: a kinetic analysis. *Comparative Biochemistry and Physiology*, **78A**: 749–54.
580. — (1984b). Reappraisal of the mucosal epithelial space associated with the surface of *Hymenolepis diminuta* and its effect on transport parameters. *Journal of Parasitology*, **70**: 516–21.
581. **Murthy, R. C. & Tayal, S.** (1978). Trehalase activity in *Stilesia globipunctata* (Cestoda). *Zeitschrift für Parasitenkunde*, **56**: 63–8.
582. **Mustafa, T., Komuniecki, R. & Mettrick, D. F.** (1978). Cytosolic glutamate dehydrogenase in adult *Hymenolepis diminuta* (Cestoda). *Comparative Biochemistry and Physiology*, **61B**: 219–22.
583. **Naash, M. T. & Al-Janabi, B. M.** (1982). Studies on some biochemical constituents of hydatid cyst fluid. *Journal of the College of Veterinary Medicine, Mosul University, Iraq*, **1**: 41–8.
584. **Nadakal, A. M. & Vijayakumaran Nair, K.** (1982). A comparative study on the mineral composition of the poultry cestode *Raillietina tetragona* Molin, 1858, and certain tissues of its hosts. *Proceedings of the Indian Academy of Sciences*, **91**: 153–8.
585. **Nakajima, K. & Egusa, S.** (1976a). [*Bothriocephalus opsariichthydis* Yamaguti (Cestoda: Pseudophyllidea) found in the gut of cultured carp, *Cyprinus carpio* (Linné). IV. Observations on the egg and

References

coracidium.] In Japanese. *Fish Pathology*, **11**: 17–22. [HA/46/5541]
586. — (1976*b*). [*Bothriocephalus opsariichthydis* Yamaguti (Cestoda: Pseudophyllidea) found in the gut of cultured carp, *Cyprinus carpio* (Linné). V. Ovicidal effects of drying, freezing, ultraviolet rays and some chemicals.] In Japanese. *Fish Pathology*, **11**: 23–5. [HA/46/5542]
587. **Nanda, S., Bhalya, A., Gairola, D., Malhotra, S. K. & Capoor, V. N.** (1987). Comparative analysis of amino acids of three species of *Gangesia* (Cestoda: Proteocephalata). *Journal of Helminthology*, **61**: 233–9.
588. **Náquira, C., Paulin, J. & Agosin, M.** (1977). *Taenia crassiceps*: protein synthesis in larvae. *Experimental Parasitology*, **41**: 359–69.
589. **Nascimento, E.** (1982). [*Taenia taeniaeformis*: aspects of the host–parasite relationship.] In Portuguese. *Memórias do Instituto Oswaldo Cruz*, **77**: 319–23. [HA/53/4144]
590. **Nelson, N. F. & Saz, H. J.** (1983). *Hymenolepis diminuta*: effects of amoscanate on energy metabolism and ultrastructure. *Experimental Parasitology*, **56**: 55–69.
591. **Németh, I. & Juhász, S.** (1980). A trypsin and chymotrypsin inhibitor from the metacestodes of *Taenia pisiformis*. *Parasitology*, **80**: 433–46.
592. — (1981). Properties of a trypsin and chymotrypsin inhibitor secreted by larval *Taenia pisiformis*. *International Journal for Parasitology*, **11**: 137–44.
593. **Németh, I., Juhász, S. & Baintner, K.** (1979). A trypsin and chymotrypsin inhibitor from *Taenia pisiformis*. *International Journal for Parasitology*, **9**: 515–22.
594. **Nieland, M. L. & Von Brand, T.** (1969). Electron microscopy of cestode calcareous corpuscle formation. *Experimental Parasitology*, **24**: 279–89.
595. **Nigam, S. C.** (1979). Effect of host diet on fatty acid composition of *Cotugnia diagonopora* [*digonopora*] (Pasquale, 1891) and *Raillietina fuhrmanni* (Southwell, 1922). *Indian Journal of Parasitology*, **3**: 67–9.
596. **Nigam, S. C. & Premvati, G.** (1980*a*). Presence of cholesterol in the neutral lipids of three sheep cestodes. *Journal of Helminthology*, **54**: 215–18.
597. — (1980*b*). Unsaponifiable lipids of *Cotugnia digonopora* and *Raillietina fuhrmanni*. (Cestoda: Cyclophyllidea). *Folia Parasitologica*, **27**: 59–61.
598. **Niyogi, A. & Agarwal, S. M.** (1983). Free and protein amino acids in *Lytocestus indicus*, *Introvertus raipurensis* and *Lucknowia indica* parasitizing *Clavias batrachus* (Linn). (Cestoda: Caryophyllidae). *Japanese Journal of Parasitology*, **32**: 341–5.
599. **Nollen, P. M.** (1975). Studies on the reproductive system of *Hymenolepis diminuta* using autoradiography and transplantation. *Journal of Parasitology*, **61**: 100–4.
600. **Normore, W. M.** (1976). Guanine-plus-cytosine (G + C) composition of the DNA of bacteria, fungi, algae and protozoa. In *CRC handbook of biochemistry and molecular biology*, 3rd edn, *Nucleic acids*, vol. 2., ed. G. D. Fasman, pp. 65–235. CRC Press: Boca Raton, FL.

References

601. **Novak, M. & Dowsett, J. A.** (1983). Scanning electron microscopy of the metacestode of *Taenia crassiceps*. *International Journal for Parasitology*, **13**: 383–8.
602. **Nugent, J. C. & O'Connor, M.** (eds.) (1984). *Mucus and mucosa*. Ciba Foundation Symposium, no. 109. Pitman Press: London.
603. **Odening, K.** (1979). Zum Erforschungsstand des 'sparganum growth factor' von *Spirometra*. *Angewandte Parasitologie*, **20**: 185–92.
604. — (1984). Zum Verhältnis zwischen Plerocercoid und adultem Bandwurm bei Pseudophyllidea (Cestoda). *Zoologischer Anzeiger*, **213**: 161–9.
605. **Odening, K., Tscherner, W. & Bockhart, I.** (1980). Experimentelle Bestimmung von Cestodeneiern (*Spirometra*) aus Goldschakal und Luchs. *Milu, Berlin*, **5**: 245–51.
606. **Ong, S. J.** (1984). *In vitro* culture of *Mesocestoides corti* (Cestoda). Ph.D. thesis, University of London.
607. **Ong, S. J. & Smyth, J. D.** (1986). Effects of some culture factors on sexual differentiation of *Mesocestoides corti* grown from tetrathyridia *in vitro*. *International Journal for Parasitology*, **16**: 361–8.
608. **Orpin, C. G., Huskisson, N. S. & Ward, P. F. V.** (1976). Molecular structure and morphology of glycogen isolated from the cestode *Moniezia expansa*. *Parasitology*, **73**: 83–95.
609. **Oshmarin, P. G. & Prokhorova, I. M.** (1978). [The biological significance of some types of reproduction in cestodes.] In Russian. *Materialy Nauchnoi Konferentsii Vsesoyuznogo Obshchestva Gel'minthologov* (Biologicheskie osnovy bor'by s gel'mintozami cheloveka i zhivotnykh), **30**: 117–25. [HA/50/2189]
610. **Osuna-Carrillo, A. & Mascaró-Lazcano, M. C.** (1982). The *in vitro* cultivation of *Taenia pisiformis* to sexually mature adults. *Zeitschrift für Parasitenkunde*, **67**: 67–71.
611. **Osuna-Carrillo, A., Mascaró-Lazcano, M. C., Guevara-Pozo, D. & Guevara-Benitez, D. C.** (1978). Cultivo *in vitro* de *Taenia hydatigena*. *Revista Ibérica de Parasitologia*, **38**: 289–99.
612. **Ovington, K. S. & Bryant, C.** (1981). The role of carbon dioxide in the formation of end-products by *Hymenolepis diminuta*. *International Journal for Parasitology*, **11**: 221–8.
613. **Owen, R. R.** (1985). Improved *in vitro* determination of the viability of *Taenia saginata* embryos. *Annals of Tropical Medicine and Parasitology*, **79**: 655–6.
614. **Owen, R. R. & Stringer, R. E.** (1984). *In vitro* screening of potential ovicides. *Annals of Tropical Medicine and Parasitology*, **78**: 252–7.
615. **Oxford, G. S. & Rollinson, D.** (eds.) (1983). *Protein polymorphism: adaptive and taxonomic significance*. The Systematics Association Special Volume no. 24. Academic Press: London.
616. **Page, C. R. III & Macinnis, A. J.** (1975). Characterization of nucleoside transport in hymneolepidid cestodes. *Journal of Parasitology*, **61**: 281–90.
617. **Page, C. R. III, Macinnis, A. J. & Griffith, L. M.** (1977). Diurnal periodicity of uridine uptake by *Hymenolepis diminuta*. *Journal of Parasitology*, **63**: 91–5.

References

618. **Pampori, N. A., Singh, G. & Srivastava, V. M. L.** (1984a). *Cotugnia digonopora*: carbohydrate metabolism and effect of anthelmintics on immature worms. *Journal of Helminthology*, **58**: 39–47.
619. — (1984b). Energy metabolism in *Cotugnia digonopora* and the effect of anthelmintics. *Molecular and Biochemical Parasitology*, **11**: 205–13.
620. — (1985). Enzymes of isolated brush border membrane of *Cotugnia digonopora*, and their insensitivity to anthelmintics *in vitro*. *Veterinary Parasitology*, **18**: 13–19.
621. **Pappas, P. W.** (1978). The inability of a trypanorhynchid cestode to utilize CO_2 produced during urea catabolism. *Ohio Journal of Science*, **78**: 152–3.
622. — (1980a). Enzyme interactions at the host–parasite interface. In *Cellular interactions in symbiosis and parasitism*, ed. C. B. Cook, P. W. Pappas & E. D. Rudolph, pp. 145–72. Ohio University Press: Ohio.
623. — (1980b). Structure, function and biochemistry of the cestode tegumentary membrane and associated glycocalyx. In *Endocytobiology*: *endosymbiosis and cell biology, a synthesis of recent research*, ed. W. Schwemmler & H. E. A. Schenk, pp. 587–603. Walter de Gruyter: Berlin.
624. — (1983a) Host-parasite interface. In *Biology of the Eucestoda*, vol. 2, ed. C. Arme & P. W. Pappas, pp. 297–334. Academic Press: London.
625. — (1983b). Glycosyl transferase activity in the brush border membrane of *Hymenolepis diminuta*. *Journal of Parasitology*, **69**: 1055–9.
626. — (1984). Kinetic analyses of the membrane-bound alkaline phosphatase activity of *Hymenolepis diminuta* (Cestoda: Cyclophyllidea) in relation to development of the tapeworm in the definitive host. *Journal of Cellular Biochemistry*, **25**: 131–7.
627. **Pappas, P. W. & Leiby, D. A.** (1986a). Variations in the sizes of eggs and oncospheres and the numbers and distributions of testes in the tapeworm *Hymenolepis diminuta*. *Journal of Parasitology*, **72**: 383–91.
628. — (1986b). Alkaline phosphatase and phosphodiesterase activities of the brush border membrane of four strains of the tapeworm *Hymenolepis diminuta*. *Journal of Parasitology*, **72**: 809–11.
629. **Pappas, P. W. & Read, C. P.** (1975). Membrane transport in helminth parasites: a review. *Experimental Parasitology*, **37**: 469–530.
630. **Pappas, P. W. & Schroeder, L. L.** (1979). *Hymenolepis microstoma*: lactate and malate dehydrogenases of the adult worm. *Experimental Parasitology*, **47**: 134–9.
631. **Pappas, P. W., Uglem, G. L. & Read, C. P.** (1973a). The influx of purines and pyrimidines across the brush border of *Hymenolepis diminuta*. *Parasitology*, **66**: 525–38.
632. — (1973b). Mechanisms and specificity of amino acid transport in *Taenia crassiceps* larvae (Cestoda). *International Journal for Parasitology*, **3**: 641–51.
633. **Parker, R. D., Jr & MacInnis, A. J.** (1977). *Hymenolepis diminuta*: isolation, purification, and reconstruction *in vitro* of a cell-free system for protein synthesis. *Experimental Parasitology*, **41**: 2–16.
634. **Parkhouse, R. M. E.** (ed.) (1984). *Parasite evasion of the immune*

References

response. Symposia of the British Society for Parasitology, vol. 21.

635. **Parshad, V. R., Guraya, S. S. & Parshad, R. K.** (1981). Biochemical and histochemical observations on the lipids of *Raillietina cesticillus* (Davaeniidae: Cestoda). *Indian Journal of Parasitology*, **5**: 229–32.

636. **Pathak, K. M. L., Gaur, S. N. S. & Verma, H. C.** (1980). Quantitative estimation of amino acids in cysticercus of *Taenia hydatigena. Veterinary Parasitology*, **7**: 375–8.

637. **Paul, J. M. & Barrett, J.** (1980). Peroxide metabolism in the cestodes *Hymenolepis diminuta* and *Moniezia expansa. International Journal for Parasitology*, **10**: 121–4.

638. **Pawlowski, Z. S.** (1982). Epidemiology and prevention of *Taenia saginata* infection. In *Cysticercosis: present state of knowledge and perspectives*, ed. A. Flisser, K. Willms, J. D. Laclette, C. Larralde, C. Ridaura, F. Beltrán & M. W. Vogt, pp. 69–85. Academic Press: New York.

639. **Pence, D. B.** (1967). The fine structure and histochemistry of the infective eggs of *Dipylidium caninum. Journal of Parasitology*, **53**: 1041–54.

640. — (1970). Electron microscope and histochemical studies on the eggs of *Hymenolepis diminuta. Journal of Parasitology*, **56**: 84–97.

641. **Phares, C. K.** (1984). A method for the solubilization of a human growth hormone analogue for plerocercoids of *Spirometra mansonoides. Journal of Parasitology*, **70**: 840–2.

642. — (1987). Plerocercoid growth factor: a homologue of human growth hormone. *Parasitology Today*, **3**: 346–9.

643. **Pietrzak, S. M. & Saz, H. J.** (1981). Succinate decarboxylation to propionate and the associated phosphorylation in *Fasciola hepatica* and *Spirometra mansonoides. Molecular and Biochemical Parasitology*, **3**: 61–70.

644. **Playfair, J. H. L.** (1984). *Immunology at a glance*. Blackwell Scientific Publications: Oxford.

645. **Podesta, R. B.** (1977). *Hymenolepis diminuta*: unstirred layer thickness and effects on active and passive transport kinetics. *Experimental Parasitology*, **43**: 12–24.

646. — (1979). Cellular Na$^+$ and ATP effects on galactose influx by tissue slices of *Hymenolepis diminuta. Journal of Parasitology*, **65**: 669–71.

647. — (1980). Concepts of membrane biology in *Hymenolepis diminuta*. In *Biology of the tapeworm* Hymenolepis, ed. H. P. Arai, pp. 505–9. Academic Press: New York.

648. — (1982*a*). Membrane physiology of helminths. In *Membrane physiology of invertebrates*, ed. R. B. Podesta, L. L. Dean, S. S. McDiarmid, S. F. Timmers, & B. W. Young, pp. 121–77. Marcel Dekker: New York.

649. — (1982*b*). Adaptive features of the surface epithelial syncytium favoring survival in an immunologically hostile environment. In *Parasites – their world and ours*, ed. D. F. Mettrick & S. S. Desser, pp. 149–55. Elsevier Biomedical Press: Amsterdam.

650. **Podesta, R. B., Mustafa, T., Moon, T. W., Hulbert, W. C. & Mettrick, D. F.** (1976). Anaerobes in an aerobic environment: role of CO_2 in

References

energy metabolism of *Hymenolepis diminuta*. In *Biochemistry of parasites and host–parasite relationships*, ed. H. Van den Bossche, pp. 81–8. North-Holland Publishing Co.: The Netherlands.

651. **Poljakova-Krusteva, O., Mizinska-Boevska, Ya., & Stoitsova, S.** (1983). A cytochemical study of some phosphatases in the tegument of two cestode species. *Khelmintologiya*, **16**: 64–7.

652. **Poljakova-Krusteva, O., Stoitsova, S. & Mizinska-Boevska, Ya.** (1984). Pinocytosis in the tegument of *Hymenolepis fraterna*. *Khelmintologiya*, **17**: 52–7.

653. **Poljakova-Krusteva, O. & Vasilev, I.** (1973). [The ultrastructure of the spermatozoan tail of *Raillietina carneostrobilata*.] In Russian. *Izvestiya na Tsentralnata Khelmintologichna Laboratoriya, Sofia*, **16**: 153–60. [HA/43/3323]

654. **Pool, D. W.** (1984). A scanning electron microscope study of the life cycle of *Bothriocephalus acheilognathi* Yamaguti, 1934. *Journal of Fish Biology*, **25**: 361–4.

655. — (1985). An experimental study of the biology of *Bothriocephalus acheilognathi* Yamaguti, 1934 (Cestoda: Pseudophyllidea). Ph.D. thesis, University of Liverpool.

656. **Pool, D. W. & Chubb, J. C.** (1985). A critical scanning electron microscope study of the scolex of *Bothriocephalus acheilognathi* Yamaguti, 1934, with a review of the taxonomic history of the genus *Bothriocephalus* parasitizing cyprinid fishes. *Systemic Parasitology*, **7**: 199–211.

657. **Potter, K. & Leid, R. W.** (1986). A review of eosinophil chemotaxis and function in *Taenia taeniaeformis* infections in the laboratory rat. *Veterinary Parasitology*, **20**: 103–16.

658. **Premvati, G. & Tayal, S.** (1978). Glycogen content and *in vitro* glycogen consumption in *Stilesia globipunctata* (Rivolta, 1874). *Indian Journal of Parasitology*, **2**: 73–5.

659. **Prokopič, J. & Jelenová, I.** (1980). Effect of fertilizers on *Taenia saginata* Goeze, 1782, egg viability *in vitro*. *Folia Parasitologica (Praha)*, **27**: 343–7.

660. **Pronina, S. V., Davydov, V. G. & Kuperman, B. I.** (1985). [Histochemical studies on some caryophyllaeid, pseudophyllid and proteocephalid cestodes.] In Russian. In *Gidrobiologiya i gidroparazitologiya Pribaĭkal'ya i Zabaĭkal'ya*, ed. A. F. Alimov & N. M. Pronin, pp. 153–67. Nauka Sibirskoe Otdelenie: Novosibirsk, USSR. [HA/55/3627]

661. **Pugh, R. E.** (1986). Effects of the development of *Dipylidium caninum* and on the host reaction to this parasite in the adult flea (*Ctenocephalides felis felis*). *Parasitology Research*, **73**: 171–7.

662. **Pugh, R. E. & Moorehouse, D. E.** (1985). Factors affecting the development of *Dipylidium caninum* in *Ctenocephalides felis felis* (Bouché, 1835). *Zeitschrift für Parasitenkunde*, **71**: 765–75.

663. **Rahaman, R. & Meisner, H.** (1973). Respiratory studies with mitochondria from the rat tapeworm *Hymenolepis diminuta*. *International Journal of Biochemistry*, **4**: 153–62.

664. **Rahman, M. S. & Bryant, C.** (1977). Studies of regulatory metabolism

References

in *Moniezia expansa*: effects of cambendazole and mebendazole. *International Journal for Parasitology*, **7**: 403–9.

665. **Rahman, M. S. & Mettrick, D. F.** (1982). Carbohydrate intermediary metabolism in *Hymenolepis microstoma* (Cestoda). *International Journal for Parasitology*, **12**: 155–62.

666. **Rahman, M. S., Mettrick, D. F. & Podesta, R. B.** (1982). 5-Hydroxytryptamine, glucose uptake, glycogen utilization and carbon dioxide fixation in *Hymenolepis microstoma* (Cestoda). *Comparative Biochemistry and Physiology*, **73B**: 901–6.

667. — (1983). Effects of 5-hydroxytryptamine on carbohydrate metabolism in *Hymenolepis diminuta* (Cestoda). *Canadian Journal of Physiology and Pharmacology*, **61**: 137–43.

668. **Rasero, F. S., Monteoliva, M. & Mayor, F.** (1968). Enzymes related to 4-aminobutyrate metabolism in intestinal parasites. *Comparative Biochemistry and Physiology*, **25**: 693–701.

669. **Rasheed, U.** (1981). Transaminase activity in *Lytocestus indicus* and its host. *Proceedings of the Indian Academy of Parasitology*, **2**: 115–16.

670. **Rausch, R. L. & Jentoft, V. L.** (1957). Studies on the helminth fauna of Alaska. XXXI. Observations on the propagation of the larval *Echinococcus multilocularis* Leuckart, 1863, *in vitro*. *Journal of Parasitology*, **43**: 1–8.

671. **Rausch, R. L. & Maser, C.** (1977). *Monoecocestus thomasi* sp.n. (Cestoda: Anoplocephalidae) from the northern flying squirrel, *Glaucomys sabrinus* (Shaw), in Oregon. *Journal of Parasitology*, **63**: 793–9.

672. **Read, C. P.** (1955). Intestinal physiology and the host–parasite relationship. In *Some physiological aspects and consequences of parasitism*, ed. W. H. Cole, pp. 27–49. Rutgers University Press: New Jersey.

673. — (1956). Carbohydrate metabolism of *Hymenolepis diminuta*. *Experimental Parasitology*, **5**: 325–44.

674. — (1973). Contact digestion in tapeworms. *Journal of Parasitology*, **59**: 672–7.

675. **Read, C. P. & Kilejian, A. Z.** (1969). Circadian migratory behaviour of a cestode symbiote in the rat host. *Journal of Parasitology*, **55**: 574–8.

676. **Reisin, I. L. & Rotunno, C. A.** (1981). Water and electrolyte balance in protoscoleces of *Echinococcus granulosus* incubated *in vitro*: general procedures for the determination of water, sodium, potassium and chloride in protoscoleces. *International Journal for Parasitology*, **11**: 399–404.

677. **Renaud, F., Gabrion, C. & Pasteur, N.** (1983). Le complexe *Bothriocephalus scorpii* (Mueller, 1776): différenciation par électrophorèse enzymatique des éspèces parasites du Turbot (*Psetta maxima*) et de la barbue (*Scophthalmus rhombus*). *Comptes Rendus des Séances de l'Académie des Sciences*, Sér. III, **296**: 127–9.

678. — (1986). Geographical divergence in *Bothriocephalus* (Cestoda) of fishes demonstrated by enzyme electrophoresis. *International Journal for Parasitology*, **16**: 553–8.

679. **Renaud, F., Gabrion, C. & Romestand, B.** (1984). Le complexe *Bothriocephalus scropii* (Mueller, 1776). Différenciation des éspèces

References

parasites du Turbot (*Psetta maxima*) et de la Barbue (*Scophthalmus rhombus*). Étude des fraction protéiques et des complexes antigéniques. *Annales de Parasitologie Humaine et Comparée*, **59**: 143–9.
680. **Reynolds, C. H.** (1980). Phosphoenolpyruvate carboxykinase from the rat and from the tapeworm *Hymenolepis diminuta*. Effects of inhibitors and transition-metals on the carboxylation reactions. *Comparative Biochemistry and Physiology*, **65B**: 481–7.
681. **Ribeiro, P. & Webb, R. A.** (1983a). The synthesis of 5-hydroxytryptamine from tryptophan and 5-hydroxytryptophan in the cestode *Hymenolepis diminuta*. *International Journal for Parasitology*, **13**: 101–6.
682. — (1983b). The occurrence and synthesis of octopamine and catecholamines in the cestode *Hymenolepis diminuta*. *Molecular and Biochemical Parasitology*, **7**: 53–62.
683. — (1984). The occurrence, synthesis and metabolism of 5-hydroxytryptamine and 5-hydroxytryptophan in the cestode *Hymenolepis diminuta*: a high performance liquid chromatographic study. *Comparative Biochemistry and Physiology, C* (*Comparative Pharmacology*), **79**: 159–64.
684. **Richards, K. S.** (1984). *Echinococcus granulosus equinus*: the histochemistry of the laminated layer of the hydatid cyst. *Folia Histochemica et Cytobiologia*, **22**: 21–3.
685. **Richards, K. S., Ilderton, E. & Yardley, H. J.** (1987). Lipids in the laminated layer of liver, lung and daughter cysts of equine *Echinococcus granulosus* (Cestoda). *Comparative Biochemistry and Physiology*, **86B**: 209–12.
686. **Rickard, M. D.** (1983). Immunity. In *The biology of the Eucestoda*, vol. 2, ed. C. Arme & P. W. Pappas, pp. 539–79. Academic Press: London.
687. — (1986). Larval taeniid cestodes – models for research on host–parasite interactions. In *Parasite lives*, ed. M. Cremin, C. Dobson & D. E. Moorehouse, pp. 151–73. University of Queensland Press: St Lucia, Australia.
688. **Rickard, M. D. & Adolph, A. J.** (1975). Vaccination of calves against *Taenia saginata* using a 'parasite-free' vaccine. *Veterinary Parasitology*, **1**: 389–92.
689. **Rickard, M. D. & Bell, K. J.** (1971). Successful vaccination of lambs against infection with *Taenia ovis* using antigens produced during *in vitro* culture of the larval stages. *Research in Veterinary Science*, **12**: 401–2.
690. **Rickard, M. D. & Lightowlers, M. W.** (1986). Immunodiagnosis of hydatid disease. In *The biology of* Echinococcus *and hydatid disease*, ed. R. C. A. Thompson, pp. 217–49. George Allen & Unwin: London.
691. **Rickard, M. D. & Williams, J. F.** (1982). Hydatidosis/cysticercosis; immune mechanisms and immunization against infection. *Advances in Parasitology*, **21**: 229–96.
692. **Rietschel, G.** (1981). Beitrag zur Kenntnis von *Taenia crassiceps* (Zeder, 1800) Rudolphi, 1810 (Cestoda, Taeniidae). *Zeitschrift für Parasitenkunde*, **65**: 309–15.

References

693. **Rietschel, P. E.** (1935). Zur Bewegungsphysiologie der Cestoden. *Zoologischer Anzeiger*, **111**: 109–11.
694. **Rishi, A. K. & McManus, D. P.** (1987a). Genomic cloning of human *Echinococcus granulosus* DNA: isolation of recombinant plasmids and their use as genetic markers in strain characterization. *Parasitology*, **94**: 369–83.
695. **Rishi, A. K. & McManus, D. P.** (1987b). DNA probes which unambiguously distinguish *Taenia solium* from *T. saginata*. *Lancet* ii: 1275–6.
696. **Roberts, L. S.** (1961). The influence of population density on patterns and physiology of growth in *Hymenolepis diminuta* (Cestoda: Cyclophyllidea) in the definitive host. *Experimental Parasitology*, **11**: 332–71.
697. — (1980). Development of *Hymenolepis diminuta* in its definitive host. In *Biology of the tapeworm* Hymenolepis diminuta, ed. H. P. Arai, pp. 357–423. Academic Press: New York.
698. — (1983). Carbohydrate metabolism. In *Biology of the Eucestoda*, vol. 2, ed. C. Arme & P. W. Pappas, pp. 343–390. Academic Press: London.
699. **Roberts, L. S. & Insler, G. D.** (1982). Developmental physiology of cestodes. XVII. Some biological properties of putative 'crowding factors' in *Hymenolepis diminuta*. *Journal of Parasitology*, **68**: 263–9.
700. **Roberts, L. S. & Mong, F. N.** (1969). Developmental biology of cestodes. IV. *In vitro* development of *Hymenolepis diminuta* in presence and absence of oxygen. *Experimental Parasitology*, **26**: 166–74.
701. **Robertson, N. P. & Cain, G. D.** (1984). Glycosaminoglycans of tegumental fractions of *Hymenolepis diminuta*. *Molecular and Biochemical Parasitology*, **12**: 173–83.
702. **Robertson, N. P., Oaks, J. A. & Cain, G. D.** (1984). Characterization of polysaccharides of the eggs and adults of *Hymenolepis diminuta*. *Molecular and Biochemical Parasitology*, **10**: 99–109.
703. **Robinson, J. M. & Bogitsh, B. J.** (1976). Cytochemical localization of peroxidase activity in the mitochondria of *Hymenolepis diminuta*. *Journal of Parasitology*, **62**: 761-5.
704. — (1978a). A morphological and cytochemical study of sperm development in *Hymenolepis diminuta*. *Zeitschrift für Parasitenkunde*, **56**: 81–92.
705. — (1978b). *Hymenolepis diminuta*: biochemical properties of peroxidase activity in mitochondria. *Experimental Parasitology*, **45**: 169–74.
706. **Rogan, M. T. & Richards, K. S.** (1986). *In vitro* development of hydatid cysts from posterior bladders and ruptured brood capsules of equine *Echinococcus granulosus*. *Parasitology*, **92**: 379–90.
707. **Rohde, K.** (1982). The nervous system of parasitic helminths. *Proceedings of the 5th International Congress of Parasitology*, Canada, pp. 70–2. Elsevier Biomedical Press: Amsterdam.
708. **Roitt, I. M.** (1984). *Essential immunology*, 5th edn. Blackwell Scientific Publications: London.
709. **Roitt, I. M., Brostoff, J. & Male, D.** (1986). *Immunology*. Churchill Livingstone: London.

References

710. **Rose, R. C.** (1981). Absorptive functions of the gall bladder. In *Physiology of the gastrointestinal tract*, ed. L. R. Johnson, pp. 1021–33. Raven Press: New York.

711. **Rosen, R. & Dick, T. A.** (1983). Development and infectivity of the procercoid of *Triaenophorus crassus* Forel and mortality of the first intermediate host. *Canadian Journal of Zoology*, **61**: 2120–28.

712. **Rosen, R. & Dick, T. A.** (1984*a*). Experimental infections of rainbow trout *Salmo gairdneri* Richardson, with plerocercoids of *Triaenophorus crassus* Forel. *Journal of Wildlife Diseases*, **20**: 34–8.

713. — (1984*b*). Growth and migration of plerocercoids of *Triaenophorus crassus* Forel and pathology in experimentally infected white fish, *Coregonus clupeaformis* (Mitchell). *Canadian Journal of Zoology*, **62**: 203–11.

714. **Rothman, A. H.** (1968). Peroxidase in platyhelminth cuticular mitochondria. *Experimental Parasitology*, **23**: 51–5.

715. **Roy, T. K.** (1979*a*). Histochemical studies on *Raillietina* (*Raillietina*) *johri* (Cestoda: Davaineidae). I. Nonspecific and specific phosphatases. *Journal of Helminthology*, **53**: 45–9.

716. — (1979*b*). Histochemical studies on *Raillietina* (*Raillietina*) *johri* (Cestoda: Davaineidae). II. Nucleoside diphosphatase and thiamine pyrophosphatase. *Journal of Helminthology*, **53**: 261–3.

717. — (1980). Histochemical studies on *Raillietina* (*Raillietina*) *johri* (Cestoda: Davaineidae). III. Esterases. *Journal of Helminthology*, **54**: 219–22.

718. — (1982). Hydrolytic enzymes and membrane digestion in parasitic platyhelminths. *Journal of Scientific and Industrial Research*, **41**: 439–54.

719. **Rubino, S., Fiori, P. L., Lubinu, G., Monaco, G. & Cappuccinelli, P.** (1983). The cytoskeleton of hydatid cyst cultured cells and its sensitivity to inhibitors. *European Journal of Cell Biology*, **30**: 182–90.

720. **Ruff, M. D. & Read, C. P.** (1973). Inhibition of pancreatic lipase by *Hymenolepis diminuta*. *Journal of Parasitology*, **59**: 105–11.

721. **Rybicka, K.** (1966). Embryogenesis in cestodes. *Advances in Parasitology*, **4**: 107–86.

722. — (1972). Ultrastructure of embryonic envelopes and their differentiation in *Hymenolepis diminuta* (Cestoda). *Journal of Parasitology*, **58**: 849–63.

723. **Safonov, N. N.** (1982). [Ecological characteristics of *Ligula intestinalis* and *Proteocephalus* sp. in Lithuanian lakes.] In Russian. *Trudy Gel'mintologicheskoi Laboratorii* **31**: 99–107. [HA/52/2068]

724. **Sagieva, A. T., Sadykov, V. M., Matchanov, N. M., Ten, V. G., Bochkarev, V. N., Yuldashev, S. Yu., Sagiev, A. G., Alimov, K. A. & Muratbaev, Zh.** (1985). [The life cycle of *Taenia multiceps* obtained from metacestodes recovered from man.] In Russian. *Meditsinskaya Parazitologiya i Parazitarnye Bolezni*, pp. 32–40. [HA/54/4920]

725. **Sakamoto, T.** (1981). Electron microscopical observations on the egg of *Echinococcus multilocularis*. *Memoirs of the Faculty of Agriculture, Kagoshima University*, **17**: 165–74.

726. **Sakamoto, T. & Sugimura, M.** (1969). Studies on echinococcosis.

References

XXI. Electron microscopical observations on general structure of larval tissue of multilocularis *Echinococcus*. *Japanese Journal of Veterinary Research*, **17**: 67–81.

727. **Salazar, P. M., de Haro, I. & Voge, M.** (1984). Hatching *in vitro* of oncospheres of *Taenia solium*. *Journal of Parasitology*, **70**: 161–2.

728. **Salminen, K.** (1973). The oxidation of external NADH by adult and plerocercoid of *Diphyllobothrium latum*. *Comparative Biochemistry and Physiology*, **44B**: 283–9.

729. — (1974). Succinate dehydrogenase and cytochrome oxidase in adult and plerocercoid *Diphyllobothrium latum*. *Comparative Biochemistry and Physiology*, **49B**: 87–92.

730. **Sanford, P. A.** (1982). *Digestive system physiology*. Edward Arnold: London.

731. **Sarciron, M. E., Azzar, G., Persat, F., Petavy, A.–F. & Got, R.** (1987). A comparative study of UTP-D-glucose-1-phosphate uridylyl transferase in the cysts of *Echinococcus multilocularis* and the livers of infected and control *Meriones unguiculatus*. *Molecular and Biochemical Parasitology*, **23**: 25–9.

732. **Sawada, I.** (1959). Experimental studies on the evagination of the cysticercoids of *Raillietina kashiwarensis*. *Experimental Parasitology*, **8**: 325–35.

733. **Schantz, P. M., Colli, C., Cruz-Reyes, A. & Prezioso, U.** (1976). Sylvatic echinococcosis in Argentina. II. Susceptibility of wild carnivores to *Echinococcus granulosus* (Batsch, 1786) and host-induced morphological variation. *Tropenmedizin und Parasitologie*, **27**: 70–8.

734. **Schantz, P. M. & Gottstein, B.** (1986). Echinococcosis (Hydatidosis). In *Immunodiagnosis of parasitic diseases*, vol. 1, ed. K. W. Walls & P. M. Schantz, pp. 69–107. Academic Press: Orlando, FL.

735. **Schantz, P. M., Van Den Bossche, H. & Eckert, J.** (1982). Chemotherapy for larval echinococcosis in animals and humans. Report of a workshop. *Zeitschrift für Parasitenkunde*, **67**: 5–26.

736. **Schiller, E. L.** (1965). A simplified method for the *in vitro* cultivation of the rat tapeworm *Hymenolepis diminuta*. *Journal of Parasitology*, **51**: 516–18.

737. — (1974). The inheritance of X-irradiated-induced effects in the rat tapeworm, *Hymenolepis diminuta*. *Journal of Parasitology*, **60**: 35–46.

738. **Schmidt, J. & Peters, W.** (1987). Localization of glycoconjugates at the tegument of the tapeworms *Hymenolepis nana* and *H. microstoma* with gold labelled lectins. *Parasitology Research*, **73**: 80–6.

739. **Schmidt, J. M. & Todd, K. S.** (1978). Life cycle of *Mesocestoides corti* in the dog (*Canis familaris*). *American Journal of Veterinary Research*, **39**: 1490–3.

740. **Schramlová, J. & Blažek, K.** (1981). Ultrastructure of the bladder tegument of cysticercus bovis in various stages of its development. *Folia Parasitologica*, **28**: 61–9.

741. — (1982). Ultrastructure of the hatched and unhatched oncospheres of *Taenia saginata*. *Folia Parasitologica*, **29**: 45–50.

742. **Schramlová, J. & Lavrov, I. L.** (1981). Scanning electron microscopic

References

studies on the scolex surface of *Hydatigera krepkogorski* (Schulz et Landa, 1934) larva. *Folia Parasitologica,* **28**: 191–2.

743. **Schroeder, L. L. & Pappas, P. W.** (1980). Trypsin absorption by *Hymenolepis diminuta. Journal of Parasitology,* **66**: 49–52.

744. **Schroeder, L. L., Pappas, P. W. & Means, G. E.** (1981). Trypsin inactivation by intact *Hymenolepis diminuta* (Cestoda): some characteristics of the inactivated enzyme. *Journal of Parasitology,* **67**: 378–85.

745. **Scott, J. S.** (1965). Evagination of the cysticercoid in *Plycercus lumbrici. Parasitology,* **55**: 421–5.

746. **Searcy, D. G. & MacInnis, A. J.** (1970). Measurements by DNA renaturation of the genetic basis of parasitic reduction. *Evolution,* **24**: 796–806.

747. **Seidel, J. S.** (1971). Hemin as a requirement in the development *in vitro* of *Hymenolepis microstoma* (Cestoda: Cyclophyllidea). *Journal of Parasitology,* **57**: 566–70.

748. — (1975). The life cycle *in vitro* of *Hymenolepis microstoma* (Cestoda). *Journal of Parasitology,* **61**: 677–81.

749. **Seidel, J. S. & Voge, M.** (1975). Axenic development of cysticercoids of *Hymenolepis nana. Journal of Parasitology,* **61**: 861–4.

750. **Sekretaryuk, K. V.** (1982). [Interrelationships between helminths in bothriocephaliasis in carp.] In Russian. *Veterinariya,* **10**: 35–7. [HA/52/950]

751. — (1984). [Ultrastructure of the zones of contact between the helminth and the intestine during bothriocephaliasis in carp.] In Russian. *Doklady̆ Vsesoyuznoi Akademii Sel'skokhozyaistvenny̆kh Nauk im. V. I. Lenina,* **4**: 35–7. [HA/53/2679]

752. **Şerban, M., Şuţeanu, M. & Lungu, T.** (1982). [Biochemical characteristics of *Coenurus cerebralis* larvae.] In Romanian. *Lucrări Ştiinţifice, Institutul Agronomic 'Nicholae Bălcescu' C,* **25**: 65–71. [HA/52/4007]

753. — (1983). [Biochemical characteristics of *Coenurus cerebralis* larvae.] In Romanian. In *Lucrările simpozionului, Probleme actuale ale diagnosticului şi profilaxiei în sistemul de creştere intensivă a animalelor, Bucureşti, 23–24 Octombrie, 1981,* pp. 153–62. Societatea de Medicină Veterinară: Bucharest, Romania. [HA/52/5189]

754. **Shepherd, J. C. & McManus, D. P.** (1987). Specific and cross-reactive antigens of *Echinococcus granulosus* hydatid cyst fluid. *Molecular and Biochemical Parasitology,* **25**: 143–54.

755. **Shield, J. M.** (1969). *Dipylidium caninum, Echinococcus granulosus* and *Hydatigera taeniformis*: histochemical identification of cholinesterases. *Experimental Parasitology,* **25**: 217–31.

756. — (1971). Histochemical localization of monoamines in the nervous system of *Dipylidium caninum* by the formaldehyde fluorescence technique. *International Journal for Parasitology,* **1**: 135–8.

757. **Shield, J. M., Heath, D. D. & Smyth, J. D.** (1973). Light microscope studies on the early development of *Taenia pisiformis* cysticerci. *International Journal for Parasitology,* **3**: 471–80.

758. **Shisov, B. A.** (1980). Biogenic amines in helminths. In *Neuro-*

353

References

transmitters. Comparative aspects, ed. J. Salánki & T. M. Turpaev, pp. 31–56. Akadémiai Kiadó: Budapest.

759. — (1984). *Biochemistry and physiology of helminths and immunity in helminthoses.* Trudy Gelan, vol. 32. Publication House 'Nauk': Moscow.

760. **Shiwaku, K. & Hirai, K.** (1982). Growth-promoting effect of *Spirometra erinacei* (Rudolphi, 1819) plerocercoids in young mice. *Japanese Journal of Parasitology*, **31**: 185–95.

761. **Shiwaku, J., Hirai, K. & Torii, M.** (1982). Growth-promoting effect of *Spirometra erinacei* (Rudolphi, 1819) plerocercoids in mature mice: relationship between number of infected plerocercoids and growth-promoting effect. *Japanese Journal of Parasitology*, **31**: 353–60.

762. **Shiwaku, K., Hirai, K., Torii, M. & Tsuboi, T.** (1983). Effects of *Spirometra erinacei* plerocercoids on the growth of Snell dwarf mice. *Parasitology*, **87**: 447–53.

763. **Siddiqui, A. A., Ahmad, M. & Nizami, W. A.** (1986). Phosphatase system of some common poultry parasites: *Ascaridia galli* and *Cotugnia digonopora*. *Indian Veterinary Journal*, **63**: 14–17.

764. **Siddiqui, A. A. & Podesta, R. B.** (1985*a*). Development regulation of protein synthesis in *Hymenolepis diminuta*: 2-dimensional electrophoretic and fluorographic analysis of polypeptide synthesis in formation of brush border membranes. *Cellular and Molecular Biology*, **31**: 209–16.

765. — (1985*b*). Developmental regulation of protein synthesis in *Hymenolepis diminuta*: two-dimensional electrophoretic and fluorographic analysis of protein synthesis in oncospheres. *Journal of Parasitology*, **71**: 119–22.

766. **Sidorov, V. S. & Smirnov, L. P.** (1980). [Fatty acid composition of some helminths of cold-blooded and warm-blooded vertebrates.] In Russian. *Zhurnal Évolyutsionnoĭ Biologii i Fiziologii*, **16**: 551–5. [HA/52/4545]

767. **Simpson, A. J. G.** (ed.) (1986). *Parasites and molecular biology: applications of new techniques. Symposia of the British Society for Parasitology*, vol. 23. [*Parasitology*, **92** (suppl.)]

768. **Simpson, A. J. G., Walker, T. & Terry, R.** (1986). An introduction to recombinant DNA technology. *Parasitology*, **91**: S7–S14 [**92** (suppl.)]

769. **Sinba, D. P., Sircar, M. & Singh, S. P.** (1978). A histochemical study of distribution of glycogen in some trematodes and cestodes. *Indian Journal of Animal Research*, **12**: 97–101.

770. **Singer, S. J. & Nicolson, G. L.** (1972). The fluid mosaic model of the structure of cell membranes. *Science*, **175**: 720–31.

771. **Singh, B. B. & Rao, B. V.** (1967). Some biological studies on *Taenia taeniaeformis*. *Indian Journal of Helminthology* (for 1966), **18**: 151–60.

772. **Singh, B. B., Singh, K. S. & Dwarkanath, P. K.** (1977). Lipase activity in *Thysaniezia giardi*. *Indian Journal of Parasitology*, **1**: 69–70.

773. **Singh, B. B., Singh, K. S., Ghosal, A. K. & Dwarkanath, P. K.** (1978). Inorganic calcium, magnesium and phosphorus in *Thysaniezia giardi*. *Indian Journal of Parasitology*, **2**: 37–8.

774. **Singh, J. P. & Baugh, S. C.** (1984). Embryos and embryonic envelopes

References

in eggs of two cyclophyllidean cestodes. *Angewandte Parasitologie*, **25**: 12–16.

775. **Sinha, D. P. & Hopkins, C. A.** (1967). The *in vitro* cultivation of the tapeworm *Hymenolepis nana* from larva to adult. *Nature*, **215**: 1275–6.

776. **Sircar, M. & Sinha, D. P.** (1978). Alkaline phosphatase activity in some tapeworms. *Indian Journal of Animal Research*, **12**: 27–30.

777. **Šišova-Kasatočkina, O. A. & Dubovskaja, A. J.** (1975). Proteinase activity in certain cestode species parasitizing vertebrates of different classes. *Acta Parasitologica Polonica*, **23**: 389–93.

778. **Šlais, J.** (1973). Functional morphology of cestode larvae. *Advances in Parasitology*, **11**: 395–480.

779. **Slepnev, N. K.** (1982). [The survival of *Echinococcus granulosus* oncospheres under the influence of chemical and physical factors.] In Russian. *Trudy Belorusskogo Nauchno-Issledovatel'skogo Instituta Èksperimental'noĭ Veterinarii. (Veterinarnaya Nauka – Proizvodstvu)*, **19**: 96–8. [HA/51/963]

780. **Smirnov, L. P.** (1982). [Lipids of helminths.] In Russian. In *Èkologiya paraziticheskikh organizmov v biogeotsenozakh severa*, pp. 128–45. Petrozavodsk, USSR: Karel'skiĭ Filial Akademi Nauk SSSR, Institut Biologii. [HA/52/2044]

781. **Smirnov, L. P. & Bogdan, V. V.** (1982). [Comparative study of the lipid composition of some cestodes and of their hosts.] In Russian. In *Èkologiya paraziticheskikh organizmov v biogeotsenozakh severa*, pp. 145–51. Petrozavodsk, USSR: Karel'skiĭ Filial Akademi Nauk SSSR, Institut Biologii. [HA/52/2043]

782. **Smirnov, L. P. & Sidorov, V. S.** (1979). [Fatty acid composition of the cestodes *Eubothrium crassum* and *Diphyllobothrium dendriticum*.] In Russian. *Parazitologyia*, **13**: 522–9. [HA/49/2581]

783. — (1984). [Comparative study of protein spectra in cestodes and their hosts by means of gel chromatography and disc electrophoresis.] In Russian: *Parazitologyia*, **18**: 430–5. [HA/54/1725]

784. **Smithers, S. R.** (1986). Vaccination against schistosomes and other systemic helminths. *International Journal for Parasitology*, **17**: 31–42.

785. **Smyth, D. H.** (1963). Intestinal absorption. In *Recent advances in physiology*, 8th edn, pp. 36–68. Churchill: London.

786. **Smyth, J. D.** (1946). Studies on tapeworm physiology. I. Cultivation of *Schistocephalus solidus in vitro*. *Journal of Experimental Biology*, **23**: 47–70.

787. — (1947). Studies on tapeworm physiology. II. Cultivation and development of *Ligula intestinalis in vitro*. *Parasitology*, **38**: 173–81.

788. — (1954*a*). Studies on tapeworm physiology. VII. Fertilization of *Schistocephalus solidus in vitro*. *Experimental Parasitology*, **3**: 64–71.

789. — (1954*b*). A technique for the histochemical demonstration of polyphenol oxidase and its application to egg-shell formation in helminths and byssus formation in *Mytilus*. *Quarterly Journal of Microscopical Science*, **95**: 139–52.

790. — (1959). Maturation of larval pseudophyllidean cestodes and strigeid trematodes under axenic conditions; the significance of nutri-

tional levels in platyhelminth development. *Annals of the New York Academy of Sciences*, **77**: 102–25.
791. — (1962). Studies on tapeworm physiology. X. Axenic cultivation of the hydatid organism *Echinococcus granulosus*; establishment of a basic technique. *Parasitology*, **52**: 441–57.
792. — (1963). The biology of cestode life-cycles. *Commonwealth Agriculture Bureaux. UK Technical Communication*, **34**: 1–38.
793. — (1964a). The biology of the hydatid organisms. *Advances in Parasitology*, **2**: 169–219.
794. — (1964b). Observations on the scolex of *Echinococcus granulosus*, with special reference to the occurrence of secretory cells in the rostellum. *Parasitology*, **54**: 515–26.
795. —(1967). Studies on tapeworm physiology, XI. *In vitro* cultivation of *Echinococcus granulosus* from the protoscolex to the strobilate stage. *Parasitology*, **57**: 111–33.
796. — (1969). *The physiology of cestodes*, 1st edn. Oliver & Boyd: Edinburgh.
797. — (1971). Development of monozoic forms of *Echinococcus granulosus* during *in vitro* culture. *International Journal for Parasitology*, **1**: 121–4.
798. — (1972). Changes in the digestive–absorptive surface of cestodes during larval/adult differentiation. *Symposia of the British Society for Parasitology*, **10**: 41–70.
799. — (1973). Some interface phenomena in parasitic protozoa and platyhelminths. *Canadian Journal of Zoology*, **51**: 367–77.
800. —(1976). *An introduction to animal parasitology*, 2nd edn. Hodder & Stoughton: London.
801. — (1979). *Echinococcus granulosus* and *E. multilocularis: in vitro* culture of the strobilar stages from protoscoleces. *Angewandte Parasitologie*, **20**: 137–47.
802. — (1982). The insemination–fertilization problem in cestodes cultured *in vitro*. In *Aspects of parasitology*, ed. E. Meerovitch, pp. 393–406. McGill University: Montreal.
803. — (1985a). The 'niche' concept in parasitology with special reference to hydatid disease. *Proceedings of the British-Scandinavian Joint Meeting in Tropical Medicine and Parasitology*, Copenhagen, September, 1985. Abstracts, p. 61.
804. — (1985b). *In vitro* culture of *Echinococcus* spp. *Proceedings of the 13th International Congress of Hydatidology*, Madrid, 1985, pp. 84–89.
805. — (1987a). Changing concepts in the microecology, macroecology and epidemiology of hydatid disease. In *Helminth zoonoses (Current topics in veterinary medicine and animal science)*, ed. S. Geerts, V. Kumar & J. Brandt, pp. 1–11. Martinus Nijhoff Publishers: Dordrecht, The Netherlands.
806. — (1987b). Asexual and sexual differentiation in cestodes: especially *Mesocestoides* and *Echinococcus*. In *Molecular paradigms for eradicating helminthic parasites, UCLA Symposium on Moleular and Cellular Biology*, new series, vol. 60, ed. A. MacInnis, pp. 19–34. Alan R. Liss Inc., New York.

References

807. **Smyth, J. D. & Barrett, N. J.** (1979). *Echinococcus multilocularis*: further observations on strobilar differentiation *in vitro*. *Revista Ibérica de Parasitología*, **39**: 39–53.
808. **Smyth, J. D. & Davies, Z.** (1974a). Occurrence of physiological strains of *Echinococcus granulosus* demonstrated by *in vitro* culture of protoscoleces from sheep and horse hydatid cysts. *International Journal for Parasitology*, **4**: 443–5.
809. — (1974b). *In vitro* culture of the strobilar stage of *Echinococcus granulosus* (sheep strain): a review of basic problems and results. *International Journal for Parasitology*, **4**: 631–44.
810. **Smyth, J. D. & Halton, D. W.** (1983). *The physiology of trematodes*, 2nd edn. Cambridge University Press: Cambridge.
811. **Smyth, J. D. & Howkins, A. B.** (1966). An *in vitro* technique for the production of eggs of *Echinococcus granulosus* by maturation of partly developed strobila. *Parasitology*, **56**: 763–6.
812. **Smyth, J. D., Miller, H. J. & Howkins, A. B.** (1967). Further analysis of the factors controlling strobilization, differentiation and maturation of *Echinococcus granulosus in vitro*. *Experimental Parasitology*, **21**: 31–41.
813. **Smyth, J. D., Morseth, D. J. & Smyth, M. M.** (1969). Observations on nuclear secretions in the rostellar gland of *Echinococcus granulosus* (Cestoda). *The Nucleus*, **12**: 47–56.
814. **Smyth, J. D. & Smyth, M. M.** (1964). Natural and experimental hosts of *Echinococcus granulosus* and *E. multilocularis*, with comments on the genetics of speciation in the genus *Echinococcus*. *Parasitology*, **54**: 493–514.
815. — (1969). Self-insemination in *Echinococcus granulosus in vivo*. *Journal of Helminthology*, **43** 383–8.
816. **Soprunov, F. F.** (1984). [Achievements in the studies of carbohydrate metabolism in helminths.] In Russian. *Trudy Gel'mintologicheskoi Laboratorii (Biokhimiya i fiziologiya gel'mintov i immunitet pri gel'mintozakh)*, **32**: 121–54. [HA/53/4233]
817. **Soulsby, E. J. L.** (ed.) (1986). *Immune responses in parasitic infections: immunology, immunopathology, and immunoprophylaxis*, vol. II *Trematodes and cestodes*. CRC Press Inc.: Boca Raton, FL.
818. **Southern, E. M.** (1975). Detection of specific sequences among DNA fragments separated by gel electrophoresis. *Journal of Molecular Biology*, **98**: 503–17.
819. **Specian, R. D., Lumsden, R. D., Ubelaker, J. E. & Allison, V. F.** (1979). A unicellular endocrine gland in cestodes. *Journal of Parasitology*, **65**: 569–78.
820. **Steelman, S. L., Glitzer, M. S., Ostlind, D. A. & Mueller, J. F.** (1971). Biological properties of the growth hormone-like factor from the plerocercoid of *Spirometra mansonoides*. *Recent Progress in Hormone Research*, **27**: 97–120.
821. **Štěrba, J. & Dyková, I.** (1979). Symptomology of taeniasis caused by *Taenia saginata*. *Folia Parasitologica*, **26**: 281–4.
822. **Štěrba, J., Šlais, J., Machnicka, B. & Schandl, V.** (1981). Development of oncospheres of *Taenia saginata* after a concomitant infection

357

References

by oral and subcutaneous routes. *Folia Parasitologica*, **28**: 353–8.
823. **Sterry, P. R. & McManus, D. P.** (1982). *Ligula intestinalis*: biochemical composition, carbohydrate utilisation and oxygen consumption of plerocercoids and adults. *Zeitschrift für Parasitenkunde*, **67**: 87–98.
824. **Stevenson, P.** (1983). Observations on the hatching and activation of fresh *Taenia saginata* eggs. *Annals of Tropical Medicine and Parasitology*, **77**: 399–404.
825. **Storey, G. W. & Phillips, R. A.** (1985). The survival of parasite eggs throughout the soil profile. *Parasitology*, **91**: 585–90.
826. **Strazhnik, L. V.** (1980). [The carbohydrate composition of fish cestodes.] In Russian. *Gidrobiologicheskiĭ Zhurnal*, **16**: 87–91. [HA/51/2079]
827. **Sukhdeo, M. V. K., Hsu, S. C., Thompson, C. S. & Mettrick, D. F.** (1984). *Hymenolepis diminuta*: behavioral effects of 5-hydroxytryptamine, acetylcholine, histamine and somatostatin. *Journal of Parasitology*, **70**: 682–8.
828. **Sukhdeo, M. V. K. & Mettrick, D. F.** (1987). Parasite behaviour: understanding platyhelminth responses. *Advances in Parasitology*, **26**: 74–144.
829. **Sulgostowska, T.** (1980). The development of organ systems in cestodes. III. Histology of *Diploposthe laevis* (Bloch, 1782) and *D. bifaria* (Siebold in Creplin, 1846) (Hymenolepididae) and histogenesis of their reproductive system. *Acta Parasitologica Polonica*, **26**: 143–52.
830. **Sultan, Sheriff D., Dar, F. K. & Kidwai, S. A.** (1984). Metallic elements in hydatid fluid. *Journal of Helminthology*, **58**: 335–6.
831. **Sun, C. N.** (1972). The fine structure of sperm tail of cotton rat tapeworm, *Hymenolepis diminuta*. *Cytobiologie*, **6**: 382–6.
832. **Suquet, C., Green-Edwards, C. & Leid, R. W.** (1984). Isolation and partial characterization of a *Taenia taeniaeformis* metacestode proteinase inhibitor. *International Journal for Parasitology*, **14**: 165–72.
833. **Swiderski, Z.** (1970). An electron microscope study of spermatogenesis in cyclophyllidean cestodes with emphasis on the comparison of fine structure of mature spermatozoa. *Journal of Parasitology*, **56**: (II, suppl.): 337–8.
834. ― (1972). La structure de l'oncosphère du cestode *Catenotaenia pusilla* (Goeze, 1782) (Cyclophyllidea, Catenotaeniidae). *La Cellule*, **69**: 207–37.
835. ― (1976a). Fine structure of the spermatozoon of *Lacistorhynchus tenuis* (Cestoda, Trypanorhyncha). *Proceedings of the 6th European Congress on electron microscopy*, Jerusalem, 1976, pp. 309–10.
836. ― (1976b). Fertilization in the cestode *Hymenolepis diminuta* (Cyclophyllidea, Hymenolepididae). *Proceedings of the 6th European congress on electron microscopy*, Jerusalem, 1976, pp. 311–12.
837. ― (1982a). *Echinococcus granulosus*: ultrastructure of the glandular regions of the infective oncosphere. *Proceedings of the 10th international congress on electron microscopy*, Hamburg, 1982, vol. 3, pp. 511–12.

References

838. — (1982*b*). *Echinococcus granulosus*: embryonic envelope formation. *Proceedings of the 10th international congress of electron microscopy*, Hamburg, 1982, vol. 3, p. 513.
839. — (1983). *Echinococcus granulosus*: hook-muscle systems and cellular organization of infective oncospheres. *International Journal for Parasitology*, **13**: 289–99.
840. **Swiderski, Z. & Eklu-Natey, R. D.** (1978). Fine structure of the spermatozoon of *Proteocephalus longicollis* (Cestoda Proteocephalidea). *Proceedings of the 9th international congress of electron microscopy*, Toronto, 1978, vol. II, pp. 572–3.
841. **Swiderski, Z., Eklu-Natey, R. D., Subilia, L. & Huggel, H.** (1978). Comparative fine structure of vitelline cells in the cestode *Proteocephalus longicollis* (Proteocephalidea). *Proceedings of the 9th international congress of electron microscopy*, Grenoble, pp. 669–70.
842. **Swiderski, Z. & Mackiewicz, J. S.** (1976*a*). Fine structure of the spermatozoan of *Glaridacris catostomi* (Cestoda, Caryophyllidea). *Proceedings of the 6th European congress of electron microscopy*, Jerusalem, pp. 307–2.
843. — (1976*b*). Electron microscope study of vitellogenesis in *Glaridacris catostomi* (Cestoidea: Caryophyllidea). *International Journal of Parasitology*, **6**: 61–73.
844. **Swiderski, Z. & Mokhtar-Maamouri, F.** (1974*a*). Vitellogenesis in *Bothriocephalus clavibothrium* Ariola, 1899 (Cestoda: Pseudophyllidea). *Zeitschrift für Parasitenkunde*, **43**: 135–49.
845. — (1974*b*). The fine structure of the coracidia of *Bothriocephalus clavibothrium* Ariola, 1899 (Cestoda: Pseudophyllidea). *Proceedings of the 3rd international congress of parasitology*, vol. I: pp. 412–13.
846. — (1980). Étude de la spermatogénèse de *Bothriocephalus clavibothrium*. *Archives de l'Institut Pasteur de Tunis*, **57**: 323–47.
847. **Swiderski, Z. & Subilia, L.** (1980). The cellular organization of the infective oncospheres of *Oochoristica* sp. (Cyclophyllidea, Anoplocephalidae). *Proceedings of the 3rd European multicolloquium of parasitology*, Cambridge, 1980, p. 187.
848. **Symons, L. E. A., Donald, A. D. & Dineen, J. K.** (eds.) (1982). *Biology and control of parasites*. Academic Press: London.
849. **Sysoev, A. V.** (1982). Composition and dynamics of invasion of the first intermediate host of *Triaenophorus nodulosus* (Pallas) (Cestoda: Triaenophoridae) under conditions in Karelia. *Helminthologia*, **19**: 249–55.
850. — (1985). On the composition of intermediate hosts of cestodes, parasitic in the nine-spined stickleback. *Angewandte Parasitologie*, **26**: 147–50.
851. **Taneya, S. K.** (1973). Morphological and cytochemical studies on the spermatogenesis of *Ophryocotyloides corvorum*, Gupta & Singh 1970 (Cestoda: Cyclophyllidea). *Zoologica Poloniae*, **22**: 197–205.
852. **Tang, Z.** (1982). [Developmental studies on *Polyonchobothrium ophiocephalina* (Tseng, 1933) and *Bothriocephalus opsariichthydis* Yamaguti, 1934.] In Chinese. *Acta Zoological Sinica*, **28**: 51–9. [HA/51/6016]

References

853. **Tayal, S. & Premvati, G.** (1982). *In vitro* estimation of glycogen content in three sheep cestodes. *Folia Parasitologica*, **29**: 259–63.
854. **Taylor, A. E. R. & Baker, J. R.** (eds.) (1978). *Methods of culturing parasites* in vitro. Academic Press: London.
855. — (eds.) (1987). In vitro *methods for parasite cultivation*. Academic Press: London.
856. **Taylor, A. E. R. & Muller, R.** (eds.) (1979). *Problems in the identification of parasites and their vectors. Symposia of the British Society for Parasitology*, **17**.
857. — (eds.) (1980). Vaccines against parasites. *Symposia of the British Society for Parasitology*, **18**.
858. **Tedesco, J. L. & Coggins, J. R.** (1980). Electron microscopy of the tumulus and origin of associated structures within the tegument of *Eubothrium salvelini* Schrank, 1790 (Cestoidea: Pseudophyllidea). *International Journal for Parasitology*, **10**: 275–280.
859. **Terenina, N. B.** (1984). Results of spectrofluorimetric determination of biogenic amines (serotonin, dopamine) in cestodes. *Helminthologia*, **21**: 275–80.
860. **Thakur, A. S., Schwabe, C. W. & Koussa, M.** (1971). Polysaccharides of the *Taenia hydatigena* cyst membrane. *Experimental Parasitology*, **30**: 94–101.
861. **Thomas, J. N. & Turner, S. G.** (1980). A reinterpretation of the evidence for contact digestion in the tapeworm, *Hymenolepis diminuta. Journal of Physiology*, **301**: 79P–80P.
862. **Thompson, C. S. & Mettrick, D. F.** (1984). Neuromuscular physiology of *Hymenolepis diminuta* and *H. microstoma* (Cestoda). *Parasitology*, **89**: 567–78.
863. **Thompson, R. C. A.** (1977). Growth, segmentation and maturation of the British horse and sheep strains of *Echinococcus granulosus* in dogs. *International Journal for Parasitology*, **7**: 281–5.
864. — (1986*a*) (ed.). *The biology of* Echinococcus *and hydatid disease.* George Allen & Unwin: London.
865. — (1986*b*). Biology and systematics of *Echinococcus*. In *The biology of* Echinococcus *and hydatid disease*, ed. R. C. A. Thompson, pp. 5–43. George Allen & Unwin: London.
866. **Thompson, R. C. A., Dunsmore, J. D. & Hayton, A. R.** (1979). *Echinococcus granulosus*: secretory activity of the rostellum of the adult cestode *in situ* in the dog. *Experimental Parasitology*, **48**: 144–63.
867. **Thompson, R. C. A. & Eckert, J.** (1982). The production of eggs by *Echinococcus multilocularis* in the laboratory following *in vivo* and *in vitro* development. *Zeitschrift für Parasitenkunde*, **68**: 227–34.
868. — (1983). Observations on *Echinococcus multilocularis* in the definitive host. *Zeitschrift für Parasitenkunde*, **69**: 335–45.
869. **Thompson, R. C. A., Hayton, A. R. & Sue, L. P. J.** (1980). An ultrastructural study of the microtriches of adult *Proteocephalus tidswelli* (Cestoda: Proteocephalidae). *Zeitschrift für Parasitenkunde*, **64**: 95–111.

References

870. **Thompson, R. C. A., Houghton, A. & Zaman, V.** (1982). A study of the microtriches of adult *Echinococcus granulosus* by scanning electron microscopy. *International Journal for Parasitology*, **12**: 579–83.
871. **Thompson, R. C. A., Jue Sue L. P. & Buckley, S. J.** (1982). *In vitro* development of the strobilar stage of *Mesocestoides corti*. *International Journal for Parasitology*, **12**: 303–14.
872. **Thompson, R. C. A. & Kumaratilake, L. M.** (1985). Comparative development of Australian strains of *Echinococcus granulosus* in dingoes (*Canis familiaris dingo*) and domestic dogs (*C. f. familiaris*), with further evidence for the origin of the Australian sylvatic strain. *International Journal for Parasitology*, **15**: 535–42.
873. **Thompson, R. C. A., Kumaratilake, L. M. & Eckert, J.** (1984). Observations on *Echinococcus granulosus* of cattle origin in Switzerland. *International Journal for Parasitology*, **14**: 283–91.
874. **Thompson, R. C. A. & Smyth, J. D.** (1975). Equine hydatidosis: a review of the current status in Great Britain and the results of an epidemiological survey. *Veterinary Parasitology*, **1**: 107–27.
875. **Thorson, R. E., Digenis, G. A., Berntzen, A. & Konyalian, A.** (1968). Biological activities of various lipid fractions from *Echinococcus granulosus* scoleces on *in vitro* cultures of *Hymenolepis diminuta*. *Journal of Parasitology*, **54**: 970–3.
876. **Threadgold, L. T.** (1962). An electron microscopic study of the tegument and associated structures of *Dipylidium caninum*. *Quarterly Journal of Microscopical Science*, **103**: 135–40.
877. — (1976). *The ultrastructure of the animal cell*. 2nd edn. Pergamon Press: Oxford.
878. —(1984). Parasitic platyhelminths. In *Biology of the integument*, ed. J. Bereiter-Hahn, A. G. Maltoltsy & K. S. Richards, pp. 132–91. Springer-Verlag: Berlin.
879. **Threadgold, L. T. & Befus, A. D.** (1977). *Hymenolepis diminuta*: ultrastructural localization of immunoglobulin-binding sites on the tegument. *Experimental Parasitology*, **43**: 169–79.
880. **Threadgold, L. T. & Dunn, J.** (1983). *Taenia crassiceps*: regional variations in ultrastructure and evidence of endocytosis in the cysticercus tegument. *Experimental Parasitology*, **55**: 121–31.
881. — (1984). *Taenia crassiceps*: basic mechanisms of endocytosis in the cysticercus. *Experimental Parasitology*, **58**: 263–9.
882. **Threadgold, L. T. & Hopkins, C. A.** (1981). *Schistocephalus solidus* and *Ligula intestinalis*: pinocytosis by the tegument. *Experimental Parasitology*, **51**: 444–56.
883. **Threadgold, L. T. & Robinson, A.** (1984). Amplification of the cestode surface: a stereological analysis. *Parasitology*, **89**: 523–35.
884. **Tkachuk, R. D., Saz, H. J., Weinstein, P. P., Finnegan, K. & Mueller, J. F.** (1977). The presence and possible function of methylmalonyl CoA mutase and propionyl CoA carboxylase in *Spirometra mansonoides*. *Journal of Parasitology*, **63**: 769–74.
885. **Torre-Blanco, A.** (1982). The collagen of *Cysticercus cellulosae*: a study in the comparative biochemistry of collagen. In *Cysticercosis: present state of knowledge and perspectives*, ed. A. Flisser, K. Willms,

References

J. P. Laclette, C. Larralde, C. Ridaura, F. Beltrán & M. W. Vogt, pp. 423–36. Academic Press: New York.

886. **Torre-Blanco, A. & Toledo, I.** (1981). The isolation, purification and characterization of the collagen of *Cysticercus cellulosae*. *Journal of Biological Chemistry*, **256**: 5926–30.

887. **Torres, P., Figueroa, L. & Franjola, R.** (1981). [Studies on Pseudophyllidea (Carus, 1813) in the south of Chile. VIII. Experimental development of *Diphyllobothrium dendriticum* Nitzsch in *Larus maculipennis* Lichenstein.] In Spanish. *Boletín Chileno de Parasitología*, **36**: 74–5. [HA/52/3721]

888. **Ubelaker, J. E.** (1980). Structure and ultrastructure of the larvae and metacestodes of *Hymenolepis diminuta*. In *Biology of the tapeworm Hymenolepis diminuta*, ed. H. P. Arai, pp. 59–156. Academic Press: New York.

889. — (1983). The morphology, development and evolution of tapeworm larvae. In *Biology of the Eucestoda*, vol. 1, ed. C. Arme & P. W. Pappas, pp. 235–96. Academic Press: London.

890. **Uglem, G. L., Dupre, R. K. & Harley, J. P.** (1983). Allosteric control of pyrimidine transport in *Hymenolepis diminuta*: an unusual kinetic isotope effect. *Parasitology*, **87**: 289–93.

891. **Uglem, G. L. & Just, J. J.** (1983). Trypsin inhibition by tapeworms: antienzyme secretion or pH adjustment? *Science*, **220**: 79–81.

892. **Uglem, G. L. & Prior, D. J.** (1980). *Hymenolepis diminuta*: chloride fluxes and membrane potentials associated with sodium-coupled glucose transport. *Experimental Parasitology*, **50**: 287–94.

893. **Valero, A., Hermoso, R. & Monteoliva, M.** (1983). [Electrophoretic studies on some species of cestodes.] In Spanish. *Revista Ibérica de Parasitología*, **43**: 89–92.

894. — (1984). [Electrofocusing of soluble proteins of *Hymenolepis fraterna* and *H. diminuta*.] In Spanish. *Revista Ibérica de Parasitología*, **44**: 213–4.

895. **Van den Bossche, H., Thienpoint, D. & Janssens, P. G.** (eds.) (1986). *Chemotherapy of gastrointestinal helminths*. Springer-Verlag: Berlin.

896. **Van Muiswinkel, W. B. & Cooper, E. L.** (1982). *Immunology and immunization of fish*. Pergamon Press: New York.

897. **Varma, T. K., Varma, V., Mohan Rao, V. K. & Ahluwalia, S. S.** (1982). Some biochemical studies on cyclophyllidean (anoplocephalid and taeniid) tapeworms of zoonotic importance. *Indian Veterinary Journal*, **59**: 343–6.

898. — (1985). Alkaline and acid phosphatase activities in cyclophyllidean (anoplocephalid and taeniid) tapeworms of zoonotic importance. *Indian Veterinary Journal*, **62**: 20–3.

899. **Vidor, E., Piens, M. A., Abbas, M. & Petavy, A. F.** (1986). Biochimie du liquide hydatique (*Echinococcus granulosus*). Influence de la localization sur la perméabilité des kystes. *Annales de Parasitologie Humaine et Comparée*, **61**: 333–40.

900. **Vinayakam, A.** (1985). Distribution of lyo- and desmo-glycogen in relation to growth and maturity of proglottids of *Moniezia benedeni*. *Foli Parasitologica*, **32**: 67–71.

901. **Vinogradov, G. A., Davydov, V. G. & Kuperman, B. I.** (1982). [Mor-

References

phological–physiological study of the mechanisms of adaptation in pseudophyllidean cestodes to different salinities.] In Russian. *Parazitologiya*, **16**: 377–83. [HA/52/434]

902. **Vlasova, T. A.** (1981). [The morpho-functional differentiation of the strobila of the cestode *Echinatrium filosomum* Spassky et Jurpalova, 1965 (Cestodoidea; Hymenolepididae).] In Russian. *Ekologiya gel'mintov 1981*: 13–17. [HA/51/6182]

903. **Voge, M.** (1973). The post-embryonic developmental stages of cestodes. *Advances in Parasitology*, **11**: 707–30.

904. — (1975). Axenic development of cysticercoids of *Hymenolepis diminuta*. *Journal of Parasitology*, **61**: 563–4.

905. — (1978). Cestoda. In *Methods of cultivating parasites in vitro*, ed. A. E. R. Taylor & J. R. Baker, pp. 193–225. Academic Press: London.

906. **Voge, M. & Berntzen, A. K.** (1961). *In vitro* hatching of oncospheres of *Hymenolepis diminuta* (Cestoda: Cyclophyllidea). *Journal of Parasitology*, **47**: 813–18.

907. **Voge, M. & Coulombe, L. S.** (1966). Growth and asexual multiplication *in vitro* of *Mesocestoides* tetrathyridia. *American Journal of Tropical Medicine and Hygiene*, **15**: 902–7.

908. **Voge, M. & Green, J.** (1975). Axenic growth of oncospheres of *Hymenolepis citelli* (Cestoda) to fully developed cysticercoids. *Journal of Parasitology*, **61**: 291–7.

909. **Voge, M. & Seidel, J. S.** (1968). Continuous growth *in vitro* of *Mesocestoides* (Cestoda) from oncosphere to fully developed tetrathyridium. *Journal of Parasitology*, **54**: 269–71.

910. **Voge, M. & Turner, J. A.** (1956). Effect of temperature on larval development of the cestode, *Hymenolepis diminuta*. *Experimental Parasitology*, **5**: 580–6.

911. **Von Brand, T.** (1979). *Biochemistry and physiology of endoparasites*. Elsevier North Holland Biomedical Press: Amsterdam.

912. **Von Brand, T. & Nylen, M. U.** (1970). Organic matrix of cestode calcareous corpuscle. *Experimental Parasitology*, **28**: 566–76.

913. **Von Brand, T., Nylen, M. U., Martin, G. N., Churchwell, F. K. & Stites, E.** (1969). Cestode calcareous corpuscles: phosphate relationships, crystallization patterns, and variations in size and shape. *Experimental Parasitology*, **25**: 291–310.

914. **Von Brand, T. & Weinbach, E. C.** (1975). Incorporation of calcium into the soft tissues and calcareous corpuscles of larval *Taenia taeniaeformis*. *Zeitschrift für Parasitenkunde*, **48**: 53–63.

915. **Vykhrestyuk, N. P., Il'yasov, I. N., Yarygina, G. V. & Nikitenko, T. B.** (1976). [Lipids in *Raillietina tetragona*, *R. echinobothrida* and in the chicken intestine.] In Russian. *Trudy Nauchno-Issledovatel'skogo Veterinarnogo Instituta Tadzhikskoĭ SSR*, **6**: 82–7. [HA/49/3663]

916. **Vykhrestyuk, N. P. & Klochkova, V. I.** (1984). [Hexokinase in the trematode *Calicophoron ijimai*, the cestode *Bothriocephalus scorpii* and the turbellarian, *Penecurva sibirica*.] In Russian. In *Parazity zhivotnykh i rastenii*, pp. 82–6, Academiya Nauk SSSR, Dal'nevostochnyi Nauchnyi Tsentr, Biologo-Poch vennyi Institut, Vladivostok. [HA/54/3002]

917. **Vykhrestyuk, N. P., Yarygina, G. V. & Il'yasov, I. N.** (1981). [Lipids of

References

Raillietina tetragona and *R. echinobothrida* from the intestine of chickens.] In Russian. *Parazitologiya*, **15**: 525–32. [HA/51/3799]

918. **Wack, M., Komuniecki, R. & Roberts, L. S.** (1983). Amino acid metabolism in the rat tapeworm, *Hymenolepis diminuta*. *Comparative Biochemistry and Physiology*, **74B**: 399–402.

919. **Wakelin, D.** (1984*a*). *Immunity to parasites. How animals control parasitic infections.* Edward Arnold: London.

920. — (1984*b*). Immunity to helminths and prospects for control. In *Critical reviews in tropical medicine*, vol. 2, ed. R. K. Chandra, pp. 209–44. Plenum Press: New York.

921. — (1986). Immunity to helminths. *EOS-Rivista di immunologia ed immunofarmacologia*, **6**: 35–46.

922. **Walker, R. W. & Barrett, J.** (1983). Mitochondrial adenosine triphosphate activity and temperature adaptation in *Schistocephalus solidus* (Cestoda: Pseudophyllidea). *Parasitology*, **87**: 307–26.

923. — (1985). Mitochondrial membrane fluorescence and temperature adaptation in *Schistocephalus solidus* (Cestoda: Pseudophyllidea). *Parasitology*, **90**: 131–5.

924. **Walker, W. A.** (1981). Intestinal transport of macromolecules. In *Physiology of the gastrointestinal tract*, ed. L. R. Johnson, pp. 1271–99. Raven Press: New York.

925. **Walker, W. A. & Isselbacher, K. J.** (1974). Uptake and transport of macromolecules by the intestine: possible role in clinical disorders. *Gastroenterology*, **67**: 531–50.

926. **Walkey, M. & Fairbairn, D.** (1973). L(+)-Lactate dehydrogenases from *Hymenolepis diminuta* (Cestoda). *Journal of Experimental Zoology*, **183**: 365–73.

927. **Walkey, M. & Körting, W.** (1985). Thermal alterations of pyruvate kinases in the fish tapeworm *Bothriocephalus acheilognathi*, Yamaguti, 1934. *Zeitschrift für Parasitenkunde*, **71**: 527–32.

928. **Walls, K. W. & Schantz, P. M.** (1986). *Immunodiagnosis of parasitic diseases*, vol. 1 *Helminthic diseases*. Academic Press: New York.

929. **Wang, S. S., Meng, X. Q., Zhou, W. Q., Ying, G. H., Li, X. & Zhad, Y. Z.** (1981). SEM observations of the membranous structure of the eggs of *Taenia solium*. *Scanning Electron Microscopy*, **3**: 183–6.

930. **Ward, C. W. & Fairbairn, D.** (1970). Enzymes of beta-oxidation and the tricarboxylic acid cycle in adult *Hymenolepis diminuta* (Cestoda) and *Ascaris lumbricoides* (Nematoda). *Journal of Parasitology*, **56**: 1009–12.

931. **Ward, P. F. V.** (1982). Aspects of helminth metabolism. *Parasitology*, **84**: 177–94.

932. **Ward, S. M., Allen, J. M. & McKerr, G.** (1986). Neuromuscular physiology of *Grillotia erinaceus* metacestodes (Cestoda: Trypanorhyncha) *in vitro*. *Parasitology*, **93**: 121–32.

933. **Wardle, R. A. & McLeod, J. A.** (1952). *The zoology of tapeworms.* University of Minnesota Press: Minneapolis.

934. **Watts, S. D. M.** (1981). Colchicine binding in the rat tapeworm, *Hymenolepis diminuta*. *Biochimica et Biophysica Acta*, **667**: 59–69.

935. **Watts, S. D. M. & Fairbairn, D.** (1974). Anaerobic excretion of

References

fermentation acids by *Hymenolepis diminuta* during development in the definitive host. *Journal of Parasitology*, **60**: 621–5.

936. **Webb, R. A.** (1976). Ultrastructure of synapses of the metacestode of *Hymenolepis microstoma*. *Experientia*, **32**: 99–100.

937. — (1977). Evidence for neurosecretory cells in the cestode *Hymenolepis microstoma*. *Canadian Journal of Zoology*, **55**: 1726–33.

938. —(1984). Intranuclear bodies in the tissues of the scolex of the cestode *Hymenolepis microstoma*. *Canadian Journal of Zoology*, **62**: 107–11.

939. — (1986). The uptake and metabolism of L-glutamate by tissue slices of the cestode *Hymenolepis diminuta*. *Comparative Biochemistry and Physiology*, **85C**: 151–62.

940. **Webb, R. A. & Davey, K. G.** (1974). Ciliated sensory receptors of the unactivated metacestode of *Hymenolepis microstoma*. *Tissue and Cell*, **6**: 587–98.

941. —(1975a). The gross anatomy and histology of the nervous system of the metacestode of *Hymenolepis microstoma*. *Canadian Journal of Zoology*, **53**: 661–77.

942. — (1975b). Ultrastructural changes in an unciliated sensory receptor during activation of the metacestode of *Hymenolepis microstoma*. *Tissue and Cell*, **7**: 519–24.

943. — (1976). The fine structure of the nervous tissue of the metacestode of *Hymenolepis diminuta*. *Canadian Journal of Zoology*, **54**: 1206–22.

944. **Webb, R. A. & Mettrick, D. F.** (1971). Patterns of incorporation of ^{32}P into the phospholipids of the rat tapeworm *Hymenolepis diminuta*. *Canadian Journal of Biochemistry*, **49**: 1209–12.

945. **Webb, R. A. & Mettrick, D. F.** (1975). The role of glucose in the lipid metabolism of the rat tapeworm *Hymenolepis diminuta*. *International Journal for Parasitology*, **5**: 107–12.

946. **Webbe, J.** (1986). Cestode infections of man. In *Chemotherapy of parasitic diseases*, ed. W. C. Campbell & R. S. Rew, pp. 457–77. Plenum Press: New York.

947. **Webster, L. A.** (1970). The osmotic and ionic effects of different saline conditions on *Hymenolepis diminuta*. *Comparative Biochemistry and Physiology*, **37**: 271–5.

948. — (1972a). Absorption of glucose, lactate and urea from the protonephridial canals of *Hymenolepis diminuta*. *Comparative Biochemistry and Physiology*, **41A**: 861–8.

949. — (1972b). Further osmotic and ionic effects of different saline conditions on *Hymenolepis diminuta*. *Comparative Biochemistry and Physiology*, **42A**: 409–13.

950. — (1972c). Succinic and lactic acids present in the protonephridial canal fluid of *Hymenolepis diminuta*. *Journal of Parasitology*, **58**: 410–11.

951. **Webster, L. A. & Wilson, R. A.** (1970). The chemical composition of protonephridial canal fluid from the cestode *Hymenolepis diminuta*. *Comparative Biochemistry and Physiology*, **35**: 201–9.

952. **Weinbach, E. C. & von Brand, T.** (1970). The biochemistry of cestode mitochondria. I. Aerobic metabolism of mitochondria from *Taenia taeniaeformis*. *International Journal for Biochemistry*, **1**: 39–56.

References

953. **Wharton, D. A.** (1983). The production and functional morphology of helminth egg-shells. *Symposia of the British Society for Parasitology*, **20**: 85–97.
954. **Whittaker, F. H., Carvajal, G. J. & Apkarian, R.** (1982). Scanning electron microscopy of the scolex of *Grillotia dollfusi* Carvajal 1971 (Cestoda: Trypanorhyncha). *Journal of Parasitology*, **68**: 1173–5.
955. **Wikgren, B.-J. P.** (1966). The effect of temperature on the cell division cycle of diphyllobothrid plerocercoids. *Acta Zoologica Fennica*, **114**: 3–27.
956. **Wilkes, J., Cornish, R. A. & Mettrick, D. F.** (1981). Purification and properties of phosphoenolpyruvate carboxykinase from *Hymenolepis diminuta* (Cestoda). *Journal of Parasitology*, **67**: 832–40.
957. **Williams, H. H.** (1960a). Some observations on *Parabothrium gadipollachii* (Rudolphi, 1810) and *Abothrium gadi* van Beneden 1870 (Cestoda: Pseudophyllidea) including an account of their mode of attachment and of variation in the two species. *Parasitology*, **50**: 303–22.
958. — (1960b). The intestine in members of the genus *Raja* and host-specificity in the Tetraphyllidea. *Nature*, **188**: 514–16.
959. **Williams, H. H. & McVicar, A.** (1968). Sperm transfer in Tetraphyllidea (Platyhelminthes: Cestoda). *Nytt Magasin for Zoologi*, **16**: 61–71.
960. **Williams, J. F.** (1982). Cestode infections. In *Immunology of parasitic infections*, ed. S. Cohen & K. Warren, pp. 676–714. Blackwell Scientific Publ.: Oxford.
961. — (1986). Prospects for prophylaxis of parasitism. *International Journal for Parasitology*, **17**: 711–19.
962. **Williams, J. F. & Sandeman, R. M.** (1982). Antigens of taeniid cestodes. In *Cysticercosis: present state of knowledge and perspectives*, ed. A. Flisser, K. Willms, J. P. Laclette, C. Larralde, C. Ridaura, F. Beltrán & M. W. Vogt, pp. 525–37. Academic Press: New York.
963. **Willis, J. M. & Herbert, I. V.** (1984). Some factors affecting the eggs of *Taenia multiceps*: their transmission onto pasture and their viability. *Annals of Tropical Medicine and Parasitology*, **78**: 236–42.
964. **Wilson, R. A. & Webster, L. A.** (1974). Protonephridia. *Biological Reviews*, **49**: 127–60.
965. **Wilson, T. H.** (1962). *Intestinal absorption*. W. B. Saunders: London.
966. **Wilson, V. C. L. C. & Schiller, E. L.** (1969). The neuroanatomy of *Hymenolepis diminuta* and *H. nana*. *Journal of Parasitology*, **55**: 261–70.
967. **Winkelman, L.** (1976). Comparative studies of paramyosins. *Comparative Biochemistry and Physiology*, **55B**: 391–7.
968. **World Health Organization** (1981). Guidelines for surveillance, prevention and control of echinococcosis/hydatidosis. WHO Geneva. Publication no. VPH/81.28.
969. — (1983). Guidelines for surveillance, prevention and control of taeniasis/cysticercosis. WHO Geneva. Publication no. VPH/83.49.
970. **Wrong, G. O. M., Edmonds, C. J. & Chadwick, V. S.** (1981). *The large*

References

intestine: its role in mammalian nutrition and homeostasis. MTP Press: Lancaster.

971. **Yakushev, V. Yu., Freze, V. I., Sysoev, A. V., Pelgunov, A. N. & Malkin, A. E.** (1985). Attempted mathematical models of seasonal dynamics of cestode infections in intermediate and final hosts. *Angewandte Parasitologie*, **26**: 139–45.

972. **Yamane, Y., Nakagawa, A., Makino, Y. & Hirai, K.** (1982). An electron microscopic study of subtegumental cells and associated structures of *Spirometra erinacei*. *Japanese Journal of Parasitology*, **31**: 487–97.

973. — (1983). *Diphyllobothrium latum*: scanning electron microscopic study on the eggshell formation. *Japanese Journal of Parasitology*, **32**: 13–25.

974. **Yamane, Y., Nakagawa, A., Makino, Y., Yazaki, S. & Fukumoto, S.** (1982). Ultrastructure of the tegument of *Diphyllobothrium latum* by scanning electron microscopy. *Japanese Journal of Parasitology*, **31**: 33–46.

975. **Yamane, Y., Yoshida, N., Nakagawa, A., Abe, K. & Fukushima, T.** (1986). Trace element content in two species of whale tapeworms, *Diphyllobothrium macroovatum* and *Diplogonoporus balaenopterae*. *Zeitschrift für Parasitenkunde*, **72**: 647–51.

976. **Yap, K. W., Thompson, R. C. A., Rood, J. I. & Pawlowksi, I. D.** (1987). *Taenia hydatigena*: isolation of mitochondrial DNA, molecular cloning, and physical mitochondrial genome mapping. *Experimental Parasitology*, **63**: 288–94.

977. **Yarygina, G. V., Vykhrestyuk, N. P. & Klochkova, V. I.** (1982). [The amino acid composition of collagenous proteins of the trematodes *Calicophoron erschowi, Eurytrema pancreaticum* and the cestodes *Bothriocephalus scorpii* and *Nybelinia* sp. larvae.] In Russian. *Zhurnal Evolyutsionnoi Biokhimii i Fiziologii*, **18**: 564–7. [HA/53/4639]

978. **Yorke, R. E. & Turton, J. A.** (1974). Effects of fasciolicidal and anti-cestode agents on the respiration of isolated *Hymenolepis diminuta* mitochondria. *Zeitschrift für Parasitenkunde*, **45**: 1–10.

979. **Zavras, E. T. & Roberts, L. S.** (1984). Developmental physiology of cestodes. XVIII. Characterization of putative crowding factors in *Hymenolepis diminuta*. *Journal of Parasitology*, **70**: 937–44.

980. — (1985). Developmental physiology of cestodes: cyclic nucleotides and the identity of putative crowding factors in *Hymenolepis diminuta*. *Journal of Parasitology*, **71**: 96–105.

981. **Zelazny, J.** (1979). Influence of some physical agents and chemical compounds on hatchability and vitality of coracidia of *Bothriocephalus gowkongensis* Yeh, 1955. *Bulletin of the Veterinary Institute in Pulawy, Poland*, **23**: 20–4.

982. **Zhuravets, S. K.** (1982). [Periods of maturation of *Echinococcus granulosus* in dogs]. In Russian. *Byulleten' Vsesoyuznogo Instituta Gel'mintologii im. K. I. Skryabina*, **32**: 28–32. [HA/52/5734]

Index

Numbers in **bold type** indicate a **figure** on that page. Numbers in *italics* indicate a *table* on that page. Well-known experimental species are sometimes quoted in an abbreviated form. ***Bold italic*** is used for easy identification of ***species names*** in entries.

Index

Index

370

Index

371

Index

Cholinesterase, cytochemistry, 22, **23**, 30
Chromalum-haematoxylin/phloxine, 28
Chymotrypsin inhibition
  by cestodes, 58
  by *Hymenolepis*, 11
  by *T. pisiformis*, 301
  by *T. taeniaeformis*, **59**
Ciliated receptors, **24**, 31, 32
Circadian migration, 236
Cirrus 'bulge', 274, **275**
Cittotaenia sp., shell formation, 183
  perplexa, chemical composition, *56*
  variabilis, egg formation, 174
Clear vesicles in nerve terminals, **26**
Cloning of specific DNA fragments
  E. granulosus, 150
  T. solium, 150
  mitochondrial genome, *T. hydatigena*,
    150
CO_2
  as trigger, 47
  in intestine of vertebrates, 46, *47*
  mucosal, **48**
CO_2-fixation
  H. microstoma, 79
  importance in cestode metabolism, 47,
    92
Cobalt chloride, effect on muscle, **32**,
  33
Coelenterates, 29
Colchicine binding
  for measuring mitosis, *215*
  H. diminuta, **116**
Collagen
  amino acid composition, *T. solium*, *115*
  in cestodes, 114
Collagenase, 212
Colon, pH, *46*
Colostrum, 44
Coomassie brilliant blue, *168*, *180*
Complements components, *294*
Complementary DNA (cDNA), definition
  154
Compression, insemination requirement,
  40, 273
Computerised image analysis, 7
Conjugation of bile salts, 51
Contact (= membrane) digestion, 10, 14,
  123
Copulatory activity, 251
Copepods
  as intermediate hosts, 197
  coracidia infections, 197, 200
Coracidium

adaptations, 195
Bothriocephalus, **197**
  defined, 170
  embryophore, **203**
  glands in, *196*, 201
  lipids, *66*
  metabolism, 205–*206*
  oxygen consumption, **205**
  penetration glands, 201
  phospholipids, *69*
  survival, **198**
  ultrastructure, SEM, **197**, **203**
Coracidium/procercoid transformation,
  202–3
Coregonus albula, *18*, *196*
Cotugnia digonopora
  egg formation, 174
  endproducts, *84*
  intermediary metabolism, 83
  lipids, 65, *66*
Covalent links, **173**
Crane's gradient hypothesis, 12
C-reactive protein, bound by *B. scorpii*,
  117
Crossbothrium squali, self-insemination,
  162
Cross-insemination, 163–**165**
Crowding effect, 243, **244**, 247
  competition for carbohydrates, 79, 247
Cryopreservation, *E. multilocularis*, 282
Crypt
  hyperplasia, *36*
  of Lieberkühn, 39, 239, 272
  size, **37**
Ctenocephalides felis felis, cysticercoid, 228
Cultivation *in vitro*, 257–82; *see individual*
  species
Culture tube, for *Schistocephalus*, **263**
Cyanide-insensitive respiration in cestodes,
  107
Cyanocobalamin
  binding to microtriches, **121**
  receptor, *Spirometra mansonoides*, **121**
Cyatocephalus truncatus, scolex glands, *18*
Cyclic nucleotide, cGMP, 247
Cyclophyllidea(n)
  egg, 156, **157**
    formation, general, **169**
    hatching, 189, **190**, 191
    reviewed, 166
    survival, 185, *186*
  embryonic development, *167*
  oncosphere, morphology, **222**
  shell/capsule, 167

Index

373

Index

Index

Index

Index

Index

Index

379

Index

Index

Index

Index

in mast cells, *290*
in oncosphere invasion, 297
IgG
  characteristics, *285*
  in cysticercosis, *299*
  in *H. diminuta* infections, 292
  in *Taenia* spp. infections, 297
IgM
  characteristics, *285*
  in cysticercosis, *299*
  in *H. diminuta* infections, 292
  in *T. crassiceps* infections, 300
Ileum, *46*
Immune response(s)
  and genotype, *294*
  'evasion of' 299–301
  to larval cestodes, 295–9
  to oncosphere, **296**
Immunisation
  against, adult cestodes, 301–2
  E. granulosus, 301
  general 301–4
  larval cestodes, 302–4
  T. pisiformis, 297
  T. taeniaeformis, 297
  see also Vaccination
Immunity
  basic concepts reviewed, **284**
  to adult cestodes, 286, **287**, 294
  to adult *E. granulosus*, 294–5
  to cestodes in general, reviewed,
    286
  to *Echinococcus* spp., 286
  to *H. nana*, 286
  to helminths, reviewed, 286
  to Taeniidae, 294
Immunobiology, 283–304
Immunocytochemical techniques
  in analysis of cestode surface **122**
  in protein and antigen characterisation,
    117
Immunocytology, in cestodes and
    turbellerians, 22, *31*
Immunodepression, *294*
Immunodiagnosis
  of *E. granulosus* in dogs, 295
  reviewed, 304
Immunofluoresence of *E. granulosus* eggs,
    304
Immunoglobulins (Ig)
  general characteristics, *285*
  in gut, *36*, **287**
  in lumen, *36*
  see individual immunoglobulins

Immunoperoxidase-antiperoxidase (PAP)
  technique, **29**, **30**
Immunosuppression, by *T. crassiceps*, 300
Inermicapsifer madagascarensis,
  spermatozoa, *159*
Inermiphyllidium pulvinatum, osmotic
  pressure, 52
Infectivity, coracidium, **199**
Inflammatory response, **287**
Inhibitor
  of protease (taeniastatin), 58, **59**
  of trypsin, 301
Inhibitory neurotransmitter, 30
Inner envelope (egg), **167**, **178**, 179, **190**
Insect, spermatozoa, 160
Insects, egg transmission, 186
Insemination
  compression requirement, 40, 262, **263**
  failure *in vitro*, 273
  in vitro, 40, 259, 262, **263**
  reviewed, 162
  tube for *in vitro* culture, **263**
Intermediary metabolism, cestodes species
  studied, 83
Intestinal mastocytosis, *294*
Intestinal mucosa
  glucose transport, **8**
  uptake of dipeptides and tripeptides, 44,
    45, 129
Intestine
  effects of parasites on, *36*
  Eh, 46
  endocytosis, **38**, 44
  immunobiology, 286, **287**
  parameters and host specificity, *37*
  pH, *37*, 45, *46*
  physiology, 41–51
Intra-epithelial leukocytes, 290
Intra-epithelial lymphocytes (IEL), **287**, 290
Intramitochondrial metabolism of malate,
  H. diminuta, **102**
Intranuclear bodies, *Hymenolepis*, 18
Introvertus raipurensis, 126
Invertebrate immunity, reviewed, 286
In vitro culture (cultivation), 257–82
  basic problems, 258
  cestode cells, 281
  criteria for growth, **252**, 259
  experiments, pitfalls, 77
  limitations, 53
  nutritional requirements, 258
  of Cyclophyllidea, 265–79
  of *Echinococcus*, 272, **273**, 274, **275**, 276
  monozoic forms, **275**, **276**

383

Index

Index

malic enzyme, 99
maturation, 235
parasitic castration, **220**
PEPCK, 92
phosphate content/calcareous corpuscles, *60*
6-phosphogluconate dehydrogenase, 112
phosphoglucomutase polymorphism, 125
phospholipids, *68*
plercocercoid, 211, **264**
  glands, *196*
  polymorphic variation, 164
  prepatent period, *242*
  procercoid/plerocerocid transformation, 212
  proteolytic activity, 132
  pyruvate kinase, 88
  scolex, 234–5
  scolex glands, adult, *18*
  sexual maturation, 214
  TCA cycle
    enzymes, *104*
    intermediates, *103*
Light
  intensity, effect on egg hatching 188
  wavelength, effect on egg hatching, **188**
Linoleic (C$_{18:2}$) acid, 69
Lipase activity, 70, *233*
Lipids
  acylglycerols, *66, 67*
  adult and larval cestodes, 65, *66, 67*
  egg, *H. diminuta*, *180*
  free fatty acids, *66, 67*
  localisation, 67
  predominant metabolic pathways, **76**
  sterols, *66, 67*
  synthesis, 74
Lipoprotein(s), 65
  antigens, 117
  scolex secretion, 18
Lithocholic acid, *49*, 51
Liver, **288**
Lota lota, *18*
Lucknowia indica, amino acids, 126
Lumen, immunoglobulins in, *36*, **287**
Lymphatics, **288**
β-Lymphocytes, 285
Lymphokines, **284**
Lysophosphatidic acid, *68*
Lysophophatidylcholine, *68*
Lysophophatidylethanolamine, *68*
Lysine, *43*
Lysosomes, **6**, **38**, 241
Lytocestus indicus

amino acids, 126
cytochemistry, sperm, 160
transaminases, 133

Macrocallista nimbosa, 29
Macrocyclops
  albidus, 200
  funcus, 200
Macrophages, *36*, **284**, **287**, 297
  activation of, *294*
Magnesium, 33
Major basic protein (MBP), 298
Malate dehydrogenase
  in carbohydrate metabolism, 93
  in isoenzyme analysis, 164
  polymorphism, *M. expansa*, 125
Malate dismutation, **91**, 93
L-Malate:NAD oxidoreductase, *206*
Malic enzyme
  in carbohydrate metabolism, 93
  in cestodes, species studied, 99
Mammalian embryo, 1
Mammalian growth hormone, *218, 219*
Mammary glands, **288**
Man
  bile composition, *49, 50*
  immune responses
    in cysticercosis, **298**, *299*
    in hydatidosis, 298
  mucosa turnover time, 39
Manganese ions, 33
Mannose
  effect on cysticercoid growth, 247
  in *H. diminuta*, *59*
Masking of host antigens, 300
Masseter muscle, 230
Mast cells
  differences, 289, *290*
  in mucosal infections, **287**
  in oncosphere invasion, 297
Mastocystosis, *36*
Maturation
  defined, 241
  stages of, 251, **252**
Meal, meat, 47
Mechanoreceptors, 31
Mediators, **284**
Medium S10E for *Echinococcus* culture, *272*
Mehlis' gland, **170**
Membrane
  α-amylase adsorption, cestodes, 123
  (=contact) digestion, 10, 14
  structure, **10**
  transport, 42–3

385

Index

Index

Index

388

Index

Index

Index

Index

Retroviruses, 300
Rhamnose, *59*
Rheoreceptors, 31
Rhinebothrium flexile, insemination, 164
Rhodoquinone, 107–9
 in *H. diminuta*, 73
Ribose, *59*
Ribosomal RNA, *E. granulosus*, 145
Ribosomes, egg, 182
RMCP I/II, *290*
RNA
 and molecular mimicry, 300-1
 characteristics and formation, *E.*
  granulosus, 143–5
 electrophoretic separation, *E. granulosus*,
  146
Roach, castrated by *Ligula intestinalis*, **220**
Roller tube culture technique, 266
Rostellar pad, 16, **240**
Rostellum, 16, **24**
Rumen, cattle, 48
Ruthenium red, 14

Sakaguchi reaction, *180*
Salivary glands, **288**
Salvelinus alpinus, *18*
Sapocholic acid, *49*
Satellite DNA in *Hymenolepis* spp., 142
Schiff test (PAS), *see* PAS, 17
Schistocephalus (sp.)
 glycogen, 19
 guanine + cytosine (G + C) content of
  DNA, *143*
 solidus
  adenine nucleotide content, *63*
  beta-oxidation enzymes, *72*
  chemical composition, *56*
  coracidia, survival, *199*
  egg shell formation, 171
  end-products, *84*
  endocytosis, 14
  enzyme electrophoresis, 125
  fertilization, 262, **263**
  glucose-6-phosphate dehydrogenase,
   112
  glycolytic enzymes, *86*, *87*
  in vitro culture 40, 262, **263**, 265
  insemination, 259, 262, **263**
  intermediary metabolism, 83
  isoenzyme analysis, 164
  lactate dehydrogenase, 88
  lipids, *67*
  malic enzyme, 99
  maturation *in vivo*, 17, **235**
  neurohormonal peptides, *31*, 251

osmotic pressure, 51
PEPCK, 92
6-phosphogluconate dehydrogenase,
 112
plerocercoid, 211, **261**
 glands in, *196*
 in fish, **221**
prepatent period, *242*
procercoid/plerocercoid
 transformation, 212
proteolytic activity, 132
lipid synthesis, 74
phosphagen (taurocyamine)
 phosphotransferase, 65
phosphofructokinase, 88, **89**
phospholipids, *68*
pyruvate dehydrogenase, 105
pyruvate kinase, 88
scolex, 17, 234, **235**
scolex glands, adult, *18*
self-insemination, 162–3, **263**
 serotonin in, 251
sexual maturation, 214
superoxide dismutase, 111
TCA cycle
 enzymes, *104*
 intermediates, *103*
pungitii
 coracidium, 196
 prevalence in copepods, 200, **201**
Schistosoma mansoni, 47
Sclerotin, chemical structure, 172, *173*
Scolex
 attachment, 235–6
 deformatus, 17
 E. granulosus, in dog gut, **240**
 evagination, *233*, 234
 general account, 15–18
 glands
  adult, 17, *18*, **19**, 240
  larval, *196*, 209, **210**, **213**
 host specificity, 36
 immunocytology, *D. dendriticum*, **29**
SDS/polyacrylamide gel electrophoresis
 (SDS/PAGE); proteins, 119
Secretion
 rostellar glands, *E. granulosus*, 18, **240**
 scolex glands, 17–**19**
Secretory antibody, 288
Secretory component (SC), **287**, 288
Secretory IgA, **288**
 in gut secretion, 297
Segmentation, 251
Segments, 5
Self and non-self tissue, 300

392

Index

Index

394

Index

Index

Index

pyruvate kinase, 88
scolex glands, adult, *18*
TCA cycle enzymes, *104*
meridionalis, scolex glands, adult, *18*
nodulosus
    chemical composition, 55, *56*
    coracidium 196, *199*
        enzymes, *206*
        metabolism, 205
    egg
        formation, 171
        hatching, 189
    glyoxalate cycle, 113
    larvae, glands, *196*
    life span, 183
    lipids, *66*
    phospholipids, *68*
    plerocercoid, 211
    procercoid glands, 209, **210**
    scolex glands, adult, *18*
    ultrastructure, 7
Tribolium confusum, *Hymenolepis* in, 202
Tricarboxylic acid (TCA) cycle,
        importance to cestodes, 102
Trickling filter, egg survival in, *186*
Trigger
    for differentation, 259
    for egg hatching, 47
    for migration, 238
Trypanorhyncha
    egg formation, **169**
    muscle, 21
    spermatozoa, *159*
Tripeptides, intestinal uptake, 42
Trypsin
    action on mucus, **41**
    adsorption, *H. diminuta*, **125**
    excystment/evagination role, *233*
    in egg hatching, *192*
    inactivation by cestodes, 58
    inhibition
        by *Hymenolepis*, 11
        by *T. pisiformis*, 301
    sexual differentiation stimulus, 250
Tryptophan, 25
Tubifex tubifex, *18*
Tubulin
    binding by benzimidazoles, 115
    isolation, *H. diminuta*, 116
Turnover rate
    glycocalyx, 9
    mucosa, 39
Two-dimensional polyacrylamide
        electrophoresis, **124**

Turbellaria, 21, *31*
Tyrosine
    egg, *H. diminuta, 180*
    in quinone tanning system, **173**
    phenol oxidase substrate, *175*

Ultrastructure
    muscle, 20, **21**
    nervous system, 22–31
    see also individual species
Unstirred water layer, 9, 42
    effect on transport kinetics, 81
Uptake
    by enterocyte, **38**
    by *H. diminuta*, methionine, **13**
    by intestine
        amino acids, 42, 44, **45**
        vitamins, 43
Urea
    in cestodes, 52
    Krebs-Henseleit cycle, **136**
    metabolism of, 52
Urease activity, in cestodes, 134
Uronic acids, 57
Ursodeoxycholic acid, *49*
Uterus, role in shell formation, 171, 183
USSR, prevalence of procercoids, **201**

Vaccination
    against adult *E. granulosus*, 302
    against cysticercosis, 302–4
    against larval cestodes in general, 302–4
    and DNA technology, 304
Vaccine(s)
    development, general comment, 283
    oncosphere secretions, 224
    sonicated oncospheres, 303
Vagina, **170**
Valine, *43*
Vasocytin, 29, *31*
Vector, definition, 154–5
Vertebrates
    lower, immunity, reviewed, 286
    neuropeptides, 29
Vesicles
    clear/dense-core, in nerve terminals, 25,
        26
    in oncospheres, **222**
Viability
    egg, 183, **184**, *185*, *186*
    oncosphere, *192*
Villous atrophy, *36*
Villus size, **37**, **229**
VIP (vaso-active intestinal polypeptide), *31*

397

Index